Networked Neighbourhoods

Networked Neighbourhood.

Patrick Purcell (Ed.)

Networked Neighbourhoods

The Connected Community in Context

 Springer

Professor Patrick Purcell
Department of Electrical & Electronic Engineering
Imperial College London
South Kensington, London SW7 2BT UK

British Library Cataloguing in Publication Data
A catalogue record for this book is available from the British Library

ISBN-13: 978-1-84996-567-5   e-ISBN-13: 978-1-84628-601-8

Printed on acid-free paper

Printed in the United States of America

9 8 7 6 5 4 3 2 1

Springer Science+Business Media, LLC
springer.com

Acknowledgements

The Editor wishes to express his appreciation to the international group of contributing authors whose unfailing cooperation throughout was vital to the successful outcome of the project.

Also, the Editor wishes to express his appreciation for the support of Mark Witkowski, Sunny Bains, Yiannis Demiris and colleagues in the Intelligent Systems & Networks Group of Imperial College London during the course of the book's development, as well as the continuing support of Ellen Haigh and colleagues of both the Departmental and Central Library Services of the campus. He also pays tribute to the collaboration of Robert Spence, Kostas Stathis and colleagues on the "Living Memory" project jointly reported in Chapter 12 of this volume.

Helen Callaghan, Catherine Brett and Felix Portnoy of Springer also deserve special thanks for their co-operation, support and intermittent forbearance throughout the process of preparing this book for publication.

The Editor also wishes to acknowledge the exceptional contribution of Jakub Wejchert of the Information Societies and Technologies programme of the European Commission who provided the original briefing and subsequent support for the research case studies reported in chapters 10, 11, 12 14 and 16 as well as support for the initial commissioning of this volume.

Contents

List of Contributors

Stefan Agamanolis
www.media.mit.edu/~stefan
stefan@media.mit.edu

Jacob Beaver
Interaction Design
Royal College of Art
Kensington Gore
London SW7 UK
j.beaver@rca.ac.uk

Kristen Berg
NetLab
Faculty of Social Work
University of Toronto
Toronto M5S 1A1 Canada
kristen.berg@utoronto.ca

Niels Ole Bernsen
Natural Interactive Systems Laboratory
University of Southern Denmark
Campusvej 55, DK-5230 Odense, Denmark
nob@nis.sdu.dk

Alberto Bianchi
Elsag Spa
Via Hamman, 102 I-53021
Abbadia San Salvatore
Siena, Italy
Alberto.Bianchi@Elsag.it

Jeffrey Boase
NetLab
Department of Sociology
University of Toronto
Toronto M5S 1A1 Canada
jeff.boase@utoronto.ca

Dr. Oscar de Bruijn
Centre for HCI design
School of Informatics
University of Manchester
O.De-bruijn@manchester.ac.uk

Juan Antonio Carrasco
Joint Program in Transportation
Dept of Civil Engineering
University of Toronto
Toronto M5S 1A1 Canada
carrasc@ecf.toronto.edu

John M. Carroll
Center for Human-Computer Interaction &
    School of Information Sciences &
    Technology
The Pennsylvania State University
University Park, PA 16802 USA
jmcarroll@psu.edu

Federico Casalegno, Ph.D.
MIT Media Lab
http://www.media.mit.edu/~federico
federico@media.mit.edu

Alan McCluskey
(writer - editor) Connected
    Magazine
Web: http://www.connected.org
alan@connected.org

Rochelle Coté
NetLab
Department of Sociology
University of Toronto
Toronto M5S 1A1 Canada
rochelle.cote@sasktel.net

Peter Day
CMIS
Faculty of Business & Information
    Sciences
Watts Building, University of
Brighton
Brighton & Hove BN2 4GJ
England BN2 4GJ UK
p.day@btinternet.com

Bernie Hogan
NetLab
Department of Sociology
University of Toronto
Toronto M5S 1A1 Canada
bernie.hogan@utoronto.ca

William Gaver
Interaction Design
Royal College of Art
London SW7 UK
w.gaver@rca.ac.uk

Antonietta Grasso
Xerox Research Centre Europe
6 chemin de Maupertuis
38240 Meylan, France
Antonietta.Grasso@xrce.xerox.com

Anu Kankainen
HIIT: Helsinki Institute for Information
    Technology>
Helsinki, Finland
anu.kankainan@ideanresearch.com

Andrea Kavanaugh
Center for Human Interaction
3160A Togerson Hall
Virginia Tech.
Blacksburg VA 24061
kavan@vt.edu

Jennifer Kayahara
NetLab
Department of Sociology
University of Toronto
Toronto M5S 1A1 Canada
jennifer.kayahara@utoronto.ca

Tracy Kennedy
NetLab
Department of Sociology
University of Toronto
Toronto M5S 1A1 Canada
tkennedy@netwomen.ca

Patrizia Marti
University of Siena,
Communication Science Department
Via dei Termini 6, 53100 Siena, Italy
marti@unisi.it

Jean-Claude Martin
Laboratoire d'Informatique et de
    Mecanique pour les
Sciences de l'Ingengenieur (LIMSI-CNRS).
Orsay, France
martin@limsi.fr

Alan McCluskey
(writer - editor) Connected Magazine
Web: http://www.connected.org
alan@connected.org

Elena Not
ITC-irst
Via Sommarive, 18
38050 Povo Trento Italy
not@itc.it

Mick O'Donnell
Escuela Politecnica Superior
Universidad Autonoma de Madrid
Calle Francisco Tomás y Valiente, 11
28049 – Madrid, Spain
micko@wagsoft.com

Michael O'Grady
Practice and Research in Intelligent
Systems & Media (PRISM) Laboratory
Department of Computer Science
University College Dublin, Ireland
michael.j.ogrady@ucd.ie

Gregory O'Hare
Practice and Research in Intelligent
Systems & Media (PRISM) Laboratory
Department of Computer Science
University College Dublin, Ireland
gregory.ohare@ucd.ie

Bill Pitkin
Research Director
United Way of Greater Los Angeles
523 West 6th Street
Los Angeles, CA 90014 USA
wpitkin@unitedwayla.org
http://www.unitedwayla.org

Patrick Purcell
Dept. Electrical & Electronic Engineering
Imperial College London
London SW7 2BT UK
purcell@imperial.ac.uk

Howard Rheingold
http://www.smartmobs.com
http://www.rheingold.com
howard@rheingold.com

Thomas Rist
University of Applied Sciences Augsburg
86161 Augsburg, Germany
tr@rz.fh-augsburg.de

Mary Beth Rosson
Center for Human-Computer Interaction
School of Information Sciences & Technology
Pennsylvania State University
University Park, PA 16802 USA
mrosson@psu.edu

Frederic Roulland
Xerox Research Centre Europe
6 chemin de Maupertuis
38240 Meylan, France
Frederic.Roulland@xrce.xerox.com

Douglas Schuler
The Evergreen State College
The Public Sphere Project (CPSR)
2202 N. 41st Street
Seattle, WA 98103 USA
douglas@cpsr.org

Reinhard Sefelin
CURE: Centre for Usability Research &
    Engineering
Hauffgasse 3-5, 1110 Vienna, Austria
sefelin@cure.at

Verena Seibert-Giller
CURE: Center for Usability Research &
    Engineering
Hauffgasse 3-5, 1110 Vienna,
    Austria
seibert-giller@cure.at

Dave Snowdon
Snowtiger Design
8 avenue de l'ilette
06600 Antibes, France
dave@snowtigerdesign.com

Robert Spence
Dept. Electrical & Electronic
    Engineering
Imperial College London
London SW7 2BT UK
.r.spence@imperial.ac.uk

Kostas Stathis
Distributed and Agent-based Systems
School of Informatics
City University London
London EC1V 0HB UK
kostas@soi.city.ac.uk

Phouc Tran
NetLab
Centre for Urban & Community Studies
University of Toronto
Toronto M5S 2G8 Canada
phuotran@chass

Manfred Tscheligi
CURE: Center for Usability Research &
    Engineering
Hauffgasse 3-5, 1110 Vienna, Austria
tscheligi@cure.at

Barry Wellman
Director, NetLab
Centre for Urban & Community Studies
University of Toronto
Toronto M5S 2G8 Canada
wellman@chass.utoronto.ca

Massimo Zancanaro
ITC-irst
Via Sommarive, 18 - 38050
Povo Trento
Italy
zancana@itc.it

Than Than Zin
Center for Human Interaction
3160A Togerson Hall
Virginia Tech.
Blacksburg VA 24061
tzin@vt.edu

Part A
Networks and Neighbours

Networked Neighbourhoods: The Purview

Patrick Purcell

1.1 Introduction

The background to this book is the pervasive networking of many aspects of civil society and the implications of this technological development for the neighbourhood and the collocated community residing therein. The *raison d'etre* of the book is the provision of a broadly based interdisciplinary forum in which some of the salient societal and technical issues flowing from this development are presented, discussed and critiqued.

This purview introduces and summarises the structure and the content of the book, together with the layout of its principal parts. The range of disciplines represented in the authorial list is broad, including sociology, ethnography, design, human factors, behavioural science and computer science. While author background is mainly research academic, some operate either in the corporate research area or as professional writers in the broad subject area based in Europe and North America.

The spectrum of values, opinions and technical insights represented in the following chapters is varied, for example where attitudes towards the dynamic of the networked society is concerned. The comparative views represented here extend on one hand from the reflective or the sceptical (even acerbic) to the unquestioning and the positive on the other.

The subject matter reported covers a broad spectrum of issues – sociological, political and technical. The topics appearing in the individual chapters range from "social capital" and the "public sphere" to "networked individualism" and "dominant cultural hegemonies".

An introduction to the successive parts of the book follows, namely Part B: "Connected Community", Part C: "Research Impetus" and Part D: "Mediated Human Communication".

This book brings together in its title the complementary triad of linked elements that together define the character of today's local social milieu, namely the physical neighbourhood, the local community residing therein, and, not least, the communication infrastructure that crucially facilitates communal organisation and operation.

In this milieu, a distinction may be drawn between neighbourhood and community. That offered by Barry Wellman seems both succinct and apt: "communities are about social relationships while neighbourhoods are about boundaries" (Wellman, 1999, p. xii).

The communication infrastructure, namely the "network", is increasingly used as a defining attribute of modern society, as expressed in the phrase "The Network Society". In the *Rise of the Network Society*, Manuel Castells asserts that "Networks constitute the new social morphology of societies" (Castells, 2000, p. 500). Similarly, in a climate of networked social structures, Peter Day in *Community Practice in the Networked Society* declares that such networks provide, inter alia, "a structural frame, within which, the practices of communities in contemporary society can be understood" (Day and Schuler, 2004, p. 3).

Also, in recognition of the strong synergy between contemporary community development and its underpinning digital communications, Barry Wellman comments that "computer-mediated communication has permitted complex social networks to become a dominant form of social organization" (Wellman, 2001, p. 228).

In the course of the book, the relationships between these three constituent elements, people, place and social informatics, are presented in a sequence of chapters, some as commentaries, others as research reports and featured case studies. The outcomes vary from trenchant critique to experimental findings and reports of innovative developments.

1.1.1 Social Informatics and the Online Community

Social Informatics is presently the focus of an active programme of research and development, a programme which currently is being reported in a substantial body of scientific and technical literature, a learned journal publication list and several ongoing conference programmes in the subject area.

Social Informatics may be defined as that "body of research and study that examines social aspects of computerisation, including the roles of information technology in social and organisational change and the ways that the social organisation of information technologies are influenced by social forces and social practices" (Indiana, 2005).

Current research and development in the field of social informatics is reported in several dedicated international conference venues, such as the Communities & Technology (C&T) biennial conferences and also the conferences of the Community Informatics Research Network (CIRN).

The editors of the *Communities & Technology 2005* (*C&T 2005*) proceedings comment on the importance of physically collocated communities as one of the basic forms of social organisation and coordination. "In this context communities provide the formulation for social practices, experience and social integration" (Van den Besselaar et al., 2005, p. ix).

In the case of the "Community Informatics Research Network" its remit "encompasses the social appropriation of information and communication technologies for local benefit, self determination and social inclusion in decision making" (CIRN, 2005). This CIRN remit is taken forward within the broad framework of a community-based approach to the implementation of systems of information and communication technology in this social application domain.

1.1.2 Community and Communities

In the context of this book, community and communities are according to Barry Wellman's definition "networks of interpersonal ties that provide sociability, support, information, a sense of belonging and social identity" (Wellman, 2001, p. 228).

The communities that feature in the following sections and chapters of this book are, in the main, physical and collocated rather than virtual and dispersed.

As a succession of case studies in the book these collocated communities are quite similar in the manner in which their members share physically adjacent lives. However, in the individual chapters in so many other respects, these featured communities are distinctively disparate in terms of their ethnicity, socio-economic levels, nationality, age and, not least, in the variety of their communal goals. These communities range from a sub-Saharan tribe to a leafy suburban community, from a remote mountain village to a close-knit urban ghetto.

1.1.3 Social Knowledge, Communal Memory and Civic Intelligence

Of the various aspects of the active community that contribute to the social capital of the neighbourhood, we find that social knowledge, communal memory and civic intelligence occur and recur in the chapters of this book.

A tribute to the significance of social knowledge and communal memory is paid by Pierre Levy in his major work *Collective Intelligence* when, in the course of discussing the social bond and its relationship to social knowledge, he asserts "through our relationship to others, mediated by processes of initiation and transmission, we bring knowledge to life" (Levy, 1997, p. 11).

Hitherto however, discussion and study of these basic aspects of community has been mainly the province of the historian, anthropologist or sociologist. From such a quarter in an interdisciplinary work *Social Memory* we find an eloquent acknowledgement of the importance of shared memory in the community:

> The memories which constitute our identity and provide the context for every thought and action are not only our own, but are learned, borrowed and inherited – in part, and part, of a common stock, constructed, sustained and transmitted by the families, communities and cultures to which we belong. (Moore, 1990)

For many of the contributing authors in the book, the attributes of community, such as social memory, communal knowledge and civic intelligence, have a core significance.

In the case studies presented in the following chapters, the reasons for creating a digital community information system were distinctly different in each case. For one village community the prospect of a system to provide a shared communal memory was the impetus to develop the local website. In the case of the "Living Memory" project (Chapter 12), the employment of advanced software agent technology was part of the rationale for the development of a system to support a living communal memory.

In a further case study, the core of the interactive community system was the development of a dynamic memory structure during community meetings that facilitated unscheduled ad hoc access by members of a widely dispersed community of remote users at different locations, at unscheduled times.

The concept of civic intelligence is the centrepiece in another chapter, presented as a nascent concept in community development, was based on the utilisation of evolving information technologies for the melioration of communal social problems.

1.1.4 Neighbourhood and the Networked Individual

As the title, subtitle and opening paragraphs indicate, the thrust of this book is concerned with communities, with neighbourhoods and, most of all, with the collective use of information and communication networks that support the functioning of both these entities. However, the following accounts of the formation of collective goals or the sharing of a communal memory should not mask the fact that much of the book's content is implicitly concerned with a prime issue in the current socio-technical "*zeitgeist*" – namely the networked individual and the degree to which the prevailing information and communication culture places a current emphasis on individualism.

On reflection, the position and significance of the individual networked user is increasingly acknowledged and proclaimed by commentators as "Networked Individualism". In the context of contemporary community

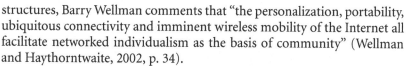

structures, Barry Wellman comments that "the personalization, portability, ubiquitous connectivity and imminent wireless mobility of the Internet all facilitate networked individualism as the basis of community" (Wellman and Haythorntwaite, 2002, p. 34).

In a somewhat broader context, the cumulative effect of changes in the contemporary culture and in current work experiences has, in the view of Manuel Castells, induced "the rise of individualism as a predominant pattern of behaviour" (Wellman and Haythorntwaite, 2002, p. xxx).

Moreover, Castells views this new focus on the individual in quite a positive light in that, for him, this individualism does not connote social isolation or even alienation. "It is a social pattern, it is a source of meaning, of meaning constructed around the projects and desires of the individual" (Wellman and Haythorntwaite, 2002).

Within the structure of this book's four parts and its list of individual chapters, there is a recurrent interest in and a current concern with the role of the individual user and especially the empowerment of the single user in a variety of social contexts employing multimodal interaction techniques and advanced software technology. While much of the research impetus is prompted by technical advance, in a number of instances, the technological imperative coexists with a philanthropic impulse. Associated research goals include progressive personalisation of the modes of person-to-person communication and computer-mediated communication generally.

In the different case studies reported in the following chapters, the individual user may be a remote islander, an urban ghetto youth, a school child or an isolated senior citizen.

1.2 Part B: Connected Community

Whereas later parts of the book report technical advance achieved by technologically inspired research, the chapters in this part are, in the main, by authors, with concerns in the area of information and communications technology and especially its ramifications in the community. The reflective mind set articulated in this section acts as a valuable counterpoise to the less questioning tone evinced by several of the chapters in the later parts.

The chapters in "Connected Community" discuss topics in the world of policy formulation for community development and also comment on the day-to-day experience of ongoing networked community practice. These authors tend to emphasise defined social benefit or citizen empowerment rather than technology advancement.

The networks referred to in the chapters of this section tend to be social networks rather than electronic networks. They represent a view where "community development" tends to take precedence over "community informatics". This part of the book "Connected Community" provides a forum where the knowledge which is acquired directly by personal contact in a familiar social setting is afforded a special value when compared with the kind of knowledge acquired from official institutional sources. Likewise, the discussion of intelligence will tend to be about the nascent concept of collectively held civic intelligence, rather than those applications in the technology of artificial intelligence, featured in later parts of the book.

1.2.1 Pathways Towards Civic Intelligence

The progressive convergence of universally accessible communication infrastructure with current digital media constitutes a powerful union of technologies which is transforming the character of social relationships. – So opines Peter Day and Doug Schuler in their chapter "Community Practice in the Network Society: Pathways Toward Civic Intelligence". They discuss the significance of these changes in social relationships from their long-standing experience in community policy formulation and community practice.

In this context they propose the emergence of "Civic Intelligence" as a nascent concept in community informatics, which putatively will play an increasingly important role in influencing social policy development and which will benefit from the appearance of social networks of enhanced awareness, scrutiny, advocacy and action.

1.2.2 Social Networks and the Nature of Communities

In "Social Networks and the Nature of Communities", Howard Rheingold reviews technology-mediated communication in the contexts of both personal and social relationships. Rheingold's measured reflections on these effects is informed by a pioneering association with this aspect of online society in the contexts of both physical, collocated communities and virtual communities.

Social Network Analysis provides Rheingold with a framework for discussing online socialising, which he appraises across a broad spectrum of strong social ties and weak social ties.

His extensive association with the field of online society also affords him the option to revisit and qualify his earlier more optimistic predictions of the beneficent effects of such technological mediation on the empowerment of the online citizen. He also revises his views on the potential of grass roots, networked society to significantly meliorate the current democratic process. A significant factor in the revision of his earlier optimism is his assessment of the increasingly baleful impact of the global hegemonic culture of powerful transnational communication and media conglomerates.

Rheingold offers his views, findings, revisions and critique against a broad context of sociological, technical, economic and cultural references from Karl Marx to Barry Wellman and Marshall Berman. His reflective, insightful, sometimes acerbic views compare favourably with some of the more unquestioning chapters expressed in the following sections of the book.

1.2.3 Community Informatics for Community Development

Next in this section "Connected Community" is the chapter "Community Informatics for Community Development: Hope or Hype?" by William Pitkin.

Pitkin's subject matter is the support technology of community informatics and the potential of this technology to contribute to community development. Pitkin is here specifically addressing the requirements of physically collocated communities of immediate neighbours.

In Pitkin's view "Community Development" can be defined broadly as a strategy to build local capacity and improve the quality of life in such physical communities. In his view, "Community Informatics" offers an approach to utilise today's information and communication technologies to further the goals of community development.

However, the promise of such benefit has to be constantly appraised against a balanced and questioning perspective. To propose a framework within which to conduct rigorous assessment, he offers a broad triadic structure where the axes of assessment are methodological, political and ideological.

His position is to maintain a constant critique of the employment of information and communication technologies, where greater social, political or economic equity is being claimed or sought.

1.2.4 Knowledge and the Community

This chapter is a reflective statement. "Knowledge and the Community" by Alan McCluskey. It presents a striking portrait of a Swiss village community, Saint Blaise, as well as an account of the origins and development of its website. The story is infused with the civic values and the philosophy of the author himself, who was, inter alia, the instigator of this community project.

The development of the website was embarked on as an experiment by the chapter's author, to record, to reveal, and primarily, to promote a wider sharing of the communal knowledge embedded in the community of Saint Blaise.

Knowledge, its acquisition and the various forms it may take in such a close-knit social milieu, has a special place in this chapter. In a collocated community, the significance and the value of knowledge gained directly from local, personal or social relationships is accorded a special value and meaning when compared with official or general institution-based knowledge. The importance of such direct experiential social knowledge in developing and sustaining the communal memory and the associated website is underlined.

In this chapter it is of interest to note the closely coupled relationship between the village Saint Blaise, the eponymous website serving this location and the author of this chapter, who effected the coupling of the physical location with its community system.

1.2.5 Connected Memories

A number of core themes relating to the social applications of information technology figure in this book. They include the related topics "social networks", "civic intelligence", "social knowledge" and "connected lives". Federico Casalegno adds "connected memories" to the core themes of the book, via his chapter "Connected Memories in a Networked Digital Era: A Moving Paradigm".

He puts forward a succession of emerging paradigms in a community context in which connected memories may be modelled, developed and disseminated using today's information and communication technologies. For this, he draws on a number of digital communal memory projects, of which he has had direct experience in both Europe and the USA.

1.2.6 Community and Communication: A Rounded Perspective

In the last chapter in this part of the book, "Community and Communication: A Rounded Perspective", Jennifer Kayahara looks at the interrelationship between community and communication, particularly in the context of the online community. In Kayahara's challenging premise, "online community" is a concept whose viability has to be investigated and rigorously validated. She carries out this investigation against a comprehensive and in-depth background of relevant contemporary and historical sources. This in-depth investigation leads to a review of how the various conceptualisations of community relate to the major arguments for and against the existence of online communities.

1.3 Part C: Research Impetus

Research is the linking theme in the following chapters of this part of the book. The research outcomes from the featured case studies take a variety of

forms, such as innovative devices, innovative applications in digital community systems or the creation of new insights into the role of social informatics in community development.

In the rapidly developing field of community-based social informatics, the three constituent areas of the domain of connected communities (namely "community practice" and "policy formulation for community development" together with "research in the domain of community practice") relate to each other as a closely coupled triad.

The research impetus reported in these chapters (and in the following chapters in Part D: "Mediated Human Communication") is driven, in the main, by sociological enquiry or alternately by the impetus of technologically focused innovation.

In a number of chapters the research aims are overtly philanthropic in character. In one case study, for example, the raison d'etre of the project was to help dispersed communities to communicate more effectively. In another case study, the project goal was to help members of disadvantaged communities to participate more effectively in the contemporary culture.

1.3.1 Connected Lives: The Project

"Connected Lives" by Barry Wellman and Bernie Hogan with their co-authors at the University of Toronto's NetLab group is an extensive, large-scale study of a community adjacent to Toronto. In this study of a collocated community, Wellman and Hogan et al. find that a subtle shift is taking place underneath the surface continuity of the social structure that constitutes the "network society". This recently perceived shift is moving progressively from group-centred structures and processes towards more individual-centred network structures and processes, giving rise to the phenomenon they term "networked individualism". Furthermore, they find that evidence in support of this phenomenon is provided by current developments in contemporary computing environments.

1.3.2 The Impact of the Internet

Technological impact on our social relations is the theme of the chapter "The Impact of the Internet on Local and Distant Social Ties" by Andrea Kavanaugh and her co-authors.

The chapter is the outcome of an empirical study of a networked community. The research goal was to determine just how local computer networking might affect the provision of different kinds of social support within various types of social ties (whether close ties, somewhat close ties or the ties of remote acquaintanceship, for example).

1.3.3 The Magic Lounge

An interesting feature of this research project was the challenge of meeting the communication needs of a distinctively different type of community with a broad range of complementary communication technologies. "Magic Lounge: Connecting Community Members Through Various Communication Services" by Thomas Rist and colleagues describe their response to just such a challenge.

Conventionally, we tend to think of physical communities as being tightly knit and local in nature, consisting of local schools, shops, sports clubs, etc. Less typical is the geographically dispersed community. A key factor in the setting up of the Magic Lounge project was the consideration of just such a dispersed community of people, living on a group of small, remote islands.

An important issue in the development of the Magic Lounge system was to investigate the feasibility of having virtual encounters on a future, reliable, high-bandwidth, low-cost network infrastructure which would allow combined spoken and text chat exchanges between multiple speakers, including spontaneous and random "drop in" members of this dispersed community.

This project also addressed the question of how emerging technologies may generate new, technically complex communication scenarios in which groups of non-professional users (both mobile and fixed location) may communicate using multiple modalities and a group of devices having very different operational capabilities.

1.3.4 The Digital Hug

"The Digital Hug: Enhancing Emotional Communication with Creative Scenarios" by Verena Seibert-Giller and colleagues is an account of the outcome of research carried out by an international multidisciplinary team known as the Maypole project consortium, which included computer scientists, designers, human factors experts, etc.

The special focus of the project team was to explore the informal patterns of communication that exist in small social groups, especially those that exist between children and in addition those patterns of communication that exist between children and other members of their families. From the outcome of such research, the Maypole research team went on to develop innovative devices which could serve or even enrich such patterns of communication between children and between children and adults.

The human factor element was very strong in the Maypole project, an impetus that was maintained throughout the course of the project, extending from the initial research effort, which sought to capture the children's patterns of behaviour and social communication, right through the various

successive stages of this project, to the final user testing of the resultant devices, prototypes and mock-ups used by the children.

1.3.5 Ambient Intelligence

The final chapter of the "Research Impetus" part of the book, "Ambient Intelligence: Human–Agent Interactions in a Networked Community" by Kostas Stathis and colleagues, studies the embedding of a networked infrastructure in the physical space of a local neighbourhood community. The prime objective was to employ this software technology in support of local social activities and the social interactions of the residents.

This infrastructure provides an environment for an interactive and distributed system whose components (acting the role of software agents) access the networked infrastructure on behalf of their individual human proprietors in the community.

In this context, human–system interactions are designed such that the electronic devices that people use (including any personal software agents that may be operating in these devices) are interconnected seamlessly. The outcome is that the physical environment of the community (local cafes, libraries, shopping malls, etc.) acquires an aura of ambient intelligence that maintains a high degree of connectivity with the local community and also acts as an infrastructure of connectivity between the individual members of the community.

1.4 Part D: Mediated Human Communication

The chapters which form this final part of the book, "Mediated Human Communication", describe a range of techniques and devices which aim to augment either human–human interaction or human–environment interaction.

1.4.1 Beyond Communication

"Humans have a fundamental need for contact with other humans. Our interactions and relations with other people form a network that supports us, makes our lives meaningful and ultimately enables us to survive". So writes Stefan Agamanolis in the opening statement to his chapter "Beyond Communication: Human Connectedness as a Research Agenda".

The research described in this chapter centres on human relationships and how such relationships may be creatively mediated by various forms of information and communication technology.

The collective research mission was "to conceive a new genre of technologies and experiences" that will meliorate the effects of imposed distance amongst members of various groups, from families to broadly based social communities generally.

Discussion on the outcomes of this research programme is collected into a number of major subthemes in the chapter under the distinctive headings "Extended Family Rooms", "Intimate Interactive Spaces", "Socially Transforming Interfaces" and "Minimization of Mediation".

1.4.2 Presence

The prevailing goal of the Presence research project is reflected in the title of this chapter "Presence: Helping Older People Engage with Their Local Communities" by William Gaver and Jacob Beaver.

The sites of the Presence project were three distinctively different localities across Europe. In each of these localities, the goal of the project was the same, namely to seek ways to establish and strengthen the social presence of the older people in their respective local communities.

The Presence project was distinctive in that its research strategy was design-inspired and also that its research impetus reflected its art campus background, in which creative, subjective and qualitative issues have a special interest and validity.

The individual communities of senior citizens that the Presence team worked amongst were quite varied in terms of location, culture, social class, ethnic background and nationality. Their single common factor was their age.

Overall, the chapter delivers a positive set of findings in terms of the strengthened engagement of the older people in their chosen communities and also in terms of the robustness of the dialogue between the researchers and the people on whose behalf the Presence project was conceived, funded and launched, namely the increasing ranks of senior citizens in modern society.

1.4.3 Enhancing Community Communication

The study reported in the chapter "Enhancing Community Communication: The Role of Interactive Displays in Comparable Settings" by Antonietta Grasso and colleagues had as its primary focus the design and functionality of community information display systems, across several operational settings.

Effective communication is a fundamental prerequisite of the sustainable community. One may point to the fact that both entities "communication" and "community" share the same etymological source. In this

project, large-scale interactive screen displays were proposed and evaluated as additional means to promote communication and information exchange amongst communities, either in public open spaces or at work sites.

Grasso and her co-authors have been involved in the design and deployment of two such systems. The first system was a multimodal interaction system conceived to promote information exchange amongst the residents of cultural heritage urban centres.

The second system was conceived and developed to exploit the earlier system's advanced interaction features in different types of community structures, such as distributed locations in large work organisations.

The authors first present the outlines of a general architecture and set of functionalities for these two comparable systems. On the basis of these common features, several different system specifications are identified and are presented for the two settings, namely the public social precinct and the distributed work locations setting.

1.4.4 Serving Visitor Communities

The final chapter in this section on mediated human communication is "Serving Visitor Communities: A Mediated Experience of the Arts", Patrizia Marti and colleagues.

Here is a distinctly different community type, a community which is essentially transient, constantly moving, continuously changing its attention focus, whose individual members are infinitely variable in terms of their prior knowledge of the complex cultural contents of their immediate physical environment, namely the museum visitor community.

The challenge for the research project, reported in this chapter, was to create a system to meet the informational needs of the ebbing and flowing activity of the ever-changing members of this dynamic community who share for just a brief period a single common interest, namely a response to the cultural content in their immediate vicinity.

An associated technological interest for the project team was the judicious and selective harvesting of state-of-the-art technologies which offered evidence of significant opportunities for enhancing the museum visitor experience.

A positive aspect of the task for the research team was the fact that museums, due to their confined nature, facilitate the close observation of visitors from which heuristics and principles could be identified for the subsequent refinement and personalisation of appropriate multimedia viewing services. Furthermore, the use of electronic positioning systems enabled the capture and interpretation of the individual visitor's physical position and orientation, thus facilitating further refinement of the user-oriented service that resulted.

The development of the resultant multimedia system offered a range of visitor friendly features. Presentation is self-paced, and the visitor can browse through the museum at will.

References

Castells, M. (2000) *The Rise of the Network Society*. Blackwell, Oxford.

CIRN (2005) <http://www.CIRN2005.org/>

Day, P. and Schuler, D. (2004) *Community Practice in the Network Society*. Routledge, London.

Indiana (2005) <http://www.slis.indiana.edu/SI/concepts.html>

Levy, P. (1997) *Collective Intelligence*. Perseus Books, Cambridge, MA, p. 11.

Moore, R.I. (1990) In Fortress, J. and Wickham, C. (eds), *Social Memory*. Blackwell, Oxford, UK, p. viii.

Turow, J. and Kavanaugh, A. (2003) *The Wired Homestead*. MIT Press, MA.

Wellman, B. (1999) In Wellman, B. (ed.), *Networks in the Global Village: Life in Contemporary Communities*. Westview Press, Boulder, CO, pp. xi–xxiv.

Wellman, B. (2001) Physical space and cyberplace: the rise of personalized networking. *International Journal of Urban and Regional Research*, **25**(2), 227–252.

Wellman, B. and Haythorntwaite, C. (2002) *The Internet in Everyday Life*. Blackwell, Oxford, UK.

Part B
Connected Community

No Man is an Island

No man is an island, entire of itself

every man is a piece of the continent, a part of the main

if a clod be washed away by the sea,

Europe is the less, as well as if a promontory were,

as well as if a manor of thy friends or of thine own were

any man's death diminishes me, because I am involved in mankind

and therefore never send to know for whom the bell tolls

it tolls for thee.

-- John Donne
(1571 - 1631)

Community Practice in the Network Society: Pathways Toward Civic Intelligence

Peter Day and Douglas Schuler

2.1 Introduction

> Climbing upwards in this way, one would reach a fork where two streams joined, and a choice had to be made. No reliable information could be obtained from the map, and no general overview was possible to guide the choice, which was based only on what could be seen within a few yards, or on any general predisposition to go towards the right or the left...Having climbed high up the side of the valley, one would pause and camp for the night...Then it was possible to feel a sense of achievement: to have climbed so high and to be able to look back over the lower country out of which one had come. And it was easy to believe that all the choices, which had been made along the way, were justified by the outcome, and were the only right choices to be made. This self-congratulation might have of course been quite unwarranted. Some other route might have led to still higher ground, and done so more easily. But if so, the knowledge was hidden, and the complacency uncontradicted. (Rosenbrock, 1990, pp. 123–124)

Although a decade and a half after Rosenbrock (1990) – drawing on his experiences in the Lushai Hills in India – first introduced this metaphor in his seminal defence of the utilisation of alternative technological systems in manufacturing industry, its central theme is as germane to the development of information and communication systems and applications in the network society as it was then. Embedded in the richness of social and cultural diversity that represents human life, newly formed networks of mutual collaboration and cooperation are emerging, which, as in the Rosenbrock metaphor, point to alternative network society pathways. Central to the purpose of such networks is the social appropriation of information and communication technology (ICT) within civil society. The increasing communication and collaboration between social movements, civil society and community networks in the information age does, we believe, possess the potential for an emerging counter-culture to the hegemony of the "space of flows" (Castells, 1996). Still very much in embryonic form these social network phenomena seek to use information and communication effectively in order to sustain communicative planning, development and activities in civil society.

The diversity of human knowledge and cultures that exist illustrates the potential for other technological pathways to be travelled, and other social landscapes to be experienced. No matter how far along the trail of technological development the knowledge society has travelled, the possibility of other routes of development must never be overlooked. Pathways that might follow other scientific and technological directions – that might enrich the human condition in deeper, more meaningful ways than we currently experience – should not be disregarded.

Evidence from around the world would appear to suggest a growing social reaction to the sanitised, top-down network society promulgated by politicians, transnational corporations and financial institutions. The dominant perspective, the dominant voice and the dominant agenda in the network society is represented by what Castells (1996) calls the "space of flows". But what is the dominant perspective, to whom do the dominant voices belong and what does the dominant agenda in this "space of flows" represent? These are questions central to understanding the nature and form of the network society as it exists today, especially as many academics, industrialists and policy makers insist that we as citizens have no alternative but to accept accommodate, and where necessary change the way we live – regardless of the social consequences (Day, 2001).

2.2 Network Society Hegemony

[D]ominant functions are organized in networks pertaining to a space of flows that links them up around the world, while fragmenting subordinate functions, and people, in the multiple space of places, made of locales increasingly segregated and disconnected from each other. (Castells, 1996, p. 476)

Castells' hypothesis provides a useful framework for describing and understanding macro level network society structures and organisation. The argument that interests of ordinary people are often subordinated to functions acting and interacting in a globalised space of flows is well understood. However, the argument that locales, i.e. communities or neighbourhoods, are increasingly segregated and disconnected is a little more problematic. Although it provides a fairly neat explanation of the situation facing many 21st century localities, it fails to take cognizance of the emerging utilisation of network technologies by many communities across the world. While still in embryonic form, community ICT initiatives are increasingly designed from the bottom-up using participatory approaches in such a way that the resultant processes and products of community dialogue and community technology contribute to community building activities in an empowering way.

Indeed, such is the growing significance of civil society to contemporary network society policy development that developing an understanding of the diversity of civil society agenda and the dialogic and networking activities, structures, organisations and processes that connect them should be an imperative for policy and academia alike.

So whilst the application of Castells' hypothesis does not necessarily assist us in developing an understanding of the functions and structures of the network society at the micro level,[1] it does highlight the socially exclusive nature of the dominant worldview and provides socio-economic context to the changing and challenging environments in which communities and neighbourhoods often struggle to exist. As MacKay illustrates, the concept of "[t]he network of flows is crucial to domination and change in society: interconnected, global, capitalist networks organize economic activity using technology and information, and are the main source of power in society" (MacKay, 2001, p. 35).

In telling his stories of the network society, Castells (1996) argues that society cannot be understood or represented without focusing on its technological tools. In the "space of flows" it is the communication technology that provides linkages and organization for dominant social functions, e.g. "flows of capital, flows of information, flows of technology, flows of organizational interaction, flows of images, sounds, and symbols" (p. 412). It should be noted though that this is the result of human agency and not a consequence of neutral evolution. The dominant functions in the "space of flows" represent the vested interests of, and are driven by, the techno-economic agenda of multinational companies and global financial institutions.

Interestingly, whereas Castells (1996) argues that locales are increasingly disconnected and segregated from one another in the "space of flows", increasing numbers of locales are utilising network technologies to forge links within and between communities. The rapacious and homogenising nature of attempts to subordinate individuals, social groups, neighbourhoods and communities to the dominant functions of globalisation and the interests of those it represents, regardless of social, cultural and environmental consequences, might, paradoxically, be sowing the seeds of its own destruction.

The remoteness and undemocratic nature of decision-making processes within the "space of flows" mean that whilst this space acts as a means for preserving positions of privilege and power for "the few", it is also becoming progressively isolated from the very structures that give meaning to the social reality of everyday life, e.g. neighbourhood and community. A consequence of this is that it risks becoming, despite the power and influence in its grasp, an irrelevance to the thinking of ordinary people as they seek to shape their own destinies and retake control of democracy. The potential

---

[1] It should be acknowledged that Castells (1997) later acknowledges the increasing civil society uses of ICT.

risk of social tensions should not be overstated but neither should it be underestimated. Already we are witnessing neighbourhoods, communities and a range of social movements, taking the first wobbly steps – through the social appropriation of ICT – along the path of developing an alternative "space of flows", a space relevant and meaningful to the everyday lives of ordinary people (Day and Schuler, 2004; Schuler and Day, 2004).

2.3 Information and Network Society: Pathways of Determinism and Partnership

In order to understand the processes at work within civil society and the actions being taken, it is necessary to acknowledge the hegemonic nature and characteristics of the network society in its current manifestation. We need to comprehend whose interests have been represented by the socio-technical developments of recent decades and whose voices have been heard in the policy-making circles that enable such developments to take place. Only then can the exclusionary nature of the current situation be expressed and the social potential of alternative pathways understood.

In order to achieve this we turn our attention momentarily to seminal moments in the sphere of recent information and communication development, starting with the emergence of the hypothesis that the world in which we live has undergone a transformation from an industrial society to post-industrial or information society. Expounded by social-forecasters such as Daniel Bell during the 1970s, notions of an information society first began to permeate the thinking of Western policy makers during the 1980s (Garnham, 1994). However, Duff et al. (1996) locate the advent of information society philosophy in Japan during the 1960s. Whether we take the philosophical or policy origins as our starting point, there is little doubt that the concept has become imbued with an almost self-fulfilling social significance during the last decade or so.

2.3.1 Determinism

This increase in significance is often attributed to the National Information Infrastructure (NII) vision of the former US Vice President Al Gore during the early 1990s (Gore, 1991; Information Infrastructure Task Force [IITF], 1993). Indeed, the Clinton/Gore administration is frequently credited with being the catalyst for contemporary global information society developments, although analysis of policy documents for this period reveals parallels in international ICT application, legislative, regulatory and policy development (Federal Trust, 1995; Niebel, 1997; Day, 2001). In fact, during this time many countries around the world adopted information society

policies with the aim of achieving economic growth and prosperity through competitive production and use of ICT by commercial interests.

Wherever responsibility for providing the initial policy impetus for these developments rests, the point is that these stimuli have precipitated major techno-economic changes worldwide. The creation of formal policy frameworks, established to steer the development of information societies through the construction of information infrastructures, is a major social development and illustrates the technologically determinist nature of policy development during this period. As Moore notes, "There can be few other examples of technological change stimulating formal policy creation in order to bring about social change" (Moore, 2000, p. 2).

The level of technological determinism at this time is illustrated by a policy action plan of the European Union in the mid-1990s. The plan provides us with a classic example of how those on the inside of the policy-making process sought to convince those excluded from it that the socio-technical developments – often promoted by politicians and bureaucrats with little or no knowledge of ICT – were inevitable and that no alternatives existed.

> The "information society" is on its way. A "digital revolution" is triggering structural change comparable to last century's industrial revolution with the corresponding high economic stakes. The process cannot be stopped and will lead eventually to a knowledge-based economy. (CEC, 1994a)

2.3.2 Public/Private Partnerships

What is interesting about this period of national and global policy development is that it was conducted almost entirely at the level of public sector/private enterprise collaboration. The purpose of this collaboration was to stimulate economic growth through the creation of markets for ICT-related goods and services (IITF, 1993; CEC, 1994b). It is worth noting, however, that whilst the intention was to stimulate economic recovery during a period of recession, the roles and benefits of/to the respective partners together with the degree of social power associated with these roles varied greatly.

Government's role in these partnerships was one of facilitation. Their purpose was to enable the development of the "information society" (a type of "code name" for information and communication *technology* – not society) by stimulating a welcoming and amenable policy climate. Through a range of advocacy activities governments promoted the use of ICT, usually quite uncritically, wherever it could. Leading by example, all levels of government, often with little planning or understanding, adopted and implemented ICT plans across its functions and services – actions extended today by e-government. Meanwhile, the actual development of an "information society" was left in the hands of private enterprise and market forces. Although some might justifiably suggest that the private sector is hardly a renowned guardian of public interest when private profit is at stake, this is precisely what happened, as acknowledged in the Bangemann Report:

> The Group believes the creation of the "information society" in Europe should be entrusted to the private sector and to market forces. (CEC, 1994b)

Since these early policy developments, there have been a number of voices calling for a more inclusive approach to policy development. It has been reasoned that global information society policy development requires effective tripartite partnership between government, private enterprise and civil society if they were to succeed (D'Orville, 1999). However, despite such calls, and with the exception of initiatives such as the World Summit on the Information Society, the majority of policy development and even more implementation is done to the exclusion of civil society.

2.3.3 The Power and Influence of Transnational Corporations

Policy makers continue to view information as the key source of competitiveness, productivity, wealth, employment and power in society; indeed it was this that gave rise to the term "information society" in the first place. However, the emergence of networks as a means of communications and as forms of societal structure and organisation has led to the increased adoption of the term "network society" to describe current socio-economic developments. Although often conflated, the two terms have related but distinct meanings. Conceptually, an "information society" is said to exist when information has become the key resource for production and the delivery of services, whereas a "network society" exists when the main social structures and modes of organisation are shaped by social and media networks.

Putting to one side any theoretical difficulties with these paraphrased definitions, it is interesting to note that the social networks with arguably the most influence in shaping modern society are the transnational corporations responsible for the development and use of ICT networks. It is this manifestation of the network society, in which information is a key source of wealth and power, owned and controlled by a miniscule percentage of the world's population that increasingly represents the globalised world in which we live (Boyd-Barrett, 2004).

As Barber illustrates, ICTs have become inextricably associated with processes of globalisation. Indeed, an argument can be made that ICTs are prerequisites of globalisation.

> In the unfettered high-tech global market, crucial democratic values become relics. Indeed, because globalisation is correctly associated with new telecommunications technologies, the globalized and privatized information economy is constructed as an inevitable concomitant of post-sovereign, postmodern society. (Barber, 2002, p. 6)

However, it is not simply the external dangers of the global marketplace that challenges national sovereignty and self-determination. As witnessed in the Bangemann Report, governments and public sector organisations have been promoting a shift in power from public to private sector through an ongoing process of public rationalisation, since the Thatcher (UK) and Reagan (USA) governments of the 1980s. It is through these processes of rationalisation and the growing bureaucratic apparatus and impenetrable policy-making authority of the G8 nations that the threat to the cornerstone of our democratic values, our communities and our neighbourhoods can be found.

There can be little doubt that whilst global markets of exchange have become the "space of flows", many locales have been forced to endure significant socio-cultural consequences. The social effects of these momentous global events have led some to herald the rise of "networked individualism" (Wellman, 2002, p. 10). Conceptually, "networked individualism" provides interesting insights into both the unquestionable socio-economic decline suffered by many localities and the dispersed environments often created by city development. It does not, however, account for the diverse range of community building, organising, developing and action that occur in, and increasingly between, communities around the world.

We believe that declarations of the end of community are unfounded. Whilst communities have been, and continue to be, challenged by the rapidly changing environments of many locales, those who claim an end to community are ignoring the enormous social value that many place on "community" as a normative social construct – in whatever form it emerges. Ignoring the wealth of collective, collaborative and cooperative actions – often interwoven with sprinklings of contested agenda, conflict and competition – emerging globally under the collective banner of civil society is, in our view, an error. Across the globe, despite significant barriers, people continue to build community – apparently they have not received the news from these pundits that community is dead! Our message on this subject is loud and clear – community exists and it matters! Understanding the developments currently taking place in civil society, and the potential challenge they pose to network society hegemony, requires a thorough consideration of the contexts in which, and for which, they exist.

2.4 Networks of Awareness, Advocacy and Action

In order to assist such considerations we frame our deliberations in the context of social network theory (Wellman, 1999). This is not only because we seek to understand these social phenomena within the broader context of the network society but because we are also investigating the conditions and communicative relationships that surround these social environments. In pursuing such an investigation we are mindful that issues of agency and political opportunities are central to understanding the evolution of social relationships. The assumption that global civil society activities and social movements will emerge automatically out of either economic globalisation or revolutions in communications technologies ignores both.

It might be too early to declare categorically that attempts at the social appropriation of technology (Surman and O'Reilly, 2003) in communities and civil society represent a transnational civil society social movement. We do believe however that they indicate an emerging grass-roots potential and willingness for social change in the network society. Understanding the way that initiatives, groups and networks emerge and are legitimized is central to understanding both the politics of civil society and successful social networking. If, as Keck and Sikkink (1998) suggest, transnational civil society relies on "dense exchanges of information and services" then the development of such an understanding is significant in cultivating knowledge of such practices in the global network society.

The dense exchanges to which Keck and Sikkink refer usually occur during the interaction of civil society groups, institutions and governments. The extent to which civil society can identify and influence targets vulnerable to material and moral leverage, in order to shape the future direction of the policy discourse, remains to be seen. Much will depend on their density, their strength and the number and size of organisations involved in civil society networks. However, network effectiveness is also dependant on the regularity and "quality" of their exchanges (Keck and Sikkink, 1998). Communicative interaction is crucial to successful networking. Similarly, the development of social network analysis (Wasserman et al., 1994; Degenne and Forse, 1999) as both participatory action research method and development tool for civil society will be of interest to the future of effective civil society networking – locally, nationally, regionally and globally.

Whilst network theory provides a framework for change through which the preferences and identities of actors engaged in social activities can be mutually transformed by interaction with others, qualitative analysis of the information flows and communication behaviour and patterns in these social networks will assist in developing knowledge of how communicative interaction can be improved and as a consequence how civil society can network more effectively.

Keck and Sikkink contest that the voluntary and horizontal nature of networks means that the motivational force for actor participation is an anticipated mutuality of learning, respect and benefits. Networking therefore has the potential to provide civil society with both a vehicle for communicative and political exchange and the potential for mutual transformation of participants.

However, not all social networking is conducted in a spirit of mutuality and reciprocity. Civil society is more often than not forced to network in conditions of power imbalance. Public and commercial sectors exercise influence that can affect the ability of civil society to shape social developments and capacity. This is especially true in the network society, where the power exercised in networks often follows from the resources that public and commercial sectors own and control.

2.5 Understanding Community: Policy and Practice

Before we can consider the practices and activities of community networks that result from the collective cognitive processes we call civic intelligence, we need to reach a common ground on what we mean by the concept of community. Of course this is easier said than done. As social constructs communities are represented by a melange of organisations, agencies, groups, networks, individuals, activities and cultures, and as such are often contested spaces that depend on the subjective and emotional loyalties of community members in order to be sustainable.

Communities vary across space and are never the same in any two locations – although, of course, similarities exist. The diversity of their composition makes classifying their characteristics, i.e. those traits that make them a community, almost impossible. They are not like organisational structures – the boundaries of which can be identified, quantified and measured – communities are messy, hard to pin down and problematic. Understanding them as social constructs requires being able to manage the dichotomous tensions between people working collaboratively and cooperatively towards common goals on the one hand and the conflict that can arise from competing values and agenda on the other. It is this level of complexity that probably explains why so many researchers studying community give up in frustration (Jewkes and Murcott, 1996).

The point about working with or seeking to understand community is that as academic researchers you must learn to leave your training in objective investigation at the door (Stoecker, 2005) – Something many find impossible to do. Interestingly though, understanding, and consequently working with, communities is no less complicated for policy makers than it is for academics. Pointing to government tendencies to use large geographical boundaries, such as city/town or electoral ward boundaries to describe community, the UK Community Development Foundation (CDF) highlight tensions between "wide and narrow definitions of community" (Chanan et al., 2000, p. 4). CDF contest that smaller areas and more localised groupings, such as neighbourhoods, provide more natural and instinctive points of reference for people. The sense of belonging that comes from being able to identify with a locality is a crucial part of being part of a community.

Hopefully, we have shown that attempting to develop hard and fast definitions of community is a complex sociological task. We do not suggest that it is entirely impossible but the literature is scattered with hundreds of such attempts and we prefer to take a different direction to reaching common ground understanding. To assist us in this task we employ three distinct but interrelated "senses of community" (Butcher, 1993). Each "sense" is broad enough and flexible enough to accommodate the subjectivity of most interpretations.

2.5.1 Sense 1: Descriptive Community

Providing our first sense of community is drawn from what Butcher calls "descriptive community". This draws on the word's etymological origins of having "something in common". The "something in common" might be location, such as neighbourhood, village, town, but it might just as easily be ideas, interest, practice or purpose – mutual activities, ethnicity, religion, sexual orientation provide examples of context. Of course, this distinction between the two groupings of community might provide us with the embryo of a conceptual taxonomy but they should not be construed as being mutually exclusive.

Indeed, geographic communities are made up of different cultures and it is not uncommon for groups and individuals to share knowledge and draw from each other's experiences, creating new forms of common interests as a consequence (Warburton, 1998). Although plagued with massive "environmental changes" (Schuler, 2007), internal disputes and occasional bouts of apathy, the development of the Seattle Community Network is an excellent example of how diversity of culture, value and belief systems can, through the synthesis of collective activities and communal communication, learn from and contribute to each other in a manner the product of which is greater than the sum of its parts (Schuler, 1996). Of course, the contested nature of community means that conflict can and does arise.

2.5.2 Sense 2: Community Values

However, healthy and sustainable communities learn from this complexity, understanding that conflicts will, from time to time, arise and that this need not be an unhealthy experience. By developing "community values" that seek to accommodate the diversity that might give rise to social tensions, solutions that respect differences can be found – although this, at times, is by no means easy. The idea here that healthy communities require certain shared values – solidarity, participation and coherence – provides the second sense of community. Of course, such values are open to interpretation but the principles upon which they are established should provide the value base of community initiatives and policies. Solidarity encourages friendliness, builds allegiances and inspires loyalty through mutual support and collaboration in relationships. Participation enables citizens to contribute to, engage in and shape the aspirations and activities of collective community life. Coherence connects individuals to the community, helping them to understand themselves, their social environments and their roles in these environments, whilst developing a communal knowledge base.

In a discussion of community networking, Schuler (1996) represents community core values in the form of a network. He highlights how each value influences and is related to others and that a weakness in any one value is a weakness in the whole. He argues that healthy communities require strong core values, which should be embedded in the planning and development of all community activities and services. However, it is important to understand that community values must also respect and be able to accommodate and value the constituent parts of the community – the communal therefore should not subsume the individual. To build and sustain healthy communities through shared *community values*, a balance is required between adequate amounts of privacy, autonomy and localism.

Shared public spaces, community associations and activities that provide the opportunity to engage with one another need to be tempered with spaces offering both privacy and respect for the diversity of cultural principles and values. The potentially contested nature of such diverse social environment, cultures and belief systems in a network society requires community members to respect and celebrate the social richness of community life if they are to coexist in the same geographical space and share social experiences.

Community values are the social product of individual and groups of citizens living in and identifying with a specific "something", often but not always a geographical space. The collective community then comprises individual community members that have developed an inherent interest in each other and in collaborating to achieve some form of common goals. In healthy communities, as well as having the ability to share the same geographical space and social experiences community members might, at times, choose not to. It is in learning to respect and celebrate the richness and diversity of human interests, needs and goals that healthy community exists. Diversity then has the potential to distinguish the individual, or group, from the community whilst at the same time possessing the potential to contribute to the richness and strength of that community life.

2.5.3 Sense 3: Active Community

It is through this sense of belonging to, and identifying with, a geographic community that people engage in community activities. The "active community" refers to collective action by community members embracing one or more communal values. In Poets Corner, a community that we are working with in Brighton and Hove in the UK, community activities are focusing on the planning of the annual festival family fun-day in Stoneham Park and the neighbouring TalkShop centre. In a multicultural and socially diverse geographic area, almost every community group in the vicinity is collaborating to ensure that the festival is a success. Such activities are normally undertaken purposively through the vehicle of groups, networks and organisations – that constitute a community's social capital – social structures that are significant to any discussion of community. Of course, as in Poets Corner, whilst the products of such community activities are important to community life, it is the dialogic processes of engaging in the planning and collaboration that builds and sustains healthy community. Community and voluntary sector groups and organisations form the bedrock of community life through the planning, organisation, provision and support of community activities and services. Although usually under resourced and overstretched, the community and voluntary sector play a significant role in building and sustaining community.

2.6 The Significance of Community Policy

Although the cornerstone of community life, the daily pressures for survival on community and voluntary sector groups often mean that enabling *active community* is a major task, and a shared value base between citizens, local civil society and community policy makers is important. However, distrust of bureaucrats and politicians often means that achieving such a shared value base can be problematic, especially between citizens/community groups and the mechanisms of local governance. Nonetheless, in the same way that community policy can create environments that block the development of healthy communities, so can it create circumstances to assist their development. By understanding what "community" means to local people at local level, it should be possible to develop policies that are meaningful and germane to people in those communities.

Although many different forms of *active community* exist – especially in more affluent areas where local resources, knowledge and expertise often exist in abundance – it should be remembered that not all communities possess the wherewithal required to facilitate, support or sustain community engagement. Marginalised or socially excluded peoples often require more direct involvement from community policy mechanisms than where healthy community already exists. Despite the need for support, community policy mechanisms should be predicated on the notion that community practices, i.e. services, activities, functions and processes need to be embedded in the aspirations, needs and culture of the people involved. Ownership and identity then are crucial elements in building healthy community. Establishing and maintaining these relationships are not accidental by-products. Dedicated engagement and dialogue between community and policy makers must focus on the development of a policy framework that supports the needs and practices of community life. Prerequisite to this are to (1) understand and meet community needs; (2) work in partnership with active community groups and organisations; (3) be based on one or more community value, i.e. solidarity, participation and coherence; (4) prioritize the needs of the community's socially excluded, marginalized, disadvantaged and oppressed; (5) valorise and celebrate cultural diversity; and (6) reflect a commitment to the objectives of community autonomy and responsibility for community initiatives.

Of course, all this is based on the normative assumptions that community is a desirable social goal, and that a function of policy is to facilitate community building, renewal and sustainability. Where healthy communities exist, the role of policy should be to support and sustain their existence and when appropriate, facilitate community renewal. Where community is being eroded or does not exist, policy should encourage community building. By this we mean policy should facilitate: capacity building, community activity and community involvement. Building capacity means stimulating the processes through which communities acquire and hone the skills to manage and develop an environment of community. Capacity building can relate to skills, knowledge and expertise among individual community members but can also apply to developing, supporting and sustaining organisational skills, knowledge and expertise in community groups, networks and institutions. Community organisations such as these – often referred to as a community's social capital – are the main driving forces in planning, organising, providing and supporting community activities and services, and in developing and nurturing community values. Community involvement is the achievement of broader participation in the activities and processes of community life and is a prerequisite to achieving healthy communities.

2.7 Community Practice

We have argued that the development of community policies is dependent on building, sustaining and renewing healthy active communities. We contend that the implementation of community policies requires changes in the mindsets of those involved in community governance – both policy makers and bureaucrats. Achieving such changes requires new and distinctive methods and techniques (Glen, 1993). These processes are known as *community practice*, which is distinct from but related to community practices. The latter relates to specific community activities and services, whereas community practice comprises the tools and techniques that support community practices. It is a method for promoting policies that encourage the planning, building and sustainability of healthy communities and usually involves some or all of the following components:

1) The sustained involvement of paid community workers; 2) A broad range of professionals who are increasingly using community work methods in their work; 3) The efforts of self-managed community groups themselves, and 4) Managerial

> attempts at reviving, restructuring and relocating services to encourage community access and involvement in the planning and delivery of services. (Glen, 1993, p. 22)

Describing the symbiotic relationship between community practice and community policies, where each is related to and promotes the other, Glen identifies three community practice approaches: (1) community services approach, (2) community development and (3) community action.

The community service approach focuses on the development of community-oriented organisations and services. It encompasses both philanthropic and compulsory forms of assistance to people in need and as such takes place in the realms of both statutory and voluntary services. Community services are usually about doing "things" to and/or for people in need whereas the other two approaches operate at grass-roots level. Community development (or community organising as it is known in the USA) concerns itself with the empowerment of communities to define and meet their own needs. It focuses on the promotion of community self-help but often, although not always, requires some degree of organisational support from outside the community. Community action emerges from community-directed planning and mobilisation. It involves campaigning for community interests and community policies in order to achieve goals set by the community themselves. Of course, this implies the occasional employment of conflict tactics in the community interest. Increasingly, community action also implies communicating with other communities. In fact it is our contention that communities adopting an introspective and insular approach to their actions often do so at their own peril.

It is important to note that due to the wide range of agencies, organisations, groups and partnerships involved in community practice, approaches can be "top-down", i.e. promoted and/or provided by local authorities, charities and voluntary bodies in a "doing to" manner. Or they may emanate from within local communities, i.e. "bottom-up" in a "being done by" manner. Usually, top-down facilities and initiatives tend to be associated with the community services approach. As community practices move toward a more action-oriented approach, they tend to adopt a more bottom-up attitude.

No matter what the composition of local partnerships or the complexion of the approach being employed, community practice should be viewed as a framework of three interrelated elements that assist in identifying, understanding and fulfilling community need. Within a network society context, community practice requires the subordination of ICT systems, artefacts and services to meeting those needs as a crucial contribution to building healthy, empowered and active community.

The acknowledgement of the importance of those conclusions does not constitute an end in itself, however. The inherent antipathy of theoretical constructs that are hegemonic and deterministic to the community practices that we are advocating in this chapter suggests the need for an intellectual counter-project. For reasons discussed above, this project must explicitly establish people and communities – not dominant institutions and technological determinism – as the *shapers*, however latent and under expressed they may now be, of the future. At the same time, this project must also take into account the *context* (including all influencing circumstances and factors) of the community practices, whether based on the perspectives of Castells or other scholars. This project, furthermore, must be flexible and dynamic; the capacity for societal *learning*, for example, must be fundamental. It is our expectation – and hope – that the concept of civic intelligence that we are incrementally developing will be valuable for these purposes.

2.8 Civic Intelligence: A New Paradigm of Thought and Action

The development of "civic intelligence" as a possible paradigm for community and civic practice is an attempt to create an orientating perspective that, unlike other theories (such as the "network society" or "information age"), provides and encourages opportunities for independent thought and action. To do so, it binds together three important threads, namely (1) recent insights, developments and experiences from civil society work, (2) potential and actual opportunities provided by the Internet and other technologies and (3) particular exigencies of our era (environmental degradation, for example). Civic intelligence, in other words, is based on an idea that has been long occluded: that humans – even "ordinary" ones – can play a role in the definition and creation of the future.

It has long been acknowledged that people have at their disposal a variety of cognitive capabilities that we call "intelligence". Intelligence describes a powerful general mechanism that individual people employ for dealing with the environment and other people as well as with abstract concepts or ideas. Less apparent is the fact that communities and other groups also possess "intelligence". This collective intelligence is played out in prosaic times and also in times of duress and dislocation. Intelligence is manifested over the long term through culture, language and institutionalisation of values and norms, but it is deeply enmeshed in transient and short-term experience as well. As Pea (1993) observes,

> Anyone who has closely observed the practices of cognition is struck with the fact that the "mind" never works alone. The intelligences revealed through these practices are distributed across minds, persons, and the symbolic and physical environments, both natural and artificial.

Because the activity is so distributed, it is often unclear "where" intelligence is "located" in a given setting, for example, a successful collaborative planning session.

Civic intelligence, as we define it, describes the capacity that organisations and society use to "make sense" of information and events and craft responses to environmental and other challenges collectively. Civic intelligence represents *potential* – not unlike Putnam's concept of social capital (Putnam, 1995). Civic intelligence is creative, active, non-deterministic and human centred. This perspective significantly contrasts with most theories of social change. It places people – not abstract systems, technology or very general historic forces – at the core. Civic intelligence combines community (or "bonding social capital"; Putnam, 1995) with civic ("bridging social capital"; Putnam, 1995) networking. The choice of the word "intelligence", moreover, was motivated by its correspondence to cognitive capabilities in individuals. Although a slavish commitment to every conceivable analogy between individual and collective intelligence is unlikely to be warranted, the supposition that useful relationships can be found seems likely. Also, it should be pointed out that although civic intelligence highlights the role of cognition, it is not intended to deny the reality of emotions and other largely non-cognitive aspects of human behaviour. At this evolving stage of the concept, it is enough to note that the non-cognitive aspects are important to individual human behaviour and to collective behaviour also. These aspects should not be ignored in a civic intelligence enterprise for long.

Civic intelligence implies the use of available tools in appropriate ways. These tools would certainly include communication and information systems including environmental monitoring systems and discussion and deliberation systems. Yet, important as it is, technology by itself will not solve humankind's problems. Civic intelligence requires intelligent – and concerned – *people.* Civic intelligence builds on and reinforces principles such as inclusivity, cooperation, justice, sustainability and other notions beyond which a simplistic measure of intelligence implies. Probably the most important aspect of civic intelligence is that it can be improved. Although we acknowledge that no system is perfect, nor totally "rational" and fundamental limits to human understanding always exist, we believe that humans can improve the situation by working together thoughtfully and applying humanistic values.

2.8.1 Requirements for the Concept of Civic Intelligence

A literature search of the term "civic intelligence" will reveal a number of very loosely related publications. One of the most relevant use was probably in "A Vision of Change: Civic Promise of the National Information Infrastructure" (Civille et al., 1993). Although uses of the term there and elsewhere were generally complementary to the usage expressed here, previous uses were generally informal characterisations rather than as a theme for serious study and focus in their own right.

Our exploration of the concept is motivated by the fact that activists, researchers and other people working on social and environmental issues are actually in some way all *working on the same project*. We are interested in developing intellectual tools that describe this phenomenon in a way that provides insight for people working in fields of social or environmental amelioration, preventing wars, for example, or repairing environmental damage. At the same time, we hope that people build on the idea that they are part of a common project and that they all can contribute to and derive strength from the common project. Ideally this would be a holistic vision that instructs and inspires. Hopefully, it would pave the way for increased collaboration and network building across boundaries wherever they exist around the world.

Although something which we claim could be called "civic intelligence" does exist independently of this exploration, we believe that the idea needs to be *socially constructed* in order to become a viable concept (or "strategic frame") intellectually for service in research, organising and integrating shared work. Hopefully, a broadly inclusive, collaborative exploration will yield models, paradigms, methodologies, projects and services to support general creation of civic intelligence throughout society. This belief, in turn, implies that (1) community processes that explore the idea should be initiated, (2) viewpoints and findings from related disciplines should be incorporated into the theory, (3) models need to be developed, tested, evaluated and reworked in the near term and (4) the practitioners and researchers who are exploring this concept must collaborate and share information. Indeed, they themselves must incorporate and simultaneously build upon the idea of civic intelligence.

2.8.2 The Significance of Civic Intelligence

While the "proof of the pudding is in the eating" there are aspects of a civic intelligence orientation that make it a hopeful approach. The first is that it is explicitly oriented towards a *dynamic* social inquiry that is *explicitly* directed towards social and environmental amelioration, not as a vague, always possible but rarely addressed or attained, *side effect* of conventional social science (Comstock, 1982). The second is that it is intended to be used both as a way to characterise past and present endeavours and as a way to *critique* current efforts and to help envision improved ones. It should hasten, in other words, social and environmental *progress*. Finally, by linking it – at least metaphorically – to models of human intelligence and learning, with attention paid to mental models, learning, communication and meta-cognition, for example, an exploration of civic intelligence can be conducted in conjunction with a variety of academic lenses including education, social psychology, political science and cognitive science and to concepts such as equity or a healthy environment that are in common usage in the non-academic world as well.

2.8.3 Emerging Civic Intelligence

Civil society historically has been at the forefront of social movements like human rights, civic rights, women's rights, environmentalism, etc. (Castells, 1997). To many observers (Barber, 1984, for example) the strong participation of civil society will be necessary if problems facing humankind in the 21st century are to be successfully addressed. There is a growing sense that communication, new modes of organising and new insights are helping civil society to address shared problems in new ways. How society uses its civic intelligence in an era marked by rapid change (propelled, for example, by new transportation and communication systems) and by daunting challenges of population growth, new diseases, environmental degradation and deadly conflict is becoming increasingly critical. Concomitant to these new circumstances, the extremely rapid growth of new civil society, business, government and scientific collaborations across traditional boundaries suggests that civic intelligence is increasing – at least within some areas of society.

Successful intelligence coexists with and reflects the world in which it inhabits (Calvin, 1996), particularly those aspects of the world that have the capacity to sustain or threaten life. The richness and complexity of modern life, ripe with threats and opportunities of all sizes and shapes, drives the need for a broad-based civic intelligence. Evidence is mounting that the new civic intelligence that our increasingly complex world seemingly requires is growing. According to the theory of civic intelligence, this would come about as a natural response to an increasingly complex and, possibly, dangerous environment. Incidentally, this conclusion would be valid regardless of whether the environment was actually more complex and dangerous or whether it was just *perceived* that way.

The number of transnational groups now number in the thousands and is still growing exponentially, providing evidence that civil society (in some sectors at least) is increasing its capacity for civic intelligence (while governments and business may, in many cases, be *causing* the problems). The diffusion of groups with a civic intelligence perspective worldwide – their agendas, strategies and tactics – is qualitatively different from their predecessors according to various observers. Keck and Sikkink (1998), for example, provide evidence that these new groups are more likely to engage in policy development and multifaceted approaches rather than simply being *for* or *against* something. They also show that many organisations are working outside of the conventional reward structures of money and power. For that reason, the importance of those groups can be undervalued by the academic disciplines that are accustomed to deal with social entities solely in these simplistic terms.

The rise in the number of transnational advocacy organisations in a loose way probably echoes the risks posed by the changed and changing environment. At the same time new technology is providing opportunities for information and communication utilisation by civil society. This new technology has helped to breach barricades that historically maintained civic ignorance while providing new venues for publishing independent viewpoints, developing shared issue frames and organising activist communities.

A multitude of new organisations are being launched, devoted to civic causes such as human rights and economic justice. The organisations often develop a networking structure that helps mobilise a critical mass. These organisations are growing in sophistication as well as in numbers. Qualitative differences in new and established organisations are emerging in ways that indicate a richer appreciation of the world and a more sophisticated and more ambitious approach to engagement. Civic intelligence organisations are creating a new "issue environment" that includes changes in number, constitution and/or diversity of issues under consideration.

Accompanying this are vocabulary changes and new framing concepts including "human rights", "sustainability" and "anti-globalism" that are all of relatively recent vintage. New active campaigns are becoming highly visible and references to their work are becoming more prevalent in educational and cultural venues such as literature, schools, museums, theatre, art, music and the mass media. Finally, increased resources, financial and contributed time resources, are flowing to civic intelligence organisations.

2.8.4 Describing Civic Intelligence

The evolving idea of civic intelligence can be used to characterise civil society (and other) organisations. Two models thus far are being developed for this purpose (Schuler, 2004) and obviously, the knowledge gleaned through further explorations will be used to adjust the models. The first model is a descriptive one, which is used to recognise, characterise and, hopefully, guide organisations or other collective enterprises. It is a naturalistic way to capture information about an enterprise that can be used to compare and contrast other organisations. A complementary model that identifies functional relationships – how organisations actually operate within an environment – is currently under development and is being used to characterise and diagnose the community network movement (Schuler, forthcoming).

In an earlier paper Schuler (2001) proposed six areas in which civic intelligence projects can be characterised and how a project that demonstrated effective civic intelligence would differ from one that did not. The six categories – orientation, organization, engagement, intelligence, products and projects and resources – are described below using language and terminology adapted from the original paper. We begin with definitions:

- *Orientation* describes the purpose, principles and perspectives that help motivate an effective deployment of civic intelligence.
- *Organization* refers to the structures, methods and roles by which people, working together, engage in civic intelligence.
- *Engagement* refers to the ways in which civic intelligence is an active force for thought, action, and social change.
- *Intelligence* refers to the ways that civic intelligence is expressed through learning, knowledge formulation and sharing, interpretation, planning, meta-cognition, etc.
- *Products and Projects* refers to some of the ways, both long-term and incremental, that civic intelligence organizations focus their efforts. This includes tangible outcomes and campaigns to help attain desired objectives.
- *Resources* refer to the types of support that people and institutions engaged in civic intelligence work need and use. (The resources they *provide* are described in the *Products and Projects* category.)

2.8.4.1 Orientation

Thriving civic intelligence stresses values that support social and environmental meliorism while acknowledging and respecting the pragmatic opportunities and challenges of specific circumstances. Central to the idea of a thriving civic intelligence is that inclusive, dedicated, long-term democratic mobilisation and strengthening of the civic sector will be necessary to address primary issues including social inequities, human suffering, environmental devastation and other collective concerns including the social management of technology. Castells (1998) describes how the civil sector is responsible for initiating the major social movements of our era, including environmentalism, the peace movement, civic and human rights movements and a wide variety of non-patriarchal causes. The civic intelligence orientation is towards moving beyond the present status, accepted norms and reward systems into more complex, nuanced and human-centred regimes. As Keck and Sikkink (1998) in their book *Activists Beyond Borders* state, networks of activists can be distinguished from other players in international, national, regional and local politics "largely by the centrality of principled ideas or values in motivating their formation".

2.8.4.2 Organisation

The civic intelligence project is global. Since the purview and resources of this project are distributed throughout the world, global "civic intelligence" will also be distributed worldwide. The global civic intelligence project likewise needs to be undertaken "everywhere at once" in order to be successful. But how should this massive effort be organised? There is no one central force or institution possessing the full set of skills, resources or authority necessary to direct the effort. Moreover, the idea of a centrally controlled hierarchical organisation is antithetical (in addition to being unrealistic) in this global project. The organisational structure of global civic intelligence of design and necessity becomes a vast network of people and institutions all communicating with each other and sharing information, knowledge, hypotheses and lessons learned. This network is necessarily composed of dissimilar institutions and individuals who cooperate with each other because they share values and commitments to similar objectives. Neither authoritarian directives nor market transactions provide the adhesive that could hold this evolving, shifting, growing ensemble together. The glue that binds it is a composite of values and commitments.

2.8.4.3 Engagement

Engagement is both a tactic and a philosophy. Engagement as a tactic means that the elements of the civic intelligence networks do not shy away from interactions with the organisations or institutions or ideas or traditions that are indifferent or opposed to the objectives of the network. These organisations may be promoting or perpetuating human rights abuses or environmental damage. They may also be thwarting civic intelligence efforts by preventing some voices and viewpoints from being heard. Engagement, of course, assumes many forms. An organisation that employs civic intelligence should, as we might expect, behave intelligently. The nature of the engagement should be principled, collective and pragmatic. Engagement represents an everyday and natural predisposition towards action; it represents a challenge and an acknowledgement that the status quo is not likely to be good enough. Engagement, ideally, is flexible and nimble and it is appropriate for the situation. Timing plays an important role in appropriate engagement. Research and study also have critical roles to play, but they must not be used as a substitute for action, postponing engagement while waiting for *all the facts to come in.*[2]

2.8.4.4 Intelligence

Intelligence implies that an appropriate view of the situation exists (or can be constructed) and that appropriate actions based on this view can be conceived and enacted on a timely basis. Clearly, the creation and dissemination

---

[2] See Rafensperger (1997) for a thoughtful approach to integrating thought and action.

of information and ideas among a large group of people is crucial. Learning is important because the situation changes and experimentation has shown itself to be an effective conceptual tool for active learning. Therefore, some of the key aspects include (1) multi-directional communication and access to information, (2) discussion, deliberation and ideating, (3) monitoring and perceiving, (4) learning, (5) experimenting, (6) adapting, (7) regulating and (8) meta-cognition.

Let us briefly touch on one aspect of intelligence – monitoring – and some examples of new civic uses. Technology ushers in both challenges and opportunities. We find, for example, that at the same time our technology (fuelled largely by economic imperatives) is creating vast problems, it is also introducing provocative new *possibilities* for the civic intelligence enterprise. Earth orbiting satellites, for example, provide data that could be used to monitor earth's vital signs from space (King and Herring, 2000). While the data itself do not specify what the earth's inhabitants will do with it, the possibility now exists for the picture of the state of the earth to be improved. This type of surveillance can expose other events to public scrutiny; it was the French "Spot" satellite that first alerted the world to the Chernobyl disasters. Also unlike previous enterprises this project makes its data readily and cheaply available to people all over the world.

2.8.4.5 Projects and Products

Projects – both campaign and product oriented – help to motivate and channel activity. An extremely wide variety of projects is important within the context of cultivating a civic intelligence. There is ample evidence that the "project" is necessary to marshal sufficient force to accomplish the desired goals (Keck and Sikkink, 1998). One of the most successful projects – at least in terms of participation and consciousness raising *worldwide* – was the rapid mobilisation opposed to the USA's invasion of Iraq. Although the Bush administration was not deterred from their plans to initiate war, it did so in full view of a disapproving worldwide audience that expressed its profound repulsion to war in no uncertain terms. While it is unclear when and where the latent antiwar sentiment will surface next, it does seem likely that the movement will build upon – and strive to reconstitute – the networks that were forged this time. It also seems likely that lessons learned during this encounter will likely result in revised tactics next time.

2.8.4.6 Resources

This category is important because all civic intelligence work requires resources. In addition to money, other common resources include labour (often volunteer), time, physical facilities, communication capabilities and focused initiatives for people and institutions. Adequate resources are necessary but not sufficient for effective civic intelligence. Although it is important to help ensure that adequate resources will be available for the current – and subsequent – projects, it is often not clear what resources will actually be needed nor available in advance of the project's launch. This of course casts doubt on any theory of social movement that relies solely on a resource perspective. In many cases, the overall project cannot wait until all the "necessary" resources are at hand before starting. Although the "adequate" resources sometimes do become available, it is also the case that needed resources do not materialise, leaving the project and the organisation that supports it at risk.

2.8.4.7 Challenges

Prudence dictates that the strength of established modalities of public and commercial sector control should not be underestimated. At the same time, the case for community and advocacy/action networks use of ICT should not be overstated (Calhoun, 2004). If the effects of community networks and other grass-roots forms of democratised communication become strong or visible enough to threaten the initiatives of powerful corporate, government or other organisations, the threatened institutions are likely to react. For this reason, it is incumbent for civil society to understand the nature of these challenges and consider how they can be effectively countered.

The first threat comes from the side effects of vast media empires going about their daily business of gaining market share and governments setting ICT policies that ignore or devalue civil society. With this scenario, community-oriented policies and institutions are incrementally degraded by "business as usual" – not as a result of a wholesale assault. One current effort in the USA illustrates how the drive for profit-taking can casually degrade social amelioration efforts by civil society. In Philadelphia, Pennsylvania, Tacoma, Washington and numerous other cities and towns, municipal governments are developing a variety of public information services that can provide access to ICT for larger populations and at lower costs than commercial providers; access for economically disadvantaged people can be provided at even lower rates. Now faced with this type of competition, large telephone and cable television companies are increasingly paying visits to representatives in state governments.

So what is the aim of these social calls? To urge the representative body to pass laws that make it *illegal* for municipalities to provide any type of ICT to its citizens, thus entering a "market" that rightfully belongs to corporations. Interestingly their ads state that it is "un-American" to compete with corporations, yet corporations did not exist in the USA in any real form until years after the constitution was written and the country was created.

Unfortunately, more profound challenges also exist. Using another example from the USA, an oft-quoted example of a "free country", the Bush Administration recently introduced the Patriot Act, which was subsequently passed by the Congress. The Act degraded across the board – in one decisive move – an entire edifice of civil liberties that were enacted over the years to protect freedom of speech, right to privacy, right to peacefully assemble, and other activities that community practices and civic intelligence rely upon in democratic nations. Sold – to the Congress and to the American people – as a tool to combat terrorism, the exceedingly rapid development of this complex and lengthy bit of legislation suggests that preparations for each aspect of the Bill were already complete and ready for nearly instant implementation once a "need" – real or perceived – was identified.

Of course not all challenges to our endeavours are so nefarious. Sometimes we are our own "worst enemy". As social innovators we have a responsibility to make our work compelling. The onus is on us to ensure that we share our vision with as many people, in as many social spheres, as possible. It is not enough to "do the right thing" – we have to do that and convince others that we are doing "the right thing" and that they have an important contribution to make in ensuring the success of the civil network society project.

2.8.4.8 Prospects for Success

Having laid out some motivations and objectives for an effective community practices movement and having discussed various factors that mitigate against such a movement, we turn our attention to assessing the likelihood of success in this endeavour. Of course, gauging what we mean by "success" is no simple task. In itself, "success" is an elusive term to define or understand. Presumably it involves attaining our goals but our goals are often culturally diffused, difficult to specify precisely and, at times, contradictory. Do we place a time limit on achieving "success"? Are we looking for success tomorrow, next week, or at some nebulous time in the distant future when "success" becomes finally established as a permanent condition? Is "success" a description of some form of normative worldview? A utopia where no wrong occurs, no violence is visited upon people or other life on earth or is a partial "success" where some pain is avoided, some scars are healed, some progress made towards reducing misery a worthwhile goal? Whether the improbability of reaching utopia is regrettable or, simply, a basic fact of human existence makes little difference here. We are concerned about trajectories, the pathways along which we are travelling as a powerful constituent of an ecosystem, and possible ways to intervene.

There exists, through the use of ICT (and within and between the social systems they are embedded in), the potential for the voices of all to travel, and be heard, over great distances. However, when information and communication environments are entirely one-directional, they amplify the worldview and the power of elites. Unfortunately, this is the situation in most technology-mediated information and communication environments today – some voices travel farther than others and are heard by greater numbers of people. Control over the content and distribution of information is an awesome power, making propaganda, for example, eminently more possible. The unremitting pounding of a single point of view can help build the necessary hatred and fear to approve the pre-emptive invasion, for example, of another country.

We believe that a public dialogue and cross-fertilisation of new ideas is essential to any democratisation of the Internet and other ICT systems. Books such as this present a diversity of new ideas and make them available for thoughtful consideration, not only in academia but in the realms of practice and policy as well. Of course, there is no geographical monopoly on these ideas; indeed, they are emerging and developing much faster than we as observers and contributors to this book can absorb. The efforts of the collaborators in this text will need to be integrated with the vast array of other activities and initiatives around the world if additional influence is to be attained globally.

Some of this integration will occur at the organisational level; groups will undoubtedly coalesce – within and across traditional borders – over shared ideas and these groups may be able to institutionalise and facilitate future thinking and actions. However, forming new organisations may not be the key idea or catalyst for social action and melioration (Keck and Sikkink, 1998). Engaging in dialogue that promotes the sharing of understanding, particularly around basic principles, may, in fact, be more immune to the challenges discussed in the previous section. And while face-to-face communication remains a rich, important venue for civil society, technology-mediated fora such as electronic mailing lists, chat rooms and the like provide additional communication platforms for the development of shared vocabularies and shared agendas necessary for group mobilisation around issues of import to civil society.

These newly emerging communities of interests, ideas, purpose and practice are players in a dynamic and complex issue-space in which they will likely need to coexist to see their long-term visions become real. Such social movements, Keck and Sikkink report, are increasing their capacities to develop more complex and insightful programmes that present in more detail the type of world they would like to see. In addition, networked communications provide the growing numbers of people involved and interested in preventing a war, protesting human rights abuses or celebrating Earth Day the potential for considerable coordinated actions.

One of the notable characteristics of the "network society" is its potential for the world's population to be connected to each other in some way or another. As participants in the "network society", active or not, we all coexist in the natural world ecosystem, the social "ecosystem' of information and communication spheres and the actions upon the social and natural environments. Connectivity between people in these "ecosystems" is not necessarily direct as we are all configured in networks of relationships, which in turn connect to other such configurations.

"Communities" in such a world web can coalesce around shared interests, values, principles, aims or other viewpoints. Working with the open source community, for example, to develop socio-technical platforms that can be distributed globally, much along the lines of the Independent Media Centers model (Morris, 2004), gives us an insight into what is possible. New overarching paradigms – like civic intelligence – that may provide the next steps in the evolution of the conscious development of ICT for the amelioration of social and other problems are emerging in ways that integrate many worldviews in a non-hierarchical network fashion. A multitude of possible paths fan out from humankind's current location. Whichever paths are chosen must be as the result of informed, conscious and democratic choices.

References

Barber, B. (1984) *Strong Democracy: Participatory Politics for a New Age.* University of California Press, Berkeley.

Barber, B. (2002) Globalizing democracy. *The American Prospect,* **11**(20). Retrieved 28 May 2005 from <http://www.prospect.org/print/V11/20/barber-b.html>

Boyd-Barrett, O. (2004) U.S. global cyberspace. In Schuler, D. and Day, P. (eds), *Shaping the Network Society.* MIT Press, Cambridge, MA, pp. 19–42.

Butcher, H. (1993) Introduction: some examples and definitions. In Butcher, H., Glen, A., Henderson, P. and Smith, J. (eds), *Community and Public Policy.* Pluto Press, London, pp. 3–21.

Calvin, W. (1996) *How Brains Think: Evolving Intelligence, Then and Now.* Basic Books, New York.

Castells, M. (1996) *The Rise of The Network Society.* Vol. I, The Information Age: Economy, Society and Culture. Blackwell Publishers, Oxford.

Castells, M. (1997) *The Power of Identity.* Vol. II, The Information Age: Economy, Society and Culture. Blackwell Publishers, Oxford.

Castells, M. (1998) *End of Millennium.* Vol. III, The Information Age: Economy, Society and Culture. Blackwell Publishers, Oxford.

Chanan, G., Garrat, C. and West, A. (2000) *The New Community Strategies: How to Involve Local People.* Community Development Foundation, London.

Civille, R., Fidelman, M. and Altobello, J. (1993) *A Vision of Change: Civic Promise of the National Information Infrastructure.* The Center for Civic Networking, Washington, DC. Retrieved 28 May 2005 from <http://www.friends-partners.org/oldfriends/telecomm/civic.promise.html>

Commission of the European Communities (1994a) *Europe's Way to the Information Society: An Action Plan.* CEC COM(94) 347 final, 19 August 1994. Commission of the European Communities, Brussels. Retrieved 28 May 2005 from <http://europa.eu.int/ISPO/infosoc/backg/bangeman.html#chap6>

Commission of the European Communities (1994b) *Europe and the Global Information Society: Recommendations to the European Council.* The Bangemann Report CD-84-94-290-EN. Retrieved 28 May 2005 from <http://europa.eu.int/ISPO/infosoc/backg/bangeman.html>

Comstock, D. (1982) A method for critical research. In Bredo, E. and Feinberg, W. (eds), *Knowledge and Values in Social and Educational Research.* Temple University Press, Philadelphia, PA, pp. 370–390.

Day, P. (2001) *The Networked Community: Policies for a Participative Information Society.* Unpublished Ph.D. thesis, University of Brighton.

Day, P. and Schuler, D. (eds) (2004) *Community Practice in the Network Society: Local Action/Global Interaction.* Sage, London.

Degenne, A. and Forse, M. (1999) *Introducing Social Networks.* Sage, London.

D'Orville, H. (1999) Towards the global knowledge and information society: the challenges for development and co-operation I. Retrieved 28 May 2005 from <http://www.undp.org/info21/public/pb-challenge.html>

Duff, A.S., David, C. and McNeill, D.A. (1996) A note on the origins of the "information society". *Journal of Information Science,* 22(2), 117–122.

Federal Trust (1995) *Network Europe and the Information Society.* Federal Trust, London.

Garnham, N. (1994) Whatever happened to the information society? In Mansell, R. (ed.), *The Management of Information and Communication Technologies.* Aslib, London, pp. 42–51.

Glenn, A. (1993) Methods and themes in community practice. In Butcher, H., Glen, A.., Henderson, P. and Smith, J. al. (eds), *Community and Public Policy.* Pluto, London, pp. 22–40.

Gore, A. (1991) Infrastructure for the global village. *Scientific American,* 265(September), 108–111.

Information Infrastructure Task Force (1993) *National Information Infrastructure [NII]: Agenda for Action.* National Telecommunications and Information Administration (NTIA), Washington, DC. Retrieved 28 May 2005 from <http://www.ibiblio.org/nii/toc.html>

Jewkes, R. and Murcott, A. (1996) Meanings of community. *Social Science and Medicine,* 43(4), 555–563.

Keck, M. and Sikkink, K. (1998) *Activists Beyond Borders: Advocacy Networks in International Politics.* Cornell University Press, Ithaca, NY.

King, M. and Herring, D. (2000) Monitoring earth's vital signs. *Scientific American,* April, 282(4), 92–97.

Mackay, H. (2001) Theories of the information society. In Mackay, H., Maples, W. and Reynolds, P. (eds), *Investigating the Information Society.* Open University/Routledge, London.

Moore, N. (2000) The international framework of information policies. In Law, D. and Elkin, J. (eds), *Managing Information.* The Open University Press, London, pp. 1–19.

Morris, D. (2004) Globalisation and media democracy: the case of indymedia. In Schuler, D. and Day, P. (eds), *Shaping the Network Society.* MIT Press, Cambridge, MA, pp. 325–352.

Niebel, M. (1997) The action plan of the European Commission. In Kubicek, H., Dutton, W. H. and Williams, R. (eds), *The Social Shaping of Information Superhighways: European and American Roads to the Information Society.* Campus Verlag, Frankfurt, pp. 61–67.

Pea, R. (1993) Practices of distributed intelligence and designs for education. In Salomon, G. (ed.), *Distributed Cognitions.* Cambridge University Press, New York.

Putnam, R.D. (1995) Bowling alone: America's declining social capital. *The Journal of Democracy,* 6(1), 65–78.

Rafensperger, L. (1997) Defining good science: a new approach to the environment and public health. In Murphy, D., Scammel, M. and Sclove, R. (eds), *Doing Community Based Research.* Loka Institute, Amherst, MA.

Rosenbrock, H. (1990) *Machines with a Purpose.* Oxford University Press, Oxford.

Schuler, D. (1996) *New Community Networks: Wired for Change.* Addison-Wesley, Reading, MA.

Schuler, D. (2001) Cultivating society's civic intelligence: patterns for a new "world brain". *Journal of Society, Information and Communication,* 4(2), 157–181.

Schuler, D. (2004) Civic intelligence functional model. Presented at On democratization of information with a focus on libraries, World Social Forum, Mumbai, India. Retrieved 28 May 2005 from <http://www.nigd.org/libraries/mumbai/reports/article-5.pdf>

Schuler, D. (2007) Community Networks and the Evolution of Civic Intelligence. *AI&Society,* Springer-Verlag, 21(2).

Schuler, D. and Day, P. (eds) (2004) *Shaping the Network Society: The New Role of Civic Society in Cyberspace.* MIT Press, Cambridge, MA.

Stoecker, R. (2005) *Research Methods for Community Change.* Sage, London.

Surman, M. and O'Reilly, K. (2003) Appropriating the Internet for social change: towards the strategic use of networked technologies by transitional civil society organizations. Retrieved 28 May 2005 from <http://commons.ca/articles/fulltext.shtml?x=336>

Warburton, D. (1998) A passionate dialogue: community and sustainable development. In Warburton, D. (ed.), *Community and Sustainable Development: Participation in the Future*. Earthscan, London, pp. 1–39.

Wasserman, S., Faust, K., Iacobucci, D. and Granovetter, M. (1994) *Social Network Analysis: Methods and Applications*. Cambridge University Press, Cambridge, UK.

Wellman, B. (ed.) (1999) *Networks in the Global Village: Life in Contemporary Communities*. Westview Press, Boulder, CO.

Wellman, B. (2002) Little boxes, glocalization, and networked individualism. In Tanabe, M., van den Besselaar, P., and Ishida, T. (eds), *Digital Cities II: Computational and Sociological Approaches*. Lecture Notes in Computer Science Series 2362. Springer-Verlag, Berlin, pp. 10–25.

Social Networks and the Nature of Communities

Howard Rheingold

3.1 Introduction

If I had encountered sociologist Barry Wellman and learned about social network analysis when I first wrote about cyberspace cultures, I could have saved us all a decade of debate by calling them "online social networks" instead of "virtual communities". Social networks predated the Internet, writing and even speech. Indeed, humans are not the only creature that makes use of social networks. I met Wellman, author of many social science journal articles about social networks; he had just written an insightful paper comparing online social networks to virtual communities. Think of the people you encounter regularly – every month, let us say – your biological family, the people from your job you hang out with, your congregation, service organisation, the people in your neighbourhood who would loan or borrow things, the people you talk with regularly on the telephone in the course of your professional or social activities, the delivery people who show up every day at your business, the people you e-mail regularly.

In a recent e-mail communication, Wellman added:

> Ever since the late 1960s, I have been arguing that community does not equal neighbourhood. That is, people usually obtain support, sociability, information and a sense of belonging from those who do not live within the same neighbourhood. They have done this through phoning, writing, driving, railroading, transiting, and flying. LA is the classic example of this, but in fact, this has been the prevalent means of connectivity in the western world at least since the 1960s.
> (Wellman, 2000)

Social networks emerge when people interact with each other continually, and they have to be useful or they would not exist. Your social network can find you a job or a husband, information you need, recommendations for restaurants and investments, babysitters and bargains, a new religion,

---

This chapter is from Howard Rheingold, *The Virtual Community*, Cambridge, MA: The MIT Press, 2000.

emotional support. Before writing letters became commonplace, social networks were confined to those people who saw each other face to face. Writing, public postal systems, telegraph, telephone and the Internet each brought new means of extending one's social network to include people who are not in the immediate geographical vicinity, who share an interest rather than a location.

It has been argued that these increasingly mediated relationships are, for the most part, increasingly superficial. As I look at the way more and more of our social communication is migrating to e-mail and cell phone, instant message and online greeting card, I tend to agree. At the same time, it certainly is possible to maintain deep relationships through regular letters, telephone calls or online chats. Like all technologies, communication tools come with a price: alienation might be the cost of the power of abstraction. We might do better by ourselves by paying more attention to how we are using the powers of abstraction. Social network analysis provides a useful framework for discussing the impact of online socialising. It counters the critique of virtual communities as alienating, dehumanising substitutes for more direct, less mediated human contact.

The notion of "strong ties and weak ties" is a useful part of that conceptual framework. The classic document explicating this idea is "The Strength of Weak Ties" (Granovetter, 1973). A weak tie is an alumnus of your alma mater, a stronger tie would be members of your college sorority or fraternity you actually lived with, an even stronger tie would be your roommate. A social network with a mixture of strong ties, familial ties, lifelong friend ties, marital ties, business partner ties is important for people to obtain the fundamentals of identity, affection, emotional and material support. But without a network of more superficial relationships, life would be harder and less fun in many ways. Weaker ties multiply people's social capital, useful knowledge, ability to get things done.

When asking questions about the impacts of any technology on community, I have learned to avoid romanticising the notion of community, of assuming a state of pastoral existence that once existed in pre-technology small towns. There is an indisputable merit to living your life in the same place, loving or hating or putting up with the same people day after day, making decisions together with people you do not necessarily like, reducing the number of your social relationships and perhaps increasing their depth. But there is a cost to this long-lost gemeinschaft of the village, the hamlet, the small town, as well. If the shadows of urban and mediated experience are alienation and superficiality, the shadows of the traditional community are narrow-mindedness and bigotry.

In *All That is Solid Melts into Air: The Experience of Modernity* (Berman, 1982), Marshall Berman claims that Goethe's Faust is a tale of the transition to modernity and includes lessons about how cruel those pastoral communities of pre-modern times could be. When Faust despoils the reputation of the maiden Gretchen, her warm, small, unmediated community of strong-tie relationships persecutes her to the point of suicide. How many people flee the idylls of small towns because they look or think or act differently than the local norm? Just as virtual communities have attractive and unattractive aspects, so do other forms of community. As Berman wrote, "So long as we remember Gretchen's fate, we will be immune to nostalgic yearning for the worlds we have lost."

I must therefore reconsider and retract the words I originally published here in 1993. I owe it to my critics Fernback and Thompson for pointing out (http://www.Well.com/user/hlr/texts/VCcivil.html) that I clearly proclaimed a nostalgia for community lost when I wrote:

> Virtual communities might be real communities, they might be pseudo-communities, or they might be something entirely new in the realm of social contracts, but I believe they are in part a response to the hunger for community that has followed the disintegration of traditional communities around the world.

The disintegration, I discovered, has been an ongoing process ever since the alphabet. And human relationships are too complex to be judged as either "deep" or "shallow" with nothing in between. But I have come to see how the benefits of communication tools have always come with a less visible cost. There is no denying that good things can be lost or destroyed by harmful use of tools and social systems, or there would be more redwoods, rainforests, town squares, convivial public transit systems and less pavement, fewer vehicles and cleaner air and water. Before we charge off to preserve those good things new media might threaten, we need to understand and agree upon what those good things are, who they are good for and why we agree they are good. Do we want cohensive societies . . . or democratic ones? Do we want warm communities, or innovative ones? Where are the spectrums of alternatives between these extremes? What are the right questions to ask about the impact of virtual communities on geographic community – the questions whose answers might improve our lives or defend against disaster?

The best attempt by social scientists to address some social critiques of virtual community is, in my opinion, "Net Surfers Don't Ride Alone: Virtual Communities as Communities". (Wellman and Gulia, 1999) Wellman and Gulia point out the vast excluded middle between virtual community utopianism and the most emphatic critics of life online, and review the findings of social science research to address seven key questions regarding virtual community:

1. Are relationships on the Net narrow and specialised or are they broadly based? What kinds of support can one expect to find in virtual community?
2. How does the Net affect people's ability to sustain weaker, less intimate relationships and to develop new relationships? Who do Net participants help those they hardly know?

3. Is support given on the Net reciprocated? Do participants develop attachment to virtual communities so that commitment, solidarity and norms of reciprocity develop?

4. To what extent are strong, intimate relationships possible on the Net?

5. What is high involvement in virtual community doing to other forms of "real-life" community involvement?

6. To what extent does participation on the Net increase the diversity of community ties? To what extent do such diverse ties help to integrate heterogeneous groups?

7. How does the architecture of the Net affect the nature of virtual community? To what extent are virtual communities solidarity groups (like traditional villages) or thinly connected Webs? Are virtual communities like "real-life" communities? To what extent are virtual communities entities in themselves or integrated into people's overall communities?

Although they do not intend to argue for definitive answers, Wellman and Gulia look at the literature of social science research, especially social network analysis, and propose ways in which the existing data can illuminate these questions.

In regard to the first question, Wellman and Gulia wrote:

> The standard pastoralist ideal of in-person, village-like community has depicted each community member as providing a broad range of support to all others. In this ideal situation, all can count upon all to provide companionship, emotional aid, information, services (such as child care or health care), money, or goods (be it food for the starving or a drill for the renovating). It is not clear if such a broadly supportive situation has ever actually been the case – it might well be pure nostalgia – but contemporary communities in the western world are quite different. Most community ties are specialized and do not form densely knit clusters of relationships. For example, our Toronto research has found that except for kin and small clusters of friends, most members of a person's community network do not really know each other. Even close relationships usually provide only a few kinds of social support. Those who provide emotional aid or small services are rarely the same ones who provide large services, companionship or financial aid. People do get all kinds of support from community members but they have to turn to different ones for different kinds of help. This means that people must maintain differentiated portfolios of ties to obtain a wide variety of resources. In market terms, they must shop at specialized boutiques for needed resources instead of casually dropping in at the general store. (Wellman, 1992)

Wellman and Gulia suggest that the nature of the Internet serves to amplify this specialisation and diversification of personal portfolios of social ties. They point out how virtual communities can organise, segment and separate kinds of social ties, from those that furnish professional information to those that provide emotional support. They note that "Emotional support, companionship, information, making arrangements, providing a sense of belonging are all non-material social resources that are relatively easy to provide from the comfort of one's computer."

In regard to their second question, Wellman and Gulia observed that people often provide information, support or favours for people they have never met – strangers. The weak ties enabled by online relationships, while perhaps reducing the depth of relationships, can help increase the diversity of relationships – the number of different kinds of people in one's social network.

To the third question, they cite evidence that there is reciprocity online and attachment to virtual communities. Indeed, the idea that cyberspace is a place where sharing is encouraged is itself a norm that influences behaviour: "Norms of generalised reciprocity and organisational citizenship are another reason for why people help others online." In most of the rest of the world of human activities, competition is the ruling norm. Building something collaboratively that creates a value for all who use it was one of the enduring values of the people who built the antecedents of today's Internet back in the 1970s and 1980s.

The ARPAnet and Internet cultures that preceded the Web by 30 years were built on norms of collaboration and cooperation. The Net was a place where informal gift economies enriched life and thought for everyone who participated. Minimal "netiquette" in social dealings, a willingness to share resources when others request them, a commitment to put value in as well as take it out of the Net are what made the Internet attractive to grow to its present status. But many question whether those norms have survived the waves of millions of newcomers and the abuse of the commons in the form of spam, chain e-mail, viruses and virus hoaxes. If, as Wellman and Gulia assert, social science data show that norms of reciprocity and organisational attachment exist online, will they continue to do so? Or are they a resource that can only live in the early stages of a network economy? The all-important and much-obviated question of whether high involvement in virtual communities removes people from involvement in their physical communities led Wellman and Gulia to several questions about the question itself. First, they questioned the certainty that "community" is a zero-sum game, that online involvement necessarily displaces offline communication.

Second, they point out that fears of virtual community in this vein indirectly "demonstrate the strength and importance of online ties, and not their weakness".

Third, they note that the question is based on a false comparison between virtual communities and unmediated, face-to-face communities. Do the critics mean communities where nobody ever uses a telephone? Citing Wellman's own research, the authors claim:

> In fact, most contemporary communities in the developed world do not resemble rural or urban villages where all know all and have frequent face to face

contact. Rather most kith and kin live further away than a walk (or short drive), so that telephone contact sustains ties as much as face-to-face get-togethers. (Wellman et al., 1988)

Fourth, they point out that most people do not divide their worlds into strictly segregated online and offline portions: online discussions are one way that people make friends offline.

Fifth, the existence of webs of personal relationships via private e-mail is not visible to most research, and "provides the basis for more multiplex relationships to develop . . . " In particular, they cite the "invisible colleges" of scholars who know each other well, meet once a year at most, but stay in touch online much more intensively via e-mails, listservs, newsgroups, web conferences.

Wellman and Gulia concluded about the relationship between online and offline community:

> In sum, the Net supports a variety of community ties, including some that are quite close and intimate. But while there is legitimate concern about whether true intimacy is possible in relationships that operate only online, the Net promotes the functioning of intimate secondary relationships and weaker ties. Nor are such weaker ties insignificant. Not only do such ties sustain important, albeit more specialized, relationships, but the vast majority of informal interpersonal ties are weak ties, whether they operate online or face-to-face. Current research suggests that North Americans usually have more than 1,000 interpersonal relations, but that only a half-dozen of them are intimate and no more than 50 are significantly strong (Kochen, 1989; Wellman, 1992.). Yet, in the aggregate, a person's other 950+ ties are important sources of information, support, companionship, and a sense of belonging.

Wellman and Gulia are less sanguine about the data regarding whether the Net increases community diversity. Noting that all people in contemporary communities belong to a number of different partial communities which expose them to a diverse set of social worlds, the authors remind us that the diversity of the online social world depends, first of all, on who is online: "Possibilities for diverse communities depend also on the population of the Net having diverse social characteristics."

I will address the issue of the "digital divide" later in this chapter. Are virtual communities "real"? Wellman and Gulia argue that

> The limited evidence available suggests that the relationships people develop and maintain in cyberspace are much like most of the ones they develop in their real life communities: intermittent, specialised, and varying strength. The net supports exchanges of information strongly, but does not hinder the exchange of communications of emotional support, as well. Specialised communities foster multiple memberships in partial communities. At the same time, the ease of group response and forwarding can foster the folding-in of formerly separate Net participants into more all-encompassing communities.

In their conclusion, Wellman and Gulia did not try to claim that social cyberspace is good or evil or that we know anywhere near enough to judge:

> It is time to replace anecdote with evidence. The subject is important: practically, scholarly, and politically. The answers have not yet been found. Indeed, the questions are just starting to be formulated.

Formulating the right questions about radically new phenomena sometimes requires thinking about old ideas in new ways. Are entirely new methodologies for social scientific inquiry required to deal with entirely new forms and media for human relationships? Wellman and Gulia point in that direction. Another social scientist makes the claim more explicitly. K.A. Cerulo, Ph.D., claims, in "Reframing social concepts for a brave new (virtual) world" (Cerulo, 1997), that

> Recent developments have touched issues at the very heart of sociological discourse – the definition of interaction, the nature of social ties, and the scope of experience and reality. Indeed, the developing technologies are creating an expanded social environment that requires amendments and alterations to ways in which we conceptualise social processes. (p. 49)

Cerulo questions the fundamental assumption made by many critics, that face-to-face communication is necessarily primary, more authentically human, than mediated communication. Cerulo proposes that social scientists and communication researchers look again, and with new eyes, at the definitions they base their studies on – definitions of social interaction, social bonding and empirical experience. Must all assumptions about social interaction be framed in terms of face-to-face communications? Do social bonds require geographic co-presence? And is it possible for ethnographers of cyberspace to do their work without becoming participant observers in virtual communities? Challenging the assumption that physical co-presence is the benchmark for social interaction, Cerulo says:

> We speak of the closeness and trust born of such mediated connections using terms such as pseudo-gemeinschaft, virtual intimacy, or imagined community. Such designations reify the notion that interactions void of the face-to-face connection are somehow less than the real thing.

The question of how virtual and geographic communities relate to each other has not been confined to theorists. In the first edition of *The Virtual Community*, I noted the birth of the "Freenet" and "Community Networking" movements in the early 1990s. Hundreds of experiments were spawned. What happened? In 1996, Douglas Schuler's book *New Community Networks: Wired for Change* provided both a manifesto and a handbook for community network building (Schuler, 1996).

Although space here does not an adequate review of community networks over the past seven years, it is worth noting that a whole class of community networks foundered on a weak business model (those that supported themselves by being Internet service providers), that well-funded and carefully designed experiments (like Blacksburg Electronic Village and Toronto's Netville) have thrived, that hundreds of others (like the Appalachian Center for Economic Networks) have managed to survive economically and slowly build social networks that bring people face to face rather than separating them by screens. I can only mention a few examples of the hundreds of community network enterprises all around the world. Blacksburg Electronic Village (BEV) is the one most journalists write about, because it had the most going for it early in the game. BEV (http://www.bev.net/) has succeeded because it grew out of a collaboration among knowledgeable community institutions that were willing to experiment, a local communications corporation, and a research university that was willing to contribute staff and resources to find out what would happen if you gave an entire community ready access to high-speed Internet connections and community services.

Virginia Tech, The Town of Blacksburg and Bell Atlantic set out in 1991 to offer Internet access to every citizen in town. BEV launched in October 1993. By the summer of 1997, more than 60% of the town's 36,000 citizens regularly used the Internet, 70% of the local businesses (more than 250) advertised online. Blacksburg senior citizens meet via listserv (http://www.bev.net/community/seniors/). Instead of watching local dollars flow through the Internet on their way out of town, hundreds of local merchants participate in main street e-commerce through an online mall (http://www.bev.net/mall/). By late 1999, more than 87% of town residents were online and more than 400 area businesses were listed on the BEV Village Mall.

Another "wired neighbourhood" in Toronto was the subject of research by sociologists Keith N. Hampton and Barry Wellman, who reported their results online: "Netville online and offline: observing and surveying a wired suburb" (Hampton and Wellman, 1999). In their abstract, the authors state:

> A connected society is more than a populace joined through wires and computers. It's a society whose people are connected to each other. For the past two years we have been looking for community online and offline, locally and globally, in the wired suburban neighbourhood of "Netville." We want to find out how living in a residential community equipped with no cost, very high speed access to the Internet affects the kinds of interpersonal relations people have with co-workers, friends, relatives, and neighbours.

In their conclusion, they write:

> Preliminary analysis suggests that the Internet supports a variety of social ties, strong and weak, instrumental, emotional, social and affiliative. Relationships are rarely maintained through computer-mediated communication alone, but are sustained through a combination of online and offline interactions. Despite the ability of the Internet to serve as a global communication technology, much

online activity is between people who live (or work) near each other, often in Netville itself. In Netville, the local network brought neighbours together to socialise, helped them to arrange in-person get-togethers – both as couples and as larger groups (barbecues, etc.) – facilitated the provision of aid, and enabled the easy exchange of information about dealing with the developer. The high rate of online activity led to increased local awareness, high rates of in-person activity, and to rapid political mobilisation at the end of the field trial.

Although many efforts in community networking failed because they lacked the funds and institutional resources available to BEV, many communities have succeeded in urban and rural middle class and lower-income communities. On 8 January 2000, the Charlotte News and Observer published a story entitled "A World Wide Web of cul-de-sacs" by staff writer Sarah Lindenfield (http://www.news-observer.com/daily/2000/01/08/tri00.html):

> John Wyman remembers the old days, before his neighbourhood went online with a Web page. Every couple of weeks, he would pay $20 to make 170 copies of a community flier. Then he would drive up to each home to deliver them. "Do you know what that does to your clutch, and do you know how long that takes?" Wyman asked. "With the Web site, I can go in there and, in 10 seconds, have it updated for everybody."

Today, Wyman is out of the printing business. Of the 170 homes in the Hardscrabble Plantation subdivision in northern Durham County, 155 have access to e-mail and the Web. Residents read neighbourhood announcements and newsletters online, and they pass the information to the 15 neighbours who are not connected.

Hardscrabble Plantation is among at least 65 Triangle neighbourhood groups that have ventured out on the Internet. Planters Walk in Knightdale, Alyson Pond in Raleigh, Park Village in Cary and Walden Pond in Durham are also among those that write newsletters, name board members, or simply list social activities on websites.

A small, but growing, group of for-profit and non-profit firms are designing sites just for neighbourhoods. And more of the 5100+ Triangle subdivisions are looking for new ways to connect.

***

Internet neighbourliness is not completely replacing knocking on doors with fresh-baked brownies, or strolling down the street and waving to people on front porches. Instead, residents see the Web as another way to communicate within the community.

Consider ACEnet, the Appalachian Center for Economic Networking (http://www.seorf.ohiou.edu/~acenet/):

> ...a community-based economic development organisation located in rural, south eastern Ohio. Our purpose is to work with others in the area to create

a healthy regional economy with many successful businesses and good jobs. Our goal is for people with low incomes to move out of poverty permanently through employment or business ownership.

As the name reveals, and as a closer inspection of their programmes makes clear, ACEnet is an economic network. Relationships with banks, businesses, educational organisations are the core of the project. It is not about technology, but technology helps it happen, and helps tie together the different parts of the community involved in the effort. In particular, ACEnet uses the Internet to link businesses with new markets and market resources such as trend information; with resources both near and far through their community network; with one another through an electronic mailing list; and to link their Appalachian community with similar communities across the country.

These examples are not offered as evidence that electronic utopia through many-to-many communications is around the corner, but as a small sample of the large number of active experiments that are still going on. Before theorists whose research is conducted primarily in libraries dismiss unequivocally the possibility that online communication can enhance rather than erode face-to-face communication in geographic communities, perhaps they should also pay attention to the results of these experiments.

One of the early enthusiastic backers of community networking, Mario Morino, now believes that community networks have failed, thus far, to live up to the promise we saw in them at the original "Ties That Bind" conference. Organized by Steve Cisler, a librarian at Apple Computer, this was the first annual face-to-face get-together of international networkers. The Morino Foundation was one of the sponsors of that first meeting, and of several efforts in the years since. Says Morino today:

> Community networking has been a movement in search of a cause and this has been its curse. It appeared that community networking never clearly articulated its purpose and this ambiguity caused some to view it from a technical perspective, others to view is as the electronic town, and while others saw it as a means of activism. Its real potential lied in bringing people together, to help people connect with one another, and more importantly, to help them toward an outcome. I wonder what would have happened had the talent and innovative minds that went into community networking focused on worker preparation programs. What would have happened if community networks had rallied around the challenge of eliminating literacy, eradicating lead poisoning for children, preparing our teachers to integrate technology into their curricula and learning delivery, or delivering health information to our most impoverished neighbourhoods? Instead, I believe the self-imposed limitation on community networking was the lack of a real vision for how it could have helped society in a focused way. The concept still holds remarkable potential. The world still hasn't grasped the potential of the online learning community. (Morino, 2000)

The foundation of a modern technological society is a population of educated individuals – humans who have been trained how to think. When new phenomena (alphabets, Internets) enable people to change the way we think, we then change the way we relate to one another. When human relationships change, human institutions change. What effects do virtual community and many-to-many communication technologies have on that fragile and precious institution, democracy? Will our grandchildren be citizens of a human-guided social system or components in a social system that guides humans? The question that trumps all the other questions is whether life online will contribute to political liberty, or diminish it.

3.2 The Prospects for the Public Sphere

Will citizens use the Internet to influence the nations of the world to become more democratic? Or will our efforts be ineffectual or even work to amplify the power of state or corporate autocracies? All other social questions about the impact of life online are secondary to this one.

Is the virtual community simply a self-hypnotising subset of the culture industry? Previously (Rheingold, 2000), I had pointed to the more sceptical worldviews of Baudrillard and Debord (global media productions as "simulacra" and "the society of the spectacle", respectively). I did not get into Adorno and Horkheimer of the "Frankfurt School" of political-cultural theory. It has been made clear to me more recently that no analysis of virtual community's political significance should ignore these thinkers who deliberated about the political implications of mass entertainment in the decades preceding the emergence of computer technology.

In the first edition of *The Virtual Community* (Rheingold, 2000), I coined the neologism "disinfotainment" to describe that sphere in which special effects, television laugh tracks, manufactured "news" programming, cross-media promotion of cultural products serve to distract and misinform a pacified population of unprotesting consumers, as well as to return profits to the owners of the cultural producers. I had also dabbled in the work of Jurgen Habermas, because his notion of the "public sphere" intuitively seemed to me like the best way to frame the political import of social cyberspaces. If I can be allowed to temporarily jack up the theoretical infrastructure for a social theory of cyberspace I started building into the clear blue sky of 1993, I need to insert Adorno and Horkheimer's ideas.

Adorno and Horkheimer were concerned with the fusion between the culture industry and mindless entertainment. Amusement is specific to the 20th century mass cultural industry and is simply another part of the cycle of routinisation. Their attack on the culture industry, first published in 1944, claimed that mass art was based on "a medicinal bath" of amusement and laughter, rather than on transcendence or happiness (Adorno and Horkheimer, 1972). People were amused and liberated from the need to think and their laughter affirmed existing society.

Are virtual communities part of a hold out from the commodification of media culture, a place of resistance and autonomy and self-empowerment?

A place where we have a chance of seeing reality for what it is, so that we can refuse to accept the present and try to change the future. Or are they disinfotainment in the guise of antidisinfotainment? Is it another way to amuse ourselves to death? These are the key questions Adorno and Horkheimer would most likely raise about the new phenomena of social cyberspaces. (Adorno died in 1969, the year the ARPAnet was born.)

Adorno and Horkheimer saw the culture industry as one that no longer tolerated autonomous thought or deviation to any degree because of the economic necessity for rapid return of capital investment. More than that, mass culture does not question the society it exists in and instead continually "confirms the validity of the system" (Adorno and Horkheimer, pp. 129). Adorno and Horkheimer saw how acceptance and reaction were permeating more and more spheres of life, and how the culture of mass society with its corporate rather than aesthetic ideology eroded cultural standards in order to quell any forms of expression which might contest the given order, producing less freedom, less individuality, and ultimately, less happiness.

Their primary concern was the transformation of society for the continuance of civilized humanity. They saw democracy and freedom to choose as diverging paths, but they found that "freedom to choose an ideology proved only to be freedom to choose what is always the same" (Adorno and Horkheimer, 167). For them, autonomy allows a conception of a different world and communicates the possibility that reason can penetrate existing barriers, allowing some to take a stance against modern culture to give the world a new direction, with a hope for the liberation of the human spirit.

Their boldest claim was that culture had become a form of domination, that the culture industry operates to diffuse oppositional consciousness and individualism. For them, the industry was selling packages of ideas and beliefs. People no longer had to think for themselves, since "the product prescribes every reaction by signals" (Adorno and Horkheimer, 137). It is characterised by a pervasive manipulation of the consumer whose intellectual capacity is continually underestimated. There is a profusion of sameness and repetition by using sets of interchangeable details, sweeping away all particularity and flattening out anything distinct, changing the nature of society as well as the way we perceive reality. One of the conclusions of the Frankfurt School was that the consumer society encouraged social and political apathy, even before the television. I offer this extended quote not as an endorsement of this rather determinist view, but because it is a sobering attack at the foundations of any utopian ideas about democratising media, from the earliest days of mass media.

The sociological theory that the loss of the support of objectively established religion, the dissolution of the last remnants of precapitalism, together with technological and social differentiation or specialisation, have led to cultural chaos is disproved every day, for culture now impresses the same stamp on everything. Films, radio and magazines make up a system which is uniform as a whole and in every part.

Even the aesthetic activities of political opposites are one in their enthusiastic obedience to the rhythm of the iron system. The decorative industrial management buildings and exhibition centres in authoritarian countries are much the same as anywhere else. The huge gleaming towers that shoot up everywhere are outward signs of the ingenious planning of international concerns, towards which the unleashed entrepreneurial system (whose monuments are a mass of gloomy houses and business premises in grimy, spiritless cities) was already hastening. Even now the older houses just outside the concrete city centres look like slums, and the new bungalows on the outskirts are at one with the flimsy structures of world fairs in their praise of technical progress and their built-in demand to be discarded after a short while like empty food cans. Yet the city housing projects designed to perpetuate the individual as a supposedly independent unit in a small hygienic dwelling make him all the more subservient to his adversary – the absolute power of capitalism.

Because the inhabitants, as producers and as consumers, are drawn into the centre in search of work and pleasure, all the living units crystallise into well-organised complexes. The striking unity of microcosm and macrocosm presents men with a model of their culture: the false identity of the general and the particular. Under monopoly all mass culture is identical, and the lines of its artificial framework begin to show through. The people at the top are no longer so interested in concealing monopoly: as its violence becomes more open, so its power grows. Movies and radio need no longer pretend to be art. The truth that they are just business is made into an ideology in order to justify the rubbish they deliberately produce. They call themselves industries; and when their directors' incomes are published, any doubt about the social utility of the finished products is removed.

Interested parties explain the culture industry in technological terms. It is alleged that because millions participate in it, certain reproduction processes are necessary that inevitably require identical needs in innumerable places to be satisfied with identical goods. The technical contrast between the few production centres and the large number of widely dispersed consumption points is said to demand organisation and planning by management. Furthermore, it is claimed that standards were based in the first place on consumers' needs, and for that reason were accepted with so little resistance. The result is the circle of manipulation and retroactive need in which the unity of the system grows ever stronger. No mention is made of the fact that the basis on which technology acquires power over society is the power of those whose economic hold over society is greatest.

A technological rationale is the rationale of domination itself. It is the coercive nature of society alienated from itself. Automobiles, bombs and movies keep the whole thing together until their levelling element shows its strength in the very wrong which it furthered. It has made the technology of the culture industry no more than the achievement of standardisation and mass production, sacrificing whatever involved a distinction between the logic of the work and that of the social system. This is the result not of a law of movement in technology as such but of its function in today's economy. The need that might resist central control has already been suppressed by the control of the individual consciousness. The step from the telephone to the radio has clearly distinguished the roles. The former still allowed the subscriber to play the role of subject, and was liberal. The latter is democratic: it turns all participants into listeners and authoritatively subjects them to broadcast programmes that are all exactly the same. No machinery of rejoinder has been devised, and private broadcasters are denied any freedom. They are confined to the apocryphal field of the "amateur" and also have to accept organisation from above. But any trace of spontaneity from the public in official broadcasting is controlled and absorbed by talent scouts, studio competitions and official programmes of every kind selected by professionals. Talented performers belong to the industry long before it displays them; otherwise they would not be so eager to fit in. The attitude of the public, which ostensibly and actually favours the system of the culture industry, is a part of the system and not an excuse for it.

Assuming that the prospect for individual liberty is as bleak as the neo-Marxist Frankfurt School portrayed it, has the technical power of the Net changed the culture machine portrayed by Adorno and Horkheimer? Is there a loophole in their critique that many-to-many communications might exploit? The broadcast technologies the Frankfurt School wrote about in 1944 centralised the power to produce and distribute cultural material, and continued to grow and consolidate as cultural products grew digital in the 1990s. The owners of the culture industry did not create the Internet, however. The Internet descended from the Atomic Bomb, not the silver screen. Does many to many turn the tables on culture monopoly, or will it be absorbed? While Sony is not the same as your desktop video on the Web, the monopolies and near-monopolies that used to control the culture industry through the high cost of culture-producing technology now are no longer alone in the mediasphere.

The Internet did not matter to the powers that be when I first wrote about virtual communities. When I interviewed executives at American, Japanese and French telecommunications companies in 1992, they were unanimous in their claim that their more serious enterprises regarded the Internet as a toy. A VP from Cap Cities, the company that owned the ABC network, told me at a lunch in 1992 that communicating via the Internet was a temporary fad, like CB radio in the 1980s. The Netscape–Yahoo–AOL–Microsoft Internet industry changed all that. When AOL bought Time-Warner in 2000, a milestone was established. At this point, Adorno and Horkheimer would remind us that when there is money to be made, when economic aspects of media culture become entrenched, then more controls will be instituted because there is more at stake.

"No machinery of rejoinder has been devised, and private broadcasters are denied any freedom." Is many to many a machinery of rejoinder that empowers media freedom for many who have been denied such power, as the printing press was? The difference, I am convinced, lies not in the nature of the technology, but in the way it is used. Manipulation, deception and marketing are not the only imaginable uses for new media. What happens when millions of people begin to create their own cultures and communicate them with each other, even giving them away to each other? The most serious critique of this book and the most serious concern about the social impact of the Internet is the challenge to my claim that many-to-many discussions could contribute to the health of democracy by making possible better communications among citizens. It seems that a great deal of the critique, although not all, is directed at one specific paragraph. I wrote this in 1993:

> We temporarily have access to a tool that could bring conviviality and under-standing into our lives and might help revitalise the public sphere. The same tool, improperly controlled and wielded, could become an instrument of tyranny. The vision of a citizen-designed, citizen-controlled world wide communications net-work is a version of technological utopianism that could be called the vision of "the electronic agora". In the original democracy, Athens, the agora was the marketplace, and more – it was where citizens met to talk, gossip, argue, size each other up, find the weak spots in political ideas by debating about them. But another kind of vision could apply to the use of the Net in the wrong ways, a shadow vision of a less utopian kind of place – the Panopticon.

Two of the criticisms directed at this paragraph have caused me to re-consider my original statement. The phrase "tool that could bring" has an implication of technological determinism that I simply let slip through be-cause I was not paying sufficient attention. Now, I pay more attention when discussing the way people, tools and institutions affect each other. It is not healthy to assume we do not have a choice. Tools are not always neutral. But neither do they determine our destinies, immune to human efforts. The rest of the book is not overly deterministic, but that paragraph is probably cited and challenged by dozens of scholarly essays over the years for reasons I have humbly come to understand.

Another flaw in my original draft is where I failed to make it clear that I was identifying, not advocating, the utopian version of an "electronic agora". I also should have mentioned that the affluent zeitgeist of Athenian democracy rested on the backs of slaves. As David Silver, one of the most thoughtful critical commentators, told me: "I'd make it clear in your new edition: neither Athens nor America nor cyberspace is a utopia."

I agree, now that critics have helped educate me. I would argue that we can still learn something from both experiments about the social nature of democracy and about the influence of public communications on political action.

Some critics have claimed that by concentrating on the Panoptic aspects of possible Internet futures as the only dystopian alternatives, I neglected to say anything about the influence of global capitalism. In fact, the Panoptic aspects of the Web today have less to do with the government spying on your every move, but with invisible and commonplace events that take place with the click of a mouse – information snooped out about your habits by websites that install information on your hard disk and read information previously stored on your computer. Capitalist competition, not totalitarian surveillance, has forced the rapid evolution of technologies that help vendors to zero in on precisely the products they need to bring to your attention.

These critics have their own evidence. It cannot be denied that over the past 5 years the Internet has been a powerful instrument of globalisation and the centralisation of great wealth in organisations like Sprint, MCI-Worldcom, Microsoft, Newscorp, AOL–Turner–Time-Warner. It is not outlandishly imaginative to think that a future Microsoft–AOL–Time-Warner–Sprint–MCI–Disney–Sony might merge into being. What kind of choice of access to the Web or variety of opinion or many-to-many broadcast capabilities citizens have then? More importantly, will it put us back in the broadcast age, where a small number of people controlled the power to inform, influence and persuade? What happens when the decentralised network infrastructure and freewheeling network economy collides with the continuing growth of mammoth, global, communication empires? We will know soon. The experiment is just beginning, but well underway.

Which citizens are going along for the ride as broadband and wireless networks enmesh every part of the human environment? Who will benefit from the replacement of physical goods by knowledge products? Who will fail to benefit?

In recent years, strong evidence has emerged of a "digital divide" between poor and middle class households, white and non-white households, in regard to ownership of personal computers and access to the Internet. As long as that divide continues to grow even while usage of new media grows explosively, no discussion of technology-assisted democracy can begin without mentioning the key question of who can afford to take advantage of the new media.

One thing all of the people online have in common today is that we have access to a computer and an Internet account. There are 200 million Internet users in a global population of 6 billion. According to a 1999 report from the US National Telecommunications and Information Administration, 40% of American families have computers, but only 8% of households earning less than $10,000 a year have PCs, and just 3% of that group have access to the Internet. The disparity between the highest and lowest income levels increased between 1997 and 1998. Urban households with incomes above $75,000 are more than 20 times more likely than low-income rural households to have home Internet access.

On 2 February 2000, President Clinton toured a Washington, DC, school with AOL CEO Steve Case and proposed spending more than $2 billion in tax breaks over 10 years as incentive for large-scale donations of computer gear and community technology centres. The idea of community technology centres where people can come to learn how to use Internet technology in their lives was a direct appropriation of an activist effort that has struggled on a shoestring for years, CTCnet (http://www2.ctcnet.org/ctcweb.asp). Clinton also called for $150 million in federal funds to train teachers and $100 million for low-income urban and rural communities.

A group of Silicon Valley executives have initiated an organisation called "ClickStart", a non-profit devoted to connecting low-income people to the Internet, ideally with the help of government subsidies. For a co-payment from the citizen of $5 per month, each participant in the proposed programme would receive a $10 monthly voucher to buy hardware and Internet access from those vendors who make their products and services available at low rates to the programme. The programme resembles in some ways the Rural Electrification Administration, a federal programme initiated in 1935 to bring electrical power and telephone service to rural regions. The millions of lower income citizens who might go online as the result of Click Start's efforts will also increase tomorrow's market for high-tech product and services. From the point of view of the socially conscious Silicon Valley libertarian, this is a simple answer to social inequality that's also good business. They might be right. It is hard to imagine a future for public–private partnerships if there isn't something to gain for the private part of the enterprise.

Henry Ford paid his workers well enough, and made his Model-T automobiles inexpensively enough that his employees could afford to buy them. In February 2000, perhaps taking a leaf from the founder, the Ford Motor Company announced that it would make a personal computer, colour printer, and Internet access available to each of its 350,000 employees for $5 per month (http://dailynews.yahoo.com/h/ap/20000203/tc/ford_computers_2. html).

Nobody will know for years to come whether or not these programmes or other government or non-profit attempts to address the problem of the digital divide will actually be implemented, or whether they will succeed. The question remains: will the advantages of online community be limited to those who can afford it at market rates? Is this a fundamental social inequality that must be addressed on all levels of community, society and nation-state? Or is the idea of a "digital divide" simply a marketing device for describing fundamental economic disparity in terms of consumption of technology products? If our concern as a society is for the welfare of our poorest citizens, perhaps food and shelter should have higher priority on taxpayer money. Or perhaps the inherent wealth captured by Moore's Law and knowledge communities can create a rising economic tide that will lift all boats, as Silicon Valley libertarians claim.

This book was not intended to be a critique of global capitalism or the problems the culture industry poses for democracy, but the power and gravitational pull of corporate power and opinion shaping cannot be ignored in any speculation about the future of media. The matter of the public sphere is a most serious one. Without liberty, all the other questions are irrelevant.

In an age where most of the journalism seen by most of the world is produced by a subsidiary of one of a few multinational entertainment companies, the question of what will remain truly "public" about communications is central today. It might not make sense tomorrow. If a theme park is all you know, you are not going to be asking where all the real parks are. In America, the idea of "public property" has grown increasingly unfashionable in the physical world of freeways, malls and skyscrapers. Is there still space in cyberspace for public property, public discourse, public opinion that emerges from informed deliberation among citizens?

Which brings us to the most serious challenge to the original draft of this book, that virtual communities might be bogus substitutes for true civic engagement or outright directly harmful to the public sphere. In 1995, two scholars, Jan Fernback and Brad Thompson, presented a "Computer-Mediated Communication and the American Collectivity" at the annual convention of the American Communication Association. With the permission of the authors, I have hosted it on my website since then, http://www.Well.com/user/hlr/texts/VCcivil.html. How certain can I be, sitting at my desk, tapping on my keyboard, about the reality and limits of the Net's political effectiveness? Would I bet my liberty on the democratising potential of the Net? I will have to say that the answer has to be "no". But that does not mean that I am convinced that we should do nothing about the way Internet media are used. To agree in theory that an action can have no consequences is to create a self-fulfilling prophecy. Therefore, I frame their critique by stating my belief that until it is proved impossible, it is important for citizens to attempt to influence the public sphere by their use of many-to-many media. Fernback and Thompson present several arguments, but to me the most serious paragraphs are these:

For the reasons stated in the preceding section, the likely result of the development of virtual communities through CMC will be that a hegemonic culture will maintain its dominance. Certainly, it cannot be assumed that the current political and technical elites would willingly cede their position of dominance or knowingly sow the seeds of their own destruction.

Indeed, it seems most likely that the virtual public sphere brought about by CMC will serve a cathartic role, allowing the public to feel involved rather than to advance actual participation. Communities seem more likely to be formed or reinforced when action is needed, as when a country goes to war, rather than through discourse alone. Citizenship via cyberspace has not proven to be the panacea for the problems of democratic representation within American society; although communities of interest have been formed and strengthened (as noted previously) and have demonstrated a sense of solidarity, they have nevertheless contributed to the fragmented cultural and political landscape of the United States that is replete with identity politics and the unfulfilled promise of a renewed "vita activa".

This research poses a larger question that has been addressed by other scholars (see Elshtain, 1995; Lasch, 1995) which emphasises the connection between the condition of fragmentation that exists within the American collectivity and democracy in theoretical terms. CMC does not, at this point, hold the promise of enhancing democracy because it promotes communities of interest that are just as narrowly defined as current public factions defined by identity (whether it be racial, sexual or religious). Public discourse ends when identities become the last, unyielding basis for argumentation that strives ideally to achieve consensus based on a common good.

If nothing else, the expressions of hope and desire for new modes of communication such as CMC speak volumes about the failures of present and past technologies to help create a just and equitable society. Perhaps these failures should prompt us to re-examine why we continue to place so much hope in technology after so many disappointments. Ultimately, we believe, the hope placed in CMC is misplaced because change will occur not by altering the technology but by reforming the political and social environment from which that technology flows.

Finally, we suggest that the term virtual community is more indicative of an assemblage of people being "virtually" a community than being a real community in the nostalgic sense that advocates of CMC would seem to be endorsing. Our comments should not be construed as protests against the corruption of a term; we recognise that community has a dynamic meaning. Our concern is that the public is more likely to forget what it means to form a true community. If, on the other hand, virtual communities can lead to action, that may be the basis for the formation of real and lasting communities of interest. But until then, any change in the communications structure, such as the widespread use of CMC, is likely to be unsettling. Therefore, we must agree with Cooley, who wrote in 1909:

> [A] rapid improvement in the means of communication, as we see in our own time, supplies the basis for a larger and freer society, and yet it may, by disordering settled relations, and by fixing attention too much upon mechanical phases of progress, bring in conditions of confusion and injustice that are the opposite of free. (Cooley, 1909, p. 55)

Have we fixed "attention too much upon mechanical phases of progress"? Are we facing "conditions of confusion and injustice" at the same time that so many people are prospering and learning in the Net-enabled environment? The questions are sobering, but are the questions alone sufficient to prevent us from investigating further the potential of many-to-many media? So many experiments are springing up on the Internet to bridge the digital divide, to make tools available to citizens, teach media literacy. I would not stop watching these experiments to see if they work simply because good critics raise good questions about the democratising potential of CMC. No theory can be any good if its effect is to prevent people from trying to improve their institutions. Are the critics themselves taking a deterministic stance that "the hegemonic culture will maintain its dominance"? Perhaps the most useful point Fernback and Thompson raise is "Ultimately, we believe, the hope placed in CMC is misplaced because change will occur not by altering the technology but by reforming the political and social environment from which that technology flows."

If any population is to succeed in this alteration, are we to do it without tools? And assuming the success of such a reformation of the political and social environment, are not we still faced with the challenge of learning how to use technology? Or are we to abandon the factories and office buildings and return to hunting and gathering? I agree, and must emphasise, that hopes placed in CMC or any technology are false hopes. Hopes must be placed in humans. I believe that knowing how to use tools is part of any successful human enterprise. Fernback and Thompson's serious challenges must also be weighed in the light of reports such as Christopher Mele, who documents the story of a group of low-income residents of public housing, all African-American women, used online communications to transform and empower the residents association in a 2-year battle with the housing authority:

> Once wired, it is difficult to predict the effects of online communication for collective action conducted by disempowered groups. For the women activists at Jervay, their connection to the Internet peeled away some of the historic and systematic layers that blanketed access to essential information. Whether it translates to long-term success is perhaps less important than the positive effect upon the activist role of the women themselves. (Mele, 1999)

The power to publish and communicate has no magical ability to make democracy happen. Only people can do that. No tool can make democracy happen without the actions of millions of people – but those millions of people would not succeed without the right tools. Most of what needs to be done has to be done face to face, person to person – civic engagement means dealing with your neighbours in the world where your body lives. But an important part of the work to be done will be mediated by new communication technologies. We need to relearn and continue to teach the communication skills necessary for maintaining healthy democracies.

Information sources and communication media that were until recently only the province of the wealthy and powerful are used daily by millions. Discourse among informed citizens can be improved, revived, restored to some degree of influence – but only if a sufficient number of people learn how to use communication tools properly, and apply them to real-world political problem-solving. Surely, this opportunity is worthy of serious consideration.

Surely we owe it to ourselves to make an effort to discover whether or not the charges of Fernback and Thompson and other critics are true in practice as well as theory. The global corporations that have consolidated control of distribution of news and entertainment will continue to command attention, reap profits and exert influence. But they are no longer the only game in town. If there is one question that lies at the foundation of the uncertainty about the Internet's future, it is whether the technical democratisation of publishing will prove to be a credible challenge to existing publishing interests.

I believe the publicness of democracy has been eroded, for the reasons Neil Postman cited in *Amusing Ourselves to Death* (Postman, 1985):

> The immense power of television as a broadcaster of emotion-laden images, combined with the ownership of more and more news media by fewer and fewer global entertainment conglomerates, has reduced much public discourse, including discussions of vital issues, to sound bites and barrages of images. In theory and a few practical examples, centralised opinion-shaping mechanisms are challenged by the decentralisation afforded by many-to-many media. But that is far from saying that the future will be less manipulated and more freely chosen by informed citizens. Much remains to be done for that rosy scenario to become a reality.

Theories and opinions about the Internet are plentiful. A good question to ask is how many real online tools exist for citizens to use today? Are there examples of successful experiments in online civic involvement that ought to be widely replicated? As a definition of "civic involvement", I suggest the one offered in Robert Wuthnow's *Loose Connections: Joining Together in America's Fragmented Communities* (Wuthnow, 1998):

> Broadly conceived, civic involvement consists in participation in social activities that either mediate between citizens and government or provide ways for citizens to pursue common objectives with or without the help of government."

The public sphere is where Kim Alexander operates when her organisation, the California Voters' Foundation (http://www.calvoter.org/aboutcvf.html), uses e-mail to organise a campaign to require political candidates to put their financial disclosures on the Internet. Civic involvement is what Paul Resnick and his students are trying to foster when they go door to door in their neighbourhoods in Ann Arbor, MI, creating web pages and e-mail lists intended to help people who live on the same block get to know one another (http://www.whothat.org). The public sphere is what Steven Clift and colleagues at the Minnesota E-Democracy project (http://www.e-democracy.org) seek to extend when they bring candidates for State office online to publish position statements and field questions from citizens. A little investigation reveals that dozens, probably hundreds, of profit-making and non-profit enterprises are experimenting with different tools for civic involvement.

Among the most notable are the following:

- CapAdvantage (http://www.capitoladvantage.com/) for communication with officials and other citizens. Their page, titled "Tools for Online Grassroots Advocacy and Mobilization", offers a comprehensive guide to Congressional publications, directories to identify state and national congressional representatives spot news and issues tracking.

- E-The People (http://www.ethepeople.com/) for petitions. "Welcome to America's Interactive Town Hall: Where Active Citizens Connect With Their Government and Each Other." "If your car is swallowed up by a pothole the size of Poughkeepsie, E-The People can help you find the person you need to tell about it. Simply come to our site, click on 'roads and transportation,' type in your address and we'll forward your note to the right officials in your city. And if your public works commissioner doesn't have Internet access, we'll convert your concern to a fax! Are you an organiser? With E-The People, you can start a petition about the same pothole and contact 10 neighbours to sign it – all on one site."

- Freedom Forum (http://www.freedomforum.org/) is a good example of vibrant discussion of political issues via message boards, along with Internet radio and news on rights. "The Freedom Forum is a non partisan, international foundation dedicated to free press, free speech and free spirit for all people."

- Civic Practices Network (http://www.cpn.org/) describes itself thus: "Born of the movement for a 'new citizenship' and 'civic revitalisation,' CPN is a collaborative and non partisan project dedicated to bringing practical tools for public problem solving into community and institutional settings across America."

- The title of the Freespeech.org page (http://www.freespeech.org/) is "Free Speech Internet Television". VolunteerMatch (http://www.volunteermatch.org) matches volunteers with opportunities and enables non-profit organisations and potential volunteers to get together. Since 1987, CompuMentor (http://www.compumentor.org/) has provided volunteer-based technology assistance to non-profits.

- National Strategy for Nonprofit Technology (http://www.nten.org/nsnt.htm) is "a leadership network of nonprofit staff members, funders, and technology assistance providers working together to analyse the technology needs of the nonprofit sector, and to develop a blueprint for how it can use technology more effectively and creatively".

- Guidestar (http://www.guidestar.org) is a clearinghouse for financial information: "Find information on the activities and finances of more than 650,000 non-profit organisations, the latest news on philanthropy, and resources for donors and nonprofits."

- While many big organisations are incorporating donation activities into their websites, smaller sites are going with a donation service like i-charity (http://www.i-charity.net/): "Free Internet fundraising service and online donations portal."

- Cause-related marketing type services like GreaterGood (http://www.greatergood.com/) provide online consumers the ability to send a portion of product purchase prices to designated organisations: "Shop where it matters."

- VoxCap (http://www.voxcap.com) aggregates tools and resources for online civic engagement as well as for "building a community of engaged citizens, where social capital can be accumulated and brought to bear", according to Jeff Fisher, VoxCap's Director of Community Development.

- Two enterprising political satirists quit their jobs in the winter of 2000 and hit the road in a van, following the early stages of the Presidential campaign from the road, updating their website daily from their own zany and well-informed angle. http://www.y2kwhistlestop.com/ is well designed and informative as well as funny. Perhaps political journalism might follow their lead and loosen up.

- The Association for Community Networking (http://www.afcn.org) is a community of interest and support for the hundreds of people working to use Internet communications to improve social capital in face-to-face communities.

- The Living Constitution Society (http://www.wethepeople.org) is dedicated to creating a continuous flow of interrelationship between government, industry, academia, citizens, and non-profit organisations.

   If the public sphere is where people act as citizens by discussing the issues that concern them, and civil society is the general name for the associations that citizens organise for social, charitable and political purposes, the name for the common wealth that they gain from acting cooperatively, in concert, rather than competitively as individuals seeking to maximise individual gain, is "social capital".

   Civic Practices Network defines social capital this way (http://www.cpn.org/sections/tools/models/social_capital.html).

   "Social capital refers to those stocks of social trust, norms and networks that people can draw upon to solve common problems". Networks of civic engagement, such as neighbourhood associations, sports clubs and cooperatives, are an essential form of social capital, and the denser these networks, the more likely that members of a community will cooperate for mutual benefit. This is so, even in the face of persistent problems of collective action (tragedy of the commons, prisoner's dilemma, etc.), because networks of civic engagement.

- foster sturdy norms of generalised reciprocity by creating expectations that favours given now will be returned later;

- facilitate coordination and communication and thus create channels through which information about the trustworthiness of other individuals and groups can flow, and be tested and verified;
- embody past success at collaboration, which can serve as a cultural template for future collaboration on other kinds of problems;
- increase the potential risks to those who act opportunistically that they will not share in the benefits of current and future transactions.

Social capital is productive, since two farmers exchanging tools can get more work done with less physical capital; rotating credit associations can generate pools of financial capital for increased entrepreneurial activity; and job searches can be more efficient if information is embedded in social networks. Social capital also tends to cumulate when it is used, and be depleted when not, thus creating the possibility of both virtuous and vicious cycles that manifest themselves in highly civic and non-civic communities.

The question of how to measure social capital is central to understanding the health of the public sphere. Indeed, as I will show later, there are those who question the idea that the social should be considered to be a form of capital. In an influential article, "Bowling Alone: America's Declining Social Capital" (*Journal of Democracy*, 6(1), January 1995, 65–78), Robert Putnam documented a broad decline in civic engagement and social participation in the United States over the past 35 years. Citizens vote less, go to church less, discuss government with their neighbours less, are members of fewer voluntary organisations, have fewer dinner parties, and generally get together less for civic and social purposes. Putnam argues that this social disengagement is having major consequences for the social fabric and for individual lives. At the societal level, social disengagement is associated with more corrupt, less efficient government and more crime.

When citizens are involved in civic life, their schools run better, their politicians are more responsive, and their streets are safer. At the individual level, social disengagement is associated with poor quality of life and diminished physical and psychological health. When people have more social contact, they are happier and healthier, physically and mentally. Putnam concluded his article prescriptively:

> In the established democracies, ironically, growing numbers of citizens are questioning the effectiveness of their public institutions at the very moment when liberal democracy has swept the battlefield, both ideologically and geo-politically. In America, at least, there is reason to suspect that this democratic disarray may be linked to a broad and continuing erosion of civic engagement that began a quarter-century ago. High on our scholarly agenda should be the question of whether a comparable erosion of social capital may be under way in other advanced democracies, perhaps in different institutional and behavioural guises. High on America's agenda should be the question of how to reverse these adverse trends in social connectedness, thus restoring civic engagement and civic trust.

The questions raised by Putnam's articles are about as serious as questions get – is the social "glue" that holds together democratic societies going to dissolve as we retreat from civic participation into more private pursuits? If, as Putnam proposed in a follow-up article, "The Strange Disappearance of Civic America" (PS, American Political Science Association, Winter 1996), the diffusion of television through the population over the past 40 years was strongly correlated with the disintegration of civic participation during that time, it is indeed important to ask now which way the Internet might push us in the future – towards or away from authentic community and deep personal ties. Or are we using the wrong assumptions and terminology when addressing the way civic practices are changing, the way Wellman and Cerulo believe we are misframing social science research in cyberspace? Another contemporary student of community, also a Harvard professor, Robert Wuthnow, recently wrote a book, *Loose Connections: Joining Together in America's Fragmented Communities* (Wuthnow, 1998), that addresses the way social affiliations seem to be changing. These paragraphs describing the book (http://hupress.harvard.edu/Fall98/catalog/loose_connect.html) summarise Wuthnow's thesis, offering an alternative to Putnam's view of the changes that seem to be taking place: It has become common to lament Americans' tendency to pursue individual interests apart from any institutional association. But to those who charge that Americans are at home watching television rather than getting involved in their communities, Wuthnow answers that while certain kinds of civic engagement may be declining, innovative new forms are taking their place. Acknowledging that there has been a significant change in group affiliations – away from traditional civic organisations – Wuthnow shows that there has been a corresponding movement towards affiliations that respond to individual needs and collective concerns. Many Americans are finding new and original ways to help one another through short-term task-oriented networks. Some are combining occupational skills with community interests in non-profit and voluntary associations.

Others use communication technologies, such as the World Wide Web, to connect with like-minded people in distant locations. And people are joining less formal associations, such as support groups and lobbying efforts, within their home communities. People are still connected, but because of the realities of daily life they form "loose connections". These more fluid groups are better suited to dealing with today's needs than the fraternal orders and ladies' auxiliaries of the past. Wuthnow looks at the challenges that must be faced if these innovative forms of civic involvement are to flourish, and calls for resources to be made available to strengthen the more constructive and civic dimensions of these organisations.

Defining, measuring, valuing, growing and preserving "social capital" is hotly disputed territory. A group from the University of Victoria, British Columbia, maintains a literature review online: "Space Between the Market and the State: A Social Capital Literature Review and Conceptual

Framework" at http://web.uvic.ca/cpss/npsri/lit_rev.html. The World Bank has a social capital website at http://www.worldbank.org/poverty/scapital/index.htm. The British Columbia group defines social capital as

> ...the intangible social features of community life – such as trust and co-operation between individuals and within groups, actions and behaviour expected from community members, networks of interaction between community members, and actions taken by community members for reasons other than financial motives or legal obligations – that can potentially contribute to the well being of that community.

In a recent online exchange, Christopher London (London, 2000) pointed out that this economic definition of one of the most uniquely human trait, the ability to cooperate, presupposes a certain economic worldview – free-market capitalism:

> Marx argued that in capitalist social relations it is possible for social life to be reduced to the mere exchange of tokens of exchange value and for the relationship between the things that people produce and the people themselves to get reversed. Rather than products existing to serve human needs, humans exist to serve products, that is, to make products come to life so that they may circulate freely in a market. Built into this monetary and exchange system is the subordination of masses of people, and their figurative and often literal degradation, so that the product of their labour may be profitably circulated by others. For capital to exist entails a host of social (institutional and cultural) arrangements to make this circulation of "surplus value" continue and for it to be characterised by continuous growth. This is a massive simplification, but I think it applies to the issue of social capital because in treating social relations as "stocks" that can be "accumulated," the relations themselves are treated as mere means to an end: that of (physical or financial) capital accumulation. Though the social capital people claim to be putting social relations at the centre of economic relations (and so supposedly illustrate that economic relations are social through and through), they do not do that at all. Rather, they reduce social relations to just another factor in an economic calculation. It's culturally premised on the idea that people come to their social relations only by thinking in terms of "what can I get out of this" and not in terms of "what do these people mean to me and me to them." (London, 2000)

The phrase "commodification of community" has been used by some social critics of virtual community. London pointed me to this passage from Marx's "Capital" in which Marx introduces (in colourful language) the notion that capitalism can turn human relations into commodities: A commodity appears, at first sight, a very trivial thing, and easily understood. Its analysis shows that it is, in reality, a very queer thing, abounding in metaphysical subtleties and theological niceties. So far as it is a value in use, there is nothing mysterious about it, whether we consider it from the point of view that by its properties it is capable of satisfying human wants, or from the point that those properties are the product of human labour. It is as clear as noonday that man, by his industry, changes the forms of the materials furnished by Nature in such a way as to make them useful to him. The form of wood, for instance, is altered by making a table out of it. Yet, for all that, the table continues to be that common, everyday thing, wood. But, so soon as it steps forth as a commodity, it is changed into something transcendent. It not only standswith its feet on the ground, but, in relation to all other commodities, it stands on its head, and evolves out of its wooden brain grotesque ideas far more wonderful than "table-turning" ever was.

***

A commodity is therefore a mysterious thing simply because in it the social character of men's labour appears to them as an objective character stamped upon the product of that labour; because the relation of the producers to the sum total of their own labour is presented to them as a social relation, existing not between themselves but between the products of their labour.

This is why the products of labour become commodities, social things whose qualities are at the same time perceptible and imperceptible by the senses. In the same way the light from an object is perceived by us not as the subjective excitation of our optic nerve, but as the objective form of something outside the eye itself. But, in the act of seeing, there is at all events, an actual passage of light from one thing to another, from the external object to the eye. There is a physical relation between physical things. But it is different with commodities. There, the existence of the things qua commodities and the value-relation between the products of labour which stamps them as commodities have absolutely no connection with their physical properties and with the material relations arising therefrom. There it is a definite social relation between men that assumes, in their eyes, the fantastic form of a relation between things. In order, therefore, to find an analogy we must have recourse to the mist-enveloped regions of the religious world. In that world the productions of the human brain appear as independent beings endowed with life, and entering into a relation both with one another and the human race. So it is in the world of commodities with the products of men's hands. This I call the Fetishism which attaches itself to the products of labour so soon as they are produced as commodities, and which is therefore inseparable from the production of commodities. This Fetishism of commodities has its origin, as the foregoing analysis has already shown, in the peculiar social character of the labour that produces them (Marx, 1987, pp. 76–77).

Outlining a programme for measuring the health of civil society and defining social capital in a way that does not transform human relationships into commodities is beyond the scope of this book. However, it is at the very heart of the kinds of discussions that must take place on a broad basis, online and offline, among millions of people. It is in the service of this broad, citizen-driven, democratic discourse that online tools for publishing and communicating hold out a hope.

If online community is *not* a commodity, it is only because people work to make it so. The hope I hold out for myself and suggest to others is that people will accomplish a task using a tool. Hope should not be vested in the tool itself. One important way of using tools wisely is informed government regulation. A tax break for corporations that donate to the public sphere, for example, might do more good than all the rhetoric and all the books decrying the deterioration of civic engagement. Consider the following scenario, not as a recipe for utopia, but a thought-experiment.

A tiny proportion of the gargantuan profits reaped by telecommunications service providers could be contributed to a well-managed fund (with its own budget and expenditures open for public inspection) that insures that every citizen has access to publicly available terminals, a free e-mail account, and free access to introductory classes on citizen use of the Net. In a world where everyone has affordable access and citizens become actively engaged in informing themselves and communicating with one another, will it be possible to make government more responsive to citizen needs – and perhaps more responsible to the public trust? All proceedings and filings at the city, state and national level could be made available to all citizens in dynamically updated databases, with easy to use web interfaces. GIS systems could enable citizens to visualise the impacts of proposed development on regional cultural and ecosystems. We could know when our legislators trade stock in companies their legislation affects.

The scenario offered in the previous paragraph is offered as an example of what I believe we should work to build, not as an unattainably ideal society expected to emerge magically from technology. There is no guarantee that the potential power of many-to-many communications will make a difference in political battles about the shape of our future. Indeed, the odds are against a media-literate population seizing the opportunities the Internet offers. But I believe the opportunity for leverage is there, waiting to be seized, ignored or mishandled. The hegemony of culture, power and capital that critics from Marx to Fernback and Thompson describe is a potent force to be reckoned with. But if we do not try to make a difference in the way tools are used and people are treated, we definitely would not make a difference. The first step in acting effectively is to know what you are acting on. Collectively, we know only a small amount about human behaviour in social cyberspaces. We need to know a lot more. I hope that this chapter, and the updated bibliography, helps inspire and orient those who pursue that knowledge, debate its meaning, and put it into action in meaningful ways.

3.3 References

Adorno, T.W. and M. Horkheimer (1972) The culture industry: enlightenment as mass deception. In Cumming, J. (trans.) *The Dialectic of Enlightenment.* Herder and Herder, New York.

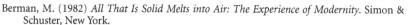

Berman, M. (1982) *All That Is Solid Melts into Air: The Experience of Modernity.* Simon & Schuster, New York.

Cerulo, K.A. (1997) Reframing social concepts for a brave new (virtual) world. *Sociological Inquiry,* **67**(1), 48–58.

Cooley, C.H. (1909) The extension of primary ideals. In *Social Organization.* Charles Scribner's Sons, New York.

Elshtain, J. (1995) *Democracy on Trial.* Basic Books, New York.

Granovetter, M. (1973) The strength of weak ties. *American Journal of Sociology,* **78**(6), 1360–1380.

Hampton, K.N. and Wellman, B. (1999) Netville online and offline: observing and surveying a wired suburb. *American Behavioral Scientist,* **43**(3), 475–492.

Kochen, M. (ed.) (1989) *The Small World.* Ablex, Norwood, NJ.

Lasch, C. (1995) *The Revolt of the Elites and the Betrayal of Democracy.* W.W. Norton, New York.

London, C. (2000) Email communication.

Marx, K. (1987) In Engels, F. (ed.), *Capital: A Critique of Political Economy,* Vol. 1. International Publishers, New York.

Mele, C. (1999) Cyberspace and disadvantaged communities: the internet as a tool for collective action. In Smith, M.A. and Peter, K. (eds), *Communities in Cyberspace.* Routledge, London.

Morino, M. (2000) Private communication.

Postman, N. (1985) *Amusing Ourselves to Death: Public Discourse in the Age of Show Business.* Viking Press, New York.

Putnam, R. (1995) Bowling alone: America's declining social capital, *Journal of Democracy.* **6**(1), John Hopkins Press, Baltimore, Maryland 21218, USA, pp. 65–78.

Rheingold, H. (2000) *The Virtual Community: Homesteading on the Electronic Frontier.* MIT Press, Cambridge, MA.

Schuler, D. (1996) *New Community Networks: Wired for Change.* ACM Press, New York.

Wellman, B. (1992) Which types of ties and networks give what kinds of social support? *Advances in Group Processes,* **9**, 207–235.

Wellman, B. (2000) Email communication with the author on 11 January.

Wellman, B., Carrington, P. and Hall, A. (1988) Networks as personal communities. In Wellman, B. and Berkowitz, S.D. (eds), *Social Structures: A Network Approach.* Cambridge University Press, Cambridge.

Wellman, B. and Gulia, M. (1999) Net surfers don't ride alone: virtual community as community. In Wellman, B. (ed.), *Networks in the Global Village,* Westview Press, Boulder, CO, USA, pp. 331–367.

Wuthnow, R. (1998) *Loose Connections: Joining Together in America's Fragmented Communities.* Harvard University Press, Cambridge, MA.

4

Community Informatics for Community Development: the "Hope or Hype" Issue Revisited

Bill Pitkin

4.1 Introduction

Community development can be defined broadly as strategies to build local capacity and improve the quality of life in geographic communities. Community informatics is a promising approach for taking advantage of information and communication technologies (ICTs) to further goals of community development. It is important, however, that proponents of this approach recognise that it is based on the assumption that technology in itself can lead to positive social development. This optimistic view of technology's role in community improvement is subject to various critiques, which can be grouped into three categories: methodological, philosophical and ideological. Reflecting on the implications of these critiques, I propose several recommendations that could serve as an ethical foundation for community informatics. In order to retain the "hope" that ICTs can help lead to greater social, political and economic equity, it is necessary to not succumb to the seductive "hype" that surrounds these technological developments.

4.2 Community Development: Can Information and Communications Technology Help?

Despite the efforts and resources of community residents, local governments, business owners and community-based organisations, there is a persistence of poverty and degrading quality of life in certain neighbourhoods and communities. Residents of these areas often lack access to good jobs or educational opportunities, live in deteriorated housing and environmental conditions and do not have access to information that would enable them to organise themselves to improve their living conditions. Do new information and communications technologies (ICTs), such as the Internet and

Geographic Information Systems (GIS), perhaps provide the opportunity to address these problems in new ways? The purpose of this essay is to explore a response to this question.

In cities throughout the world, low-income residents are cordoned off into deteriorated neighbourhoods, be they squatter settlements in the developing world or inner-city slums in the industrialised world. Efforts to ameliorate these social and economic divisions at the local community can be referred to broadly as part of a strategy of community development. In the USA, the beginning of the field of community development is usually traced to progressive social reform movements in the late 19th century, being formalised in federal legislation to promote urban housing reform in the 1940s and 1950s (Halpern, 1995). Community development can also take place in rural contexts, such as in poor rural areas of the USA,[1] and especially in the rural sectors of the developing world, where non-governmental organisations work to provide social and economic development strategies for rural peasants and subsistence farmers.

For some, community development implies a narrow focus on an issue like affordable housing or employment creation. The editors of a recent book on community development in US cities, however, take a broad view of the field:

> community development is asset building that improves the quality of life among residents of low- to moderate-income communities, where communities are defined as neighbourhoods or multineighbourhood areas. (Ferguson and Dickens, 1999, p. 5)

Community development, then, encompasses building various kinds of capital – not only financial, but also physical, intellectual, social and political – to enhance how people live. I find this broad definition of community development preferable to more narrow perspectives because it encompasses social – as opposed to merely economic – aspects of life. Therefore, community development is more than just building physical structures or increasing economic opportunities; it also includes efforts to build local capacity, educate and organise community residents and increase their access to local policy making that affects their lives. Moreover, this broad definition is equally applicable in both rural and urban contexts.

With the growing prevalence of ICTs in everyday life, a number of scholars, policy makers and community activists have begun to ask how these new technologies may play a role in furthering the goals of community development. In essence, advocates of this linkage between ICTs and community development seek to utilise tools such as the Web or GIS to gather, share and analyse information that can help them deal with local

---

[1] See, for example, the US Department of Agriculture's Office of Community Development Web site: http://www.rurdev.usda.gov/ocd/.

concerns and even influence policy decisions. Community networks, for example, provide a range of information services for neighbourhoods, cities or even rural areas (Schuler, 1996; Hecht, 1998). City governments have also become involved in these initiatives by developing "public access computer networks" to improve service delivery and increase citizen participation (Guthrie and Dutton, 1992). This type of response is a burgeoning field of practice and research, which has been termed "community informatics":

> Community Informatics is a technology strategy or discipline which links economic and social development efforts at the community level with emerging opportunities in such areas as electronic commerce, community and civic networks and telecentres, electronic democracy and on-line participation, self-help and virtual health communities, advocacy, cultural enhancement, and others. (Gurstein, 2000, p. 1)

Do these efforts represent a new hope for improving the quality of life in low-income communities, or are their proponents merely subsumed in the hype of the information superhighway? In an attempt to help answer this question, I first address a fundamental assumption of a community informatics approach to community development, relying both on literature review as well as a case study of a community informatics project with which I have been closely involved. Next, I consider several critiques that could be levied against a community informatics approach, grouped in three categories: methodological, philosophical and ideological. Finally, I reflect on the implications of these critiques by proposing four responses that could serve as ethical foundations for those who hope to further the goals of community development through ICTs. As both a community informatics researcher and practitioner, this essay in many ways reflects my own self-critique and struggles with these questions. I trust, however, that this personal reflection will be useful to others contemplating or working on ways to use ICTs to improve the plight of low-income communities. In the end, it is my hope that this exercise will help proponents of community-based ICT projects retain a realistic hope without succumbing to unbridled hype.

4.3 The Inherent Optimism of Community Informatics

With the growing influence of the Web and other ICTs in all aspects of life over the past couple of decades, researchers, activists and policy makers have begun to ask how are these new technologies being designed and used, and what are their impacts? Some scholars, especially in computer and information sciences, have focused on how computer networking changes the ways in which individuals communicate and organisations conduct business, as reflected in the wide body of research on "computer-mediated communication" (Lea et al., 1999; Turoff et al., 1999). Many activists and policy makers have been concerned with the economic impacts of these new technologies, inquiring whether their adoption will lead to more corporate downsizing, job loss and decrease in sales tax revenues. Other researchers – from fields such as information studies and communications – have focused on the social aspects of ICTs: "'Social informatics' is the new working name for the interdisciplinary study of the design, uses, and consequences of information technologies that takes into account their interaction with institutional and cultural contexts" (Kling, 2000). Social informatics researchers focus on the social settings of how ICTs develop and get utilised in various aspects of life, using empirical research methods to inform design and usage of ICTs.

Community informatics is an emerging area of research and practice in the broad inquiry into the interaction between people, society and ICTs, focusing specifically on the implications of developments in information technologies for communities. In this setting, "community" can have at least two meanings: (1) a *geographic community* such as a neighbourhood or region and (2) *a community of interest* such as persons of a certain ethnicity or some specific interest in common. Following several pioneers in the field of community informatics, I focus on the territorial aspect of community and how ICTs intersect with efforts to improve local communities. For Gurstein (2000), community informatics projects are inherently tied to physical, as opposed to virtual, places. Likewise, researchers from the Community Informatics Research and Applications Unit[2] concentrate on how ICTs connect to local space:

> Community informatics is an approach which offers the opportunity to connect cyberspace to community place: to investigate how ICTs can be geographically embedded and developed by community groups themselves to support networks of people who already know and care about each other. (Loader et al., 2000, p. 81)

Much of the current literature on community informatics tends to be speculative, reflecting optimistic assumptions about the potential role that information technology can play in improving life and society. Futurists view ICTs as providing solutions to a wide range of social and economic problems. For example, Nicholas Negroponte, Director of the MIT Media Lab, foresees a completely digital world in which information technology "can be a natural driving force drawing people into greater world harmony" (Negroponte, 1995). William Mitchell, Dean of MIT's School of Architecture and Planning, develops a similarly futuristic outlook of these possibilities in two books, *City of Bits* and *E-topia*, that more specifically speculate how urban communities are changing. In *City of Bits*, Mitchell outlines the many ways in which information technology will increasingly influence all aspects

---

[2] For more information, see the CIRA Web site: http://www.cira.org.uk/.

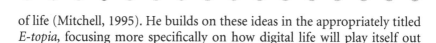
of life (Mitchell, 1995). He builds on these ideas in the appropriately titled *E-topia*, focusing more specifically on how digital life will play itself out in an increasingly urban world. He envisions "lean, green cities that work smarter, not harder" as a result of the digital revolution:

> In the twenty-first century, then, we can ground the condition of civilized urbanity less upon the accumulation of things and more upon the flow of information, less upon geographic centrality and more upon electronic connectivity, less upon expanding consumption of scarce resources and more upon intelligent management. (Mitchell, 1999)

In translating the ideas of these information-age sages to practical application in community development projects, leaders of community informatics projects retain the utopian assumption that information technology can aid in effecting social change. Community networks are among the best known of community informatics applications, and their proponents hail the potential of these networks for creating and strengthening personal relationships, as well as for providing access to vital community information and resources (Schuler, 1996). This movement is diverse in origin and function, but the networks tend to share a common commitment to serving the local needs of a neighbourhood, city or region, by providing online information sharing and networking. Many community networks hope to move forward an agenda of electronic democracy that will increase public access to information, increase political deliberation and heighten community participation in decision making (Tsagarousianou, 1999). Gurstein (1999) argues that ICTs can provide the engine for local economic development, especially for remote rural areas like Nova Scotia, helping small businesses market their products and network and coordinate their efforts more efficiently.

Even within traditional community development practice, ICTs are increasingly being employed to improve local communities. Internet-based neighbourhood information systems provide community residents with the opportunity to conduct online property and neighbourhood research and enter into local policy debates, thus pushing the boundaries of participation in urban planning (Krouk et al., 2000).[3] Organisations that support the community development field are increasingly looking to provide information and data, training materials and networking opportunities through the Web (Schwartz, 1998). Even in the traditionally centralised world of GIS, organisations are taking advantage of decreasing computing costs and user-friendly desktop software to utilise GIS to increase community participation in community development processes (Harris et al., 1995; Howard, 1998; McGarigle, 1998).

---

[3] In this chapter from *Community Informatics: Enabling Community Uses of Information Technology* (Gurstein, 2000), I compared NKLA with similar projects in Chicago and Seattle.

In an article on the potential for using information technologies in community development, Samuel Nunn presents a litany of reasons why organisations working in this field should take advantage of these new technologies:

> [to] become more efficient and lower labor costs, offer better information management, provide more intelligence about clientele and other stakeholders, empower individual employees, improve service delivery, and present the organization as a progressive user of advanced technology. (Nunn, 1999, pp. 13–14)

Although the article is interspersed with examples of how groups are using technology to achieve some of these goals, Nunn's argument is in general speculative and futuristic, presenting what information technology *can* do for community organisations without necessarily questioning the risks or potential problems with this approach. It is this type of language that in general characterises common approaches to community informatics.

4.4 A Community Informatics Case Study: Neighbourhood Knowledge Los Angeles

As noted in the introduction, my reflection in this essay on the assumptions and potential of locally-based ICT projects for effecting social change in physical communities comes from both a review of community informatics literature and my own intimate involvement with one such project. The Neighbourhood Knowledge Los Angeles (NKLA) (http://nkla.ucla.edu) project was initiated in the mid-1990s within the context of struggles against processes of neighbourhood disinvestment and slum housing in Los Angeles, CA. Working with tenants of slum buildings and affordable housing advocates, Danny Krouk, a graduate student in the University of California at Los Angeles (UCLA) Department of Urban Planning, discovered that local government possesses property-level data that can serve as early warning indicators of neighbourhood deterioration (Krouk, 1996).

Recognising that these indicators could serve as more than just data for a traditional academic research study, Krouk proposed following the lead of the Center for Neighbourhood Technology's Neighbourhood Early Warning System (http://www.newschicago.org/) project in Chicago and providing dynamic access to these data over the Web. Working closely with Neal Richman, a member of the UCLA Urban Planning faculty and one of his thesis advisors, Krouk raised the financial and political support necessary to launch NKLA in the spring of 1996. Their innovative site quickly grew in utilisation and scope, vividly demonstrating through interactive data queries and mapping how early warning indicators such as property tax delinquency,

utility liens and building code complaints could be used to address issues of disinvestment.[4]

I came on board the NKLA team while studying in the UCLA Urban Planning Department, assisting in the technical and content development of the site. In the fall of 1997, I began to work full-time on the project and took over many of the day-to-day responsibilities of developing and maintaining the site. We were soon able to secure a 3-year implementation grant from the US Department of Commerce's National Telecommunication and Infrastructure Agency,[5] allowing us to complete a major redesign of the site in the summer of 1999. Whereas in the previous system users were required to query each of the property databases separately, the new design contains an integrated database that permits users to view all of the attribute data for a property at one time. Moreover, the redesign allowed for complex querying of the data in the "Policy Room" and began to experiment with gathering community-created data through a Web form and displaying it in interactive maps on the website (see screen shot of NKLA in Figure 4.1).

Finally, the new funding helped us to greatly expand our community outreach and training to encourage people working to improve Los Angeles neighbourhoods to use NKLA in their community development efforts.

As a community informatics project, NKLA began with the assumption that an ICT, namely the Web, could play a role in improving physical communities. In developing NKLA, we have promoted the usefulness of the Web as a tool for timely data dissemination and analysis, as well as for communication among stakeholders, which can aid in improving the quality of life for residents of Los Angeles neighbourhoods. This "inherent optimism" is evident in how we have pitched our strategy in a grant funding proposal:

> Enthusiasm for NKLA's new comprehensive neighbourhood information initiative is certainly spurred by hopes that new technology can help unify a city torn apart by urban unrest only five years ago and now facing rising threats of political secession... NKLA is being challenged to demonstrate that technology can help connect and mobilize neighbourhoods across this multi-ethnic, automobile-dependent municipal region.[6]

From the beginning, we believed that the Web could help bring about positive social change in Los Angeles neighbourhoods. In order to reflect

---

[4] For more background on NKLA, see Krouk et al. (2000). There is additional information in the "NKLA History" page and the "NKLA How-To Kit" on the NKLA website.

[5] The funding program was the Telecommunications and Information Infrastructure Assistance Program (since renamed Technology Opportunities Program; more information is available at: http://www.ntia.doc.gov/otiahome/top/index.html. Federal funding for TOP was eliminated in 2005.

[6] This excerpt is from a grant proposal submitted by NKLA to the US Department of Commerce's Telecommunications and Information Infrastructure Assistance Program (now called the Technology Opportunities Program) in 1999. The project narrative from the proposal is available at: http://nkla.sppsr.ucla.edu/tiiap.htm.

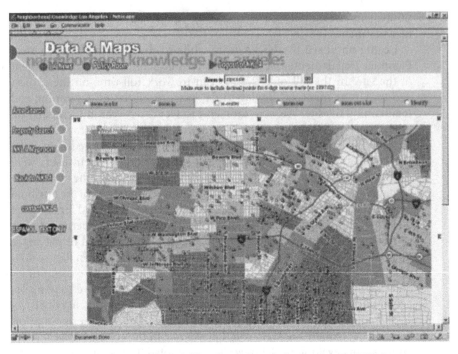

Figure 4.1 Interactive Map on the Neighbourhood Knowledge Los Angeles Website.

on the implications of the inherent optimism of community informatics projects like NKLA, I next consider several critiques of this type of approach to community development.

4.5 Critiquing Community Informatics

I believe that there are important critiques that can be made towards a community informatics approach to community development. It is crucial that people and groups working on community informatics projects reflect on these critiques in order to develop thoughtful, grounded theory on which they base their work. I have grouped these critiques into three general categories: methodological, philosophical and ideological.

4.5.1 Methodological Critique

The first series of critiques centres on the tendency of community informatics advocates to be non-critical about their work. This is obvious in the utopian, futuristic language of much community informatics literature, reflecting a fascination with the hype surrounding emerging developments

in information technologies. Many proponents of community informatics seem to be enamoured by indiscriminate claims of futurists such as Alvin Toffler, who foresaw many of the sweeping impacts that information technology would have on economic and social life in his 1981 book *The Third Wave*. Though he has tempered his overwhelming optimism somewhat, Toffler still holds that information technology can "make possible the substantial alleviation of poverty" (Harris and Gold, 1999, p. 16).

Much of the source of this hype is undoubtedly mass media, as advertising banners fill the screens of many websites and companies post their website URLs in television and print advertising.

> Most large media, computing and telecommunications companies, for example, are involved in some way in encouraging the current frenzy of debate and hype over the "information superhighways", which are often being cast as some sort of technological panacea for all the social, economic and environmental ills of society. (Graham and Marvin, 1996)

Therefore, we are bombarded with images of these new networking technologies breaking down physical, cultural, economic and interpersonal borders, implying that we cannot be truly fulfilled employees, students, parents, lovers, or citizens if we do not jump on the technology bandwagon. Judging by the largely uncritical language of the community informatics literature, it appears that many well-intentioned persons have been seduced by what Kellner calls the "new infotainment society" (Kellner, 1997a, b), without recognising that it has been promoted largely to serve commercial interests.

The inherent danger in this lack of critical reflection is that it stunts historical memory and does not acknowledge possible unintended consequences. Constantly in search of "today's next big something" or "killer app", even community informatics advocates can forget the lessons of yesterday. Mosco (1998) parallels the current fascination with the so-called information highway with that among early users of radio, who hoped this new technology would bring about positive social change. This lack of historical understanding leads to the prevalence of what Mosco refers to the myth of the information highway, one that encourages people to forget that technological development is always part of a social and political context: "the denial of history is central to understanding myth as depoliticised speech because to deny history is to remove from discussion active human agency, the constraints of social structure, and the real world of politics" (Mosco, 1998, p. 60). In paralleling Internet hype with the hope that many community activists placed in public access television in the early 1970s, Stephen Doheny-Farina points out that "because we are increasingly afflicted with that particularly postmodern disability, acontextuality, we tend to forget failed dreams" (Doheny-Farina, 1996). Even early experiments in promoting political dialogue via computer-mediated communication – such as the PEN system in Santa Monica, CA – have encountered serious problems because of uninhibited "flaming" (i.e. sending messages with personal attacks) that is more likely in electronic than in face-to-face communication (Dutton, 1999).

Because of this tendency to overlook past failures, community informatics proponents are unlikely to consider the possible negative consequences of their actions, no matter how well meaning they may be. In promoting community networks, for example, are they turning people inward, away from face-to-face interaction, thus undermining "the public, civic sense of cities as physical and cultural spaces of social interaction" (Graham and Marvin, 1996)? Might the promotion of digital interaction "induce democratically unpromising psychopathologies, ranging from escapism to passivity, obsession, confusing watching with doing, withdrawal from other forms of social engagement, and psychological distancing from moral consequences" (Sclove, 1995)? By promoting e-commerce economic development strategies, could they be heightening social and economic divisions and contributing to the development of what Borja and Castells (1997) call the "dual city"? Finally, what are the implications of networking projects for privacy concerns in the light of current threats to individual privacy as commerce and information increasingly become digitised and integrated (Agre and Rotenberg, 1998)?

These questions should force community informatics proponents to reconsider their methodologies and be more self-critical in their approach. In the case of NKLA, we have at times been tempted to view the Web as providing such a revolutionary way of accessing data and information that will greatly enhance the ability of community developers to improve the quality of life in neighbourhoods. While there may be some element of truth in this, it is crucial that we reflect on the failed dreams of the past and consider potential unintended consequences. For example, one possible consequence of which I have become aware is that by using NKLA to access information about properties, it is possible that neighbourhood activists may de-emphasise the face-to-face community organising and relationship building that is necessary for positively developing their communities.

4.5.2 Philosophical Critique

A second set of critiques questions the very philosophical assumptions of a community informatics approach that places hope in the power of technology to catalyse positive social change. According to Leo Marx, the historical roots of current utopian views of technology are found in the 19th-century Enlightenment ideals of social progress, determinism and positivist epistemology (Marx, 1999). This modernist heritage has led to linear, simplistic interpretations of how technology impacts society, thus leaving out the complexities of these relationships. According to a current philosopher of technology, this determinism is based on two premises: (1) that technology

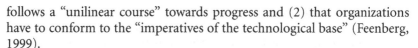
follows a "unilinear course" towards progress and (2) that organizations have to conform to the "imperatives of the technological base" (Feenberg, 1999).

This type of deterministic thinking is obviously embedded in the current hype around the Web and other ICTs and undoubtedly seeps into community informatics thinking and practice. These tenets are likely popular with techno-utopists because they seem to jibe with common sense and justify efforts to solve social problems with the latest technical fix. As historians of technology note, however, seemingly innocuous technical innovations such as mechanical looms, automobiles and piped water systems have always had negative, as well as positive, impacts (Sclove, 1995; Marx, 1999). History tells us that there is not a simple, linear relationship between technological innovation and social progress. This should produce caution in proponents of community informatics, especially given our discussion of unintended consequences in the previous section.

While the debate over determinism may seem little more than an argument of semantics – and writers such as Castells and Kranzberg characterise it as a "false problem" – it is important to guard against determinism for two major reasons. First of all, given the many negative impacts of technological innovations throughout history, determinism can just as easily lead to a bleak dystopian outlook as it can to a utopian interpretation. Therefore, it sets up the common dualism between optimists and pessimists, technophiles and technophobics, who share this deterministic outlook despite their divergent conclusions. The second reason to reject technological determinism is that – as I have already mentioned – it tends to oversimplify causal relationships and – as I will explore more deeply below – denies the complex social and cultural contexts in which technologies develop.

Besides the debate over determinism, several of the most prominent philosophers and social critics of the 20th century have questioned the role of technology in society on more substantive grounds. Writers such as Heidegger, Weber, and Ellul represent "a grand tradition of romantic protest against mechanization", which argues that "technology is not neutral but embodies specific values" (Feenberg, 1999). Likewise, for Habermas media co-optation and the "technization of the lifeworld" are impediments to democratic discourse through communicative action. For Jean-Francois Lyotard, technology is "a game pertaining not to the true, the just, or the beautiful, etc., but to efficiency" (Lyotard, 1984). This ontological argument has more recently been taken up by Albert Borgmann, who laments the social distancing impact of "hyper intelligence":

> Plugged into the network of communications and computers, they seem to enjoy omniscience and omnipotence; severed from their network, they turn out to be insubstantial and disoriented. They no longer command the world as persons in their own right. Their conversation is without depth and wit; their attention is roving and vacuous; their sense of place is uncertain and fickle. (Borgmann, 1992)

These substantive critiques of technology should likewise produce caution and reflection for community informatics advocates, as they again question the unintended consequences of increasing reliance on ICTs. Instead of "increasing democratic participation" as many community informatics proponents naively assume, it is important to consider how these new technologies might deepen current divisions and create new ones.

Implicit in the NKLA strategy is the assumption that providing more data about properties and neighbourhoods will lead to enhanced community participation, improved decision making and – ultimately – improved quality of life for neighbourhood residents. This assertion opens us up to charges of determinism. Does this website lead directly to the improvement of neighbourhood conditions? Do the data themselves? The philosophical critique also forces us to consider possible negative consequences of our work. Are we increasing a division between NKLA users and non-users?

4.5.3 Ideological Critique

I call the final set of critiques ideological, though they could perhaps just as easily fit within the philosophical category. Therefore, there is substantial overlap between the critiques in these two areas. Perhaps the defining feature of what I call ideological critiques is that they tend to come from a "postmodern" or "poststructuralist" perspective. For example, one of the basic questions raised by postmodernism is that of agency and expertise. Whereas traditional modernist views of technology led to a form of "technocracy", popular social movements and postmodern writers of the 1960s began to question the foundational positivism and determinism of science and technology. Building on the substantive critiques of Heidegger, writers such as Marcuse and Foucault analysed the power relations inherent in the technocracy. As Feenberg explains,

> It is not easy to explain the dramatic shift in attitudes toward technology that occurred in the 1960s. By the end of the decade early enthusiasm for nuclear energy and the space program gave way to technophobic reaction. But it was not so much technology itself as the rising technocracy that provoked public hostility. (Feenberg, 1999)

These doubts bring into question the very role of technical experts, something that has prompted crises in many professions. More fundamentally, postmodern writers such as Alcoff challenge the right of experts to "speak for others", as all forms of communication and representation are biased by the contexts of both the giver and the receiver. Speaking for others perhaps becomes even more problematic with electronic forms of communication because, as Alcoff (1995) points out, it is "increasingly difficult to know

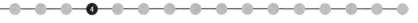
anything about the context of reception". Obviously, technical experts appeal to the supposed neutrality of technology in order to retain their "privileged space", but postmodernism's critique reveals the prominence of biases in the technocratic tradition.

While, in general, advocates of community informatics may not be aloof technocrats, it is crucial that they consider the critiques outlined above, as there is likely some degree of "representation" or "speaking for others" in any conceivable community-based technology innovation. Probably the great majority of community informatics projects involve some type of information provision, often to information that would not otherwise be readily accessible to community residents. One common vehicle for providing community data is through GIS software, and geographers with a postmodern sensibility have raised serious ethical concerns about the use of GIS. As Curry (1995) warns, for example, any data used for policy decisions projects some type of representation of the "other", thus insinuating that those with access to the information are "empirically better able to make decisions" than those who do not.

This should produce reflection for those who purport to enhance policy making or community involvement through community informatics, as there is generally some type of filtering process before providing information in maps or on a website. The question of how that representation takes place is critical for ensuring that community informatics projects move towards their goal of improving communities by mobilising residents through information technology. As Curry points out,

> It is helpful here to distinguish between knowing *how* and knowing *that*. Knowing *how* refers to the ability to do something, the ability of the average person, say, to use a computer, to enter data, or to do analysis using simple, perhaps menu-driven routines. Knowing *that* refers to knowledge about how something works. When we look at the role of technology in society, we find a somewhat complicated story. In the last few years some technological systems – and computers are a prime example – have obviously become much more complicated. Hence, for the average person, the ability to operate a system is increasingly a matter of knowing *how*, and decreasingly a matter of knowing *that*. It is less and less related to the understanding of how a system works, and technological systems look more and more like black boxes. (Curry, 1995)

If the leaders of community informatics projects do not recognise their own privileged position as experts and how this role is challenged by these ideological critiques, they will limit the political viability of their work and, ultimately, their impact on communities.

Again, the experience of NKLA is illustrative. The primary designers of NKLA were not residents of disinvested communities, but rather professionals working to improve those neighbourhoods. While we have tried to involve a wide range of people – including residents – in the design and implementation of NKLA, we cannot deny that our work contains a degree of "speaking for others". Our maps, for example, provide specific representations of neighbourhoods, ones that certainly do not contain all important data and potentially contain erroneous data. While I do not view myself as a "technocrat" by any means, I believe it is important that in this work I recognise the privileged position from which I represent communities as a designer of NKLA.

4.6 Towards an Ethics for Community Informatics

The critiques outlined above provide severe challenges to the optimistic assumptions of a community informatics approach to community development. Reflecting on the implications of these critiques, I next offer four recommendations directed to community informatics advocates. In one respect, they are pragmatic suggestions for avoiding potential criticisms of community informatics. On a more foundational level, they could serve as ethical guidelines for utilising ICTs in community development.

4.6.1 Enter Wholeheartedly the Debate over the Role of ICT in Society

Lest community informatics leaders be discouraged by the critiques of their utopian assumptions, they should be admonished to not retreat from these criticisms, but rather to meet them head on. Several writers with a sympathetic, yet critical, view of the community informatics approach present the conflictual nature of information age struggles. Graham and Marvin liken the struggle to fights over neighbourhood planning issues:

> The battle is on over the future complexion of electronic spaces. Many of the new telematics networks on the Internet for special interest groups are 'places' that they defend from incursions in similar ways as physical neighbourhoods of cities. (Graham and Marvin, 1996)

For Kellner, many different forces are positioning themselves for a place in the information-age power structure:

> The Internet is thus a contested terrain, used by Left, Right, and Center to promote their own agendas and interests. . . . Those interested in the politics and culture of the future should therefore be clear on the important role of the new public spheres and intervene accordingly. (Kellner, 1997a, b)

Rather than fleeing from the critiques that can be made against a community informatics approach, it is possible to take advantage of them in developing a more coherent approach. Thus, Feenberg combines a radical Foucauldian "counter-hegemony" to technocracy that will help realise a Habermasian vision of "communicative action" in developing what he terms a process of "democratic rationalization". This approach looks to technology to break down power structure and barriers that have excluded social movements from participation in public decision making.

In NKLA outreach work, we have often come across people who are sceptical that a website like ours can help in the messy, nitty-gritty world of community development while others view the website as the answer to their prayers. We try to convey a middle ground between these two extremes when we train people to use NKLA and other resources on the Web. Perhaps the most pragmatic advice comes from Doheny-Farina:

> Just as it is naive to trust the design of the net to the technotopists, it is equally naive to assume that by turning off our televisions and boycotting the net, we can somehow recapture something we've lost. The only long-term option is to work to use the technologies for the local good... What communities need are people who have some technical skills, a willingness to examine how electronic communication technologies can enhance the community, some drive, and *a healthy dose of constructive skepticism. Bring doubt to every claim about the net, but be committed to moving forward.* (Doheny-Farina, 1996)

Advocates of community informatics would do well to heed these balanced words. It is precisely those who are sceptical about the potential of community informatics projects that need to be involved in designing and implementing them. This will allow them to avoid the methodological errors of succumbing to information-age hype and losing historical memory of past failures. Understanding why this approach is important, as well as the critiques against it, will only help increase the ability of community informatics advocates to impact the political process.

4.6.2 Recognise That Information Technology Is Part of a Social, Political and Cultural Context

Community informatics advocates should anticipate philosophical critiques by rejecting technological determinism, instead of basing their understanding of how technology relates to society in what Graham and Marvin call a "recursive relationship". Melding together a political economy approach with social constructivism, these theorists of electronic urban space contend that technical innovation in cities is "socially, politically, and culturally shaped rather than being purely technical" (Graham and Marvin, 1996). Commenting more generally on technology, Feenberg argues similarly that technological innovation is not due simply to efficiency – be it technical or economic – but rather part of a social construction: "What singles out an artefact is its relationships to the social environment, not some intrinsic property" (Feenberg, 1999).

This constructivist perspective certainly makes analysing impacts and potential impacts of information technologies more complicated, but it will also enrich the potential for community informatics projects. They are tools that can be adapted and used by residents to assist in their community building efforts:

> Technologies, then, are not value free but neither do they determine our future. They are shaped and developed through social relations and thereby offer the potential of being used as a tool by the disadvantaged and excluded to challenge entrenched positions and structures. (Loader et al., 2000, p. 87)

Instead of merely setting up computer centres in low-income neighbour-hoods or providing information over the Web, projects developed with this outlook will have to take into account the "human element" in both planning and implementation. Though not in itself an objective agent of change and innovation, technology is malleable and therefore can be adapted and evolved into innovative uses by opening up information design to a broader group of users (Vaughan and Schwartz, 1999). As Beale explains,

> One of the precepts of community informatics is that information systems *enable* and support human endeavors, rather than replace them. In concrete terms, this means supporting projects or initiatives devised by people in the community, rather than defining them. (Beale, 2000, p. 66)

The experience of NKLA shows how community informatics projects are socially constructed. Despite being part of "cyberspace", NKLA relies heavily on social networks and interpersonal feedback. Since the inception of NKLA, focus groups and training of users have played an important role in its design, as well as in the identification of new data sets and applications for NKLA. For example, the idea of adding community-created data to the site was suggested by participants in a planning meeting, and their insistence on using NKLA to allow residents to present their own "bottom-up" perspectives of their neighbourhoods led to the Asset Mapping component of the site. From surveys and interviews of NKLA users, it is clear that people use the site not because it is "cool" but rather for the data and information they can get from it. NKLA has a substantive and geographic focus, namely housing and community development in Los Angeles. While we get many visitors from outside the Los Angeles area, it is clear that those who use the site most regularly do so because they are working on housing and community development issues in Los Angeles neighbourhoods. Therefore, it is not the technology itself, but rather how this technological tool has been designed and employed as part of a social and political process, that has led to the success of NKLA.

4.6.3 Preserve Public Access to Information

A critical battlefield on which community informatics proponents should position themselves is the debate over the privatisation of public information. As Graham and Marvin explain,

> Telematics and computing allow information to be controlled, processed and managed with unprecedented sophistication and precision. This means that highly individualistic market solutions become possible where previously services often had to be offered at generalised charges or as free public services. (Graham and Marvin, 1996)

This leads to what Mosco calls the "Pay-per Society", in which consumers are forced to pay per phone call, per television show, or even "per bit or screenful of material in the information business" (Mosco, 1988). The "commodification" of information poses a serious threat to access, as only those with the necessary financial resources are able to purchase it. This happens even with supposedly public data sets, such as property records, that are sold by local governments to third party vendors, which then repackage the data and resell it, thus effectively limiting access to supposedly public data.

Onsrud (1998) refers to the privatisation of public information as the "tragedy of the information commons", likening the costs of letting private interests dominate information to that of market interests contaminating the natural environment. The information commons has shrunk as information and data have become increasingly digitised, creating "threats ranging from the elimination of the public's ability to read copyrighted works in public library–like arrangements in our digital future to the creation of scarcities in personal information privacy" (Onsrud, 1998, p. 143). For Schroeder (1999), public-minded professionals – such as librarians, planners, educators and GIS experts – need to take leadership in ensuring that information remains in the public domain. By preserving access to local data and information, communities got through a process of empowering "self-discovery":

> Key to the information strategy suggested here is the formation of what might be termed cooperative lateral information sharing networks. Information that is needed is obtained as raw data in many forms, ranging from hearsay and gossip through the contents of sophisticated databases. This is often fragmentary and colored by contexts of origin and integrated into local frameworks of interpretation. In receiving and interpreting this knowledge, communities have an opportunity to learn about themselves; with this learning comes the potential for change. (Schroeder, 1999, p. 48)

Schroeder's description of local networks that gather and distribute information about communities parallels in many ways the experience of NKLA and similar projects. Website projects such as the Right-to-Know Network (http://www.rtk.net), the Center for Neighbourhood Technology's

Neighbourhood Early Warning System (http://www.newschicago.org/) and NKLA are bucking the privatisation trend by providing free public access to environmental and housing information that community residents and organisations might otherwise have to purchase. Community informatics advocates should support right-to-know legislation and involve themselves in fighting to keep information public. This will ensure that, despite the growing influence of market relations determining trends on the Web, community informatics will retain relevancy by democratising access to data in society.

4.6.4 Open Up Space for Complementing Representational Democracy with Direct Democracy

The final recommendation provides a response to the ideological critique, which questions the privileged role of community informatics professionals. Instead of responding with syrupy rhetoric of "community empowerment", leaders of community informatics projects should acknowledge that they do have their own biases that stem from their role as "technical experts". At the same time, they should resist the development of any type of community informatics technocracy by ensuring that their projects become truly participatory in every stage.

After Alcoff (1995) elucidates the danger of experts speaking for others, she provides four "interrogatory practices" that can help professionals avoid paternalism:

1. The *impetus to speak* must be carefully analysed and, in many cases (certainly for academics), fought against.
2. We must also interrogate the *bearing of our location and context* on what we are saying.
3. Speaking should always carry with it an *accountability and responsibility* for what an individual says.
4. Analyse the *probable or actual effects of the words on the discursive and material context.*

Community informatics advocates should consider the potential implications of representation in their work, critically analysing their own context and taking responsibility for the impact of their actions and words. By considering and heeding these warnings, community informatics proponents can avoid harmfully speaking for others and increase the potential for democratising their work.

As Sclove argues, this democratic process should begin in the design phase and be much more than just broadening participation. Feenberg posits a "deep democratization" of technology that would provide a "popular agency" as an alternative to the dominant technocratic ideology. This is precisely the type of democratic process to which community informatics should strive, and one for which information technologies such as the Internet are particularly suited. With the potential for multiple-way communication in computer networking technologies, community informatics projects can enhance representational democracy with applications that provide residents the opportunity to directly enter policy discussions.[7] A simple example of this might be that a community network hosts an online discussion of city housing policy, in which tenants, homeowners, business owners, municipal staff and elected officials all participate.

A final of caution on this matter must be mentioned. As noted by researchers who conducted a participatory GIS project in South Africa, "the question of 'who' participates will be central to the outcome of a participatory process" (Harris et al., 1995). This statement should serve as a mantra for all leaders of community informatics projects. Just when they think that they have included all necessary "stakeholders" in the design and implementation of their project, they should ask themselves what interests may still not be represented. At NKLA, this self-questioning has encouraged us to expand our outreach among youth organisations and to experiment with providing unedited, community-created content on the website. This constant questioning of assumptions forces us to escape the bounds of technical expert and make our community informatics projects more participatory and democratic.

4.7 Conclusion

The purpose of the essay has been to analyse the assumptions inherent in community informatics in order to chart a balanced course between indiscriminate hope and careless hype. Basing this reflection on both the community informatics literature and my own experience, I have tried to be openly self-critical of a community informatics approach; not because I wish to diminish it, but on the contrary because I hope to support it. Keeping in mind the danger of uncritical methodologies, simplistic philosophies and elitist ideologies, it is my desire that activists and researchers will collaborate in constructing truly participatory, transformative and ethical community informatics applications that support community development. I believe that the four recommendations offered in this essay provide a good framework for measuring how much these projects meet these goals.

---

[7] I should be clear that I do not purport to be an expert on political theory. I am not arguing for "direct democracy" as a superior model to "representative democracy"; rather I am simply suggesting that ICTs can help introduce experiments in direct democracy within a representational system. There is a vibrant discussion on these issues in several recently published books. For example, see Tsagarousianou et al. (1998), Hague and Loader (1999), Hacker and van Dijk (2000) and Wilhelm (2000).

I believe that community informatics advocates would all do well to heed Doheny-Farina's call to constructive scepticism, bringing doubt to the hype but with a hopeful disposition to moving forward. Community informatics appears to be a promising approach to supporting community development. Technically minded people with a social conscience and long-time community activists who have discovered the potential of using ICTs to advance their causes are coming together to develop innovative local informatics applications that serve as tools for community development. It is impossible to tell at this point whether these relatively new applications will profoundly change community development practice for the better. I believe, however, that this likelihood will increase if community informatics practitioners deliberately address possible critiques of their work and apply ethical principles to their work. The needs in low-income communities are substantial, and their residents cannot afford to be disappointed once again. Let us, who know the needs in these communities and see the potential for ICTs to address these needs, be honest and forthright about our work in order to provide true hope without mere hype.

4.8 Acknowledgements

I would like to thank the following persons for their comments on earlier versions of this chapter, namely Neal Richman, Howard Besser, Mike Gurstein and Wal Taylor. Any shortcomings in the chapter are of course my own.

4.9 References

Agre, P. and Rotenberg, M. (eds.) (1998) *Technology and Privacy: The New Landscape.* MIT Press, Cambridge, MA.

Alcoff, L.M. (1995) The problem of speaking for others. In Roof, J. and Wiegman, R. (eds), *Who Can Speak?* University of Illinois Press, Illinois.

Beale, T. (2000) Requirements for a regional information infrastructure for sustainable communities: the case for community informatics. In Gurstein, M. (ed.), *Community Informatics: Enabling Community Uses of Information Technology.* Idea Group Publishing, Hershey, PA, pp. 52–80.

Borgmann, A. (1992) *Crossing the Postmodern Divide.* University of Chicago Press, Chicago.

Borja, J. and Castells, M. (1997) *Local and Global: Management of Cities in the Information Age.* Earthscan Publications Ltd., London.

Curry, M.R. (1995) GIS and the inevitability of ethical inconsistency. In Pickles, J. (ed.), *Ground Truth: The Social Implications of Geographic Information Systems.* The Guilford Press, New York.

Doheny-Farina, S. (1996) *The Wired Neighbourhood.* Yale University Press, New Haven.

Dutton, W. (1999) *Society on the Line: Information Politics in the Digital Age.* Oxford University Press, Oxford, UK.

Feenberg, A. (1999) *Questioning Technology.* Routledge, London.

Ferguson, R.F. and Dickens, W.T. (1999) Introduction. In Ferguson R.F. and Dickens, W.T. (eds.), *Urban Problems and Community Development.* Brookings Institution Press, Washington, DC, pp. 1–31.

Graham, S. and Marvin, S. (1996) *Telecommunications and the City: Electronic Spaces, Urban Places.* Routledge, London.

Gurstein, M. (1999) Flexible networking, information and communications technology and local economic development. *First Monday,* **4**(2). Retrieved from <http://www.firstmonday.org/issues/issue4_2/gurstein/index.html>

Gurstein, M. (2000) Introduction. In Gurstein, M. (ed.), *Community Informatics: Enabling Community Uses of Information Technology.* Idea Group Publishing, Hershey, PA, pp. 1–30.

Guthrie, K.K. and Dutton, W.H. (1992) The politics of citizen access technology: the development of public information utilities in four cities. *Policy Studies Journal,* **20**(4), 574–597.

Hacker, K. and van Dijk, J. (2000) *Digital Democracy: Issues of Theory and Practice.* Sage Publications, London.

Hague, B. and Loader, B. (1999) *Digital Democracy: Discourse and Decision Making in the Information Age.* Routledge, London.

Halpern, R. (1995) *Rebuilding the Inner City: A History of Neighbourhood Initiatives to Address Poverty in the United States.* Columbia University Press, New York.

Harris, B. and Gold, B. (1999) Interview with Alvin Toffler. *Government Technology,* **12**(15), 11–17.

Harris, T.M., Weiner, D., Warner, T. and Levin, R. (1995) Pursuing social goals through participatory GIS: redressing South Africa's historical political ecology. In Pickles, J. (ed.), *Ground Truth: The Social Implications of Geographic Information Systems,* edited by New York: The Guilford Press.

Hecht, L. (1999) *U.S. Community Networks and the Services they Offer,* Master's Thesis, Department of Public Policy, Georgetown University.

Howard, D. (1998) Geographic information technologies and community planning: spatial empowerment and public participation. Paper presented at Project Varenius Specialist Meeting on Empowerment, Marginalization, and Public Participation GIS. Retrieved 5 April 2006 from <http://www.ncgia.ucsb.edu/varenius/ppgis/papers/howard.html>

Kellner, D. (1997a) Crossing the postmodern divide with borgmann: adventures in cyberspace. Retrieved 5 April 2006 from <http://www.gseis.ucla.edu/courses/ed253a/newDK/borg.htm>

Kellner, D. (1997b) Theorizing new technologies. Retrieved 5 April 2006 from <http://www.gseis.ucla.edu/courses/ed253a/newDK/theor.htm>

Kling, R. (2000) Learning about information technologies and social change: the contribution of social informatics. *The Information Society,* **16**(3), 217–232.

Krouk, D. (1996) *Tax Delinquency and Urban Disinvestment in Los Angeles.* Master's Thesis, UCLA Department of Urban Planning.

Krouk, D., Pitkin, B. and Richman, N. (2000) Internet-based neighbourhood information systems: a comparative analysis. In Gurstein, M. (ed.), *Community Informatics: Enabling Community Uses of Information Technology.* Idea Group Publishing, Hershey, PA, pp. 275–297.

Lea, M., O'Shea, T. and Fung, P. (1999) Constructing the networked organization. In DeSanctis, G. and Fulk, J. (eds.), *Shaping Organization Form: Communication, Connection, and Community.* Sage, Newbury Park, CA, pp. 295–324.

Loader, B., Hague, B. and Eagle, D. (2000) Embedding the net: community empowerment in the age of information. In Gurstein, M. (ed.), *Community Informatics: Enabling Community Uses of Information Technology.* Idea Group Publishing, Hershey, PA, pp. 81–102.

Lyotard, J.F. (1984) *The Postmodern Condition: A Report on Knowledge.* University of Minnesota Press, Minneapolis, MN.

Marx, L. (1999) Information technology in historical perspective. In Schön, D.A., Sanyal, B. and Mitchell, W.J. (eds.), *High Technology and Low-Income Communities: Prospects for the Positive Use of Advanced Information Technology.* MIT Press, Cambridge, MA.

McGarigle, B. (1998) Democratizing GIS. *Government Technology,* **14–15**, 48.

Mitchell, W.J. (1995) *City of Bits: Space, Place, and the Infobahn.* MIT Press, Cambridge, MA.

Mitchell, W.J. (1999) *E-topia.* MIT Press, Cambridge, MA.

Mosco, V. (1988) Introduction: information in the Pay-per Society. In Mosco, V. and Wasko, J. (eds.), *The Political Economy of Information*. The University of Wisconsin Press, Madison, WI.

Negroponte, N. (1995) *Being Digital*. MIT Press, Cambridge, MA.

Nunn, S. (1999) The role of information technologies in Community Development Organizations. *Journal of Urban Technology*, 6(2), 13–37.

Onsrud, H. (1998) Tragedy of the information commons. In Fraser Taylor, D.R. (ed.), *Policy Issues in Modern Cartography*. Pergamon, Oxford, UK, pp. 141–158.

Richman, N. and Waldman, J. (1999) Publicizing privatized information: a new role for university-based planners. Paper presented at 1999 Meeting of the Association of Collegiate Schools of Planning, Chicago, IL.

Sawicki, D.S. and Craig, W.J. (1996) The democratization of data: bridging the gap for community groups. *Journal of the American Planning Association*, 62(4), 512–523.

Schroeder, P. (1999) Changing expectations of inclusion, toward community self-discovery. *URISA Journal*, 11(2), 43–51.

Schuler, D. (1996) *New Community Networks: Wired for Change*. ACM Press, New York.

Schwartz, Ed. (1998) An Internet resource for neighbourhoods. In Tsagarousianou, R., Tambini, D. and Bryan, C. (eds.), *Cyberdemocracy: Technology, Cities and Civic Networks*. Routledge, London, pp. 110–124.

Sclove, R.E. (1995) *Democracy and Technology*. The Guilford Press, New York.

Tsagarousianou, R. (1999) Electronic democracy: rhetoric and reality. *Communications: The European Journal of Communication Research*, 24(2), 189–208.

Tsagarousianou, R., Tambini, D. and Bryan, C. (eds.) (1998) *Cyberdemocracy: Technology, Cities and Civic Networks*. Routledge, London.

Turoff, M., Hiltz, S.R., Bieber, M., Fjermestad, J., and Rana, A. (1999) Collaborative discourse structures in computer mediated group communications. *Journal of Computer Mediated Communication*, 4(4).

Vaughan, M. and Schwartz, N. (1999) Jumpstarting the information design for a community network. *Journal of the American Society for Information Science*, 50(7), 588–597.

Wilhelm, A. (2000) *Democracy in the Digital Age: Challenges to Political Life in Cyberspace*. Routledge, New York.

5

Knowledge and the Local Community

Alan McCluskey

5.1 The Global and the Local

When talking about the local community, it is not possible to ignore the influence of the global marketplace and the impact of worldwide networks. Often the global and the local are depicted as being antagonists. From certain perspectives this is clearly the case: globalisation makes local jobs precarious; its disrespect for frontiers tends to steamroller local culture; etc. Taking up arms and opposing globalisation and global networks in the defence of the local community is, to my mind, not a very fruitful strategy. In this text I aim to show that once you have set local knowledge building on the right footing with the measured and appropriate use of such tools as the Internet, you reinforce that community and make the relationship with the rest of the world worthwhile and enriching. When the local community is thriving and healthy, globalisation does not need to be an incurable illness.

5.2 The Example of Universal Access

The question of universal access to Internet-based services is revealing when it comes to the relationship between the global and the local. For that reason, I begin by questioning the notion of "universal access" before exploring knowledge and the local community in relation to Internet use.

5.3 Initial Assumptions

Our calls for action and our discussions about how to embark on that action are generally based on a number of assumptions that often go unchallenged. That is normal. Assumptions are made to save us from having to revisit basic issues every time we act. However, situations change and assumptions cease to be valid especially in a fast-changing world. In addition, the beliefs behind certain assumptions may well not have been questioned at the outset. Masquerading as good ideas full of promise, some assumptions are acclaimed by those whose vested interest consists of building their empire on our unquestioning acceptance of those ideas. The faster they act, the more likely their ideas are to take root and lead to a multitude of ramifications that make undoing what has been done almost impossible. In the mean time, these ideas have been taken up and championed by other honest folk whose only aim is to do good.

5.4 The Battle for Universal Access

Such is the case with the battle for universal access to networks and related services. It would be pretty unpopular to insist on less. In fact, discussion about the subject rarely concerns the desirability of such universality, but rather concentrates on how to make it come about. I would like to take the risk of challenging the assumption that the drive to make networked services universally accessible is desirable or even feasible. Let me hasten to add that I am not arguing for elitism or programmed exclusion. The point I am trying to make is that, paradoxically, in seeking to attain universality, exclusion is the foregone outcome.

5.5 Desire . . .

One of the phenomena to be addressed in any discussion of the use of technology is the impact of the desire created by the availability of a new tool. Let us take the example of mobility. The extent of mobility afforded by motorised transport would have been almost unthinkable prior to its invention. The advent of the train, the car and the plane has modified our perception of the world and created desires even in people not having access to these means of transport. Mobility has become a human right.

5.6 . . . and Exclusion in the Shopless Village

Embodied in the concept of "haves-and-havenots" is both a generalised desire to use a given tool and the judgement that that use is good for one and all. Let us go back to our example of mobility. If there are no buses or trains to your village and you have no car – and what's more, walking, cycling or horse riding is out of the question – when the last shop closes in your village for lack of custom, you could be in a very difficult situation. Note that it was the presence of motorised transport that enabled the development of supermarkets and subsequent hypermarkets that led to the decline of local shops. Going even further back in time, industrialisation produced an ever-increasing dependence of families on shopkeepers for their everyday

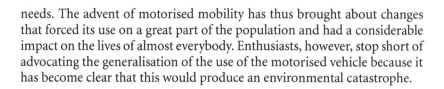

needs. The advent of motorised mobility has thus brought about changes that forced its use on a great part of the population and had a considerable impact on the lives of almost everybody. Enthusiasts, however, stop short of advocating the generalisation of the use of the motorised vehicle because it has become clear that this would produce an environmental catastrophe.

5.7 Without Access in the Networked Society

With electronic networks and related services, advocates of universality are arguing that all those who do not have access (presumably both physical access as well as the necessary know-how to use it) will be seriously penalised in the future heavily-networked society. How will they be penalised? Rather like the shopless village requires cars or public transport, so essential services provided initially both online and "offline" will increasingly be provided only online because it is economically more efficient to do so, leaving those who do not have access out in the lurch. Think of Amazon.com. It is a "place" with an apparently very large collection of books, a round-the-clock personalised advisory system and a fast delivery service that local bookshops cannot compete with. Despite the fact that the corner bookstore has a good number of books for you to browse, that may not be enough to stop a good many of them from going out of business. So you can see how Amazon (despite its very positive and enticing features) could have a lasting negative effect even on those who never use it.

5.8 The Driving Forces

Whether or not those online services like Amazon are really essential will not matter. Many of them will become essential, like the Sunday family outing in the car or the weekly drive to the local supermarket. Between the strongly desired and the absolutely necessary, exclusion comes when these new tools are the only perceived means of access. Social pressure plays a key role. Your friend, for example, urges you to get an e-mail account because it is easier to exchange messages with him because he is already using the Net. Not to mention the ever-present billboards and other sirens that invite you to take the plunge, the commercial forces driving the generalisation of the Internet have a vested interest in making their services appear indispensable to as many people as possible. The same applies to administrations that measure the success of their services in terms of the number of people using them. In addition, the hope for saving in costs on certain services can come only if everybody uses the system. The numerical success of a given service in terms of income or numbers of users is, however, not necessarily an indication of the fundamental need for such a service or of its long-term good for society.

5.9 The Paradox

Against this backdrop, why insist that the idea of universal access necessarily leads to exclusion? Because the concept embodies the assumption that using the network is necessary for everybody, whereas it is far from proven that everybody needs or wants access. What's more, experience with the telephone has shown that – despite enormous efforts – it is not at all easy to provide access to everybody. We could also question the so-called role of the telephone or the Internet in improving communication. Those that advocate "universal access" are cautious about how they approach individual freedom. Their aim is to give everybody the possibility to choose to have access rather than obliging the individual to use the network. Yet the insistence on access is more than enough to create exclusion. The more we insist on the necessity and desirability of access, the greater the feeling of exclusion will be in those who do not have it, even if they do not want it or need it.

5.10 Will People Be Free Not to Have Access?

In asking ourselves the question "Do I really need to use the network for this activity?" we are already tacitly giving the network priority over our own individual and collective fundamental needs. Of course, the network is marketed as satisfying many of our needs: for communication and exchange, for community, for learning, for work, for health, etc., not to mention a lot of exciting new needs and satisfactions we had never even dreamed of. Faced with all these promises, it might be propitious to draw back a moment, before we decide, and ask ourselves what we really need, especially in our day-to-day life in the local community. Doing so is no easy task, if it is at all possible. It requires us to free ourselves from those desires inculcated in us by a host of vested interests whether they be commercial, political or personal. A possible solution is the "alien" perspective: how would someone from another planet "understand" what we are doing. Such a "Martian" estrangement coupled with the naïve, almost child-like approach to the world, in which the self-evident is questioned, might prove useful.

5.11 What About Knowledge?

Is not the attention given to universal access symptomatic of a belief that possessing the tools is in itself sufficient? "Just plug-in and away you go!" says the slogan. Strangely enough for a proclaimed "knowledge society", emphasis continues to be placed on the possession of mass produced tools rather than on the individual development and use of knowledge. Access to

the Global Information Infrastructure without the knowledge to use that access is meaningless?

5.12 Empowerment?

Access to the Global Information Infrastructure and (occasionally) the related knowledge are seen as a source of empowerment for those excluded from power. Is this hope justified? When portable video came onto the market in the 1960s, militant organisations were full of hope. Here was a tool that would democratise mass media. Events turned out quite different. For one, militants and would-be artists had overlooked the problem of language. Although the Net is not really comparable, there has been a similar belief in the empowerment of tools forgetting that most disempowerment springs from social systems and the lack of human, rather than technological, skills.

5.13 Beyond Universal Access

My discomfort with texts advocating widespread access to information, however, goes beyond the idea of universal access. I have come to believe that a good part of my discomfort is due to the fact that the information to be accessed is almost invariably created by others and comes from elsewhere. We have been dispossessed as creators of knowledge (at least in the minds of those advocates) and our personal and collective experience has been relegated to the distant background. The underlying paradigm is that of a society of knowledge consumers rather than a community of knowledge creators. This is rather unfortunate, to say the least, because following the latter paradigm confers a much greater significance and importance to the use of the Internet in the local community. In what follows, I explore such a creative use of the Internet in a project called Saint-Blaise. Net in a small Swiss village (Figure 5.1).

5.14 Creative Approach to Knowledge

The beginnings of my experience with Saint-Blaise.Net – integrating the use of the Internet in a village of 3000 inhabitants in Switzerland – indicate that the local community must first build on its own knowledge and experience. Saint-Blaise.Net is a creative approach to knowledge rather than one of consumption. The accent elsewhere seems to be on a passive model in which most people are seen and see themselves as consumers rather than as creators of knowledge. At best, people can play around with information created by others.

Figure 5.1   Saint Blaise: The environs.

5.15 Models of Learning as Consumption of Knowledge

Despite a more experiential approach to learning in early school years, young people are rapidly channelled into a way of learning centred on consuming the knowledge of others rather than building on personal experience. What's more, they learn that the criteria for judging their knowledge can only be set by others rather than by assessing it themselves in relationship to their needs and those of the situation they are in. The very principle of educational institutions is that they are privileged places for learning in which knowledge is a pre-packed commodity dispensed by appointed experts (teachers or professors) outside the learners' daily experience. Adult society, given the complexity of the world, has delegated much knowledge generation to experts. By experts, I mean those people we pay to develop knowledge for us. You might call them professional learners, but that is only a part of their role. Their work consists of providing or developing knowledge about a particular problem that directly concerns us. This is seen as an economical approach as it avoids the costly task of taking all those concerned through a possibly lengthy learning process. The difficulty arises when the experts attempt to transfer the knowledge they have gained to the rest of us "ordinary people" so that we can integrate it into our everyday lives and work. Unfortunately, what experts have learnt by working their way through the problem cannot be communicated so easily to those ordinary mortals who have not been down that road.

5.16 Knowledge Based on Experience

What sort of knowledge are we talking about? First of all, there is an ongoing confusion between information and knowledge. I too have treated the two here almost as if they were synonymous. When we read about access to knowledge, often the authors mean information – that is to say: news, weather reports, prices, stock values, statistics, etc., as well as scientific facts. When they talk about creating knowledge, they generally imply "filtering, sorting and prioritising" what is already available elsewhere (to quote one of them). Have you noticed how dreary people can be when they recite ill-digested facts they read somewhere? Why should that be so? Because those ideas are disembodied and lifeless. That is why, I stress the link between knowledge and experience. When people talk to others about their experience, what they have to say is anchored in their being, in their lives, even if their memories of what they relate have become distorted with time. Their tale is generally less abstract and more present than the more general, disembodied considerations that are often put forward as knowledge. What people have to say not only creates a link between them and us and weaves the threads of community around us, but it invariably leads to subjects from which we can learn much for ourselves. It is this knowledge based on experience that I am talking about. This is the knowledge which when shared serves to build and reinforce the community.

5.17 Against the Backdrop of the Past

In this local knowledge, there is both memory and accumulated knowledge in the form of experience. Memory as understood here is not the recall of what has been stored in the brain, but rather a more or less structured past experience about people, places and events, and about how and why things are done. Recognising the role of memory is particularly important in talking about knowledge and the local community because memory represents a good part of the background against which current activities take place. Without this backdrop, what we do does not really make full sense. Without it, our acts are cut off from their roots.

5.18 Recognising the Value of Personal Experience

In reality, much of the knowledge available in the local community is not widely known or shared, even within the community. The loss of the oral tradition and the notion of apprenticeship or companionship have – maybe – contributed to the decrease in the circulation of such experience. The ever-increasing size of local communities has no doubt also had its impact as has people's increased mobility and the resulting ever-changing local population. In addition, many people are reticent about sharing their experience or are convinced they have nothing to offer. One of the major tasks in opening up and sharing experience and knowledge locally consists of encouraging individuals to see that their experience could be of value to others. This can be achieved by providing suitable forms of recognition for that experience. Note that sharing experience implies a high level of respect for others that includes respecting the individual's right to withhold his or her experience should the person wish to. It also means that listening to what others have to say is our free choice. Publishing on the Web makes such a choice possible. Creating mutual confidence is, as a result, extremely important.

5.19 Communicating Experience

Once the barriers to expression have been overcome, there is still a need to communicate that experience and knowledge. How good it can be to share our experience with others – but talking to people about what you know has its limits if you want to do so more widely without losing something of the heart-felt direct contact. It is here that the Internet can play a key role both as a less rigid form of publication (the Web) allowing people to evolve

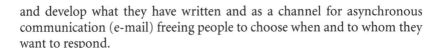
and develop what they have written and as a channel for asynchronous communication (e-mail) freeing people to choose when and to whom they want to respond.

5.20 Writing

Bringing people to write about their experience can be difficult. Many people have given up on writing as a form of expression, daunted as they were by earlier unpleasant experiences. Much of the difficulty people had in writing at school was probably due to the fact that what they were obliged to write was not based on their own experience but on imposed themes and ill-understood styles. If the success of teaching were to be measured in its statistical frequency in getting across a message, then teaching would have been very successful in having people learn that they cannot write. At the same time, many people can tell a pretty good story, given the right conditions. How about getting people to tell their story to someone in whom they have confidence, someone who has less problems with writing it down?

5.21 Interview

So one way of approaching the writing of experience is to use the interview. Listening carefully (full of care) to what others have to say is a first step towards helping them recognise the value of their experience. Writing what they say and then submitting those texts to them for correction is a further step, this time towards the written formulation of their ideas. This is not always easy however as difficulties can be encountered with the choice of words and style. I have always tried to remain close to spoken language in the belief that the reader will retain something of the experience of listening to a tale told. Unfortunately, most people have been taught that the written word cannot be like spoken language. They stifle the natural expression of their words by trying to make them what they imagine to be formally acceptable. Publishing the corrected text of the interview is not only an additional step towards the wider recognition of the value of their experience but also a first move towards widespread sharing of that experience within the community. A further difficulty can be encountered here as some people are hesitant about making their words public. On the one hand, there are those who, as already mentioned, feel they have nothing of value to say. On the other hand, there are those who do not want to be criticised for placing themselves in the spotlight. This may be a reaction to interview writing in the press where the motivation is often to make a star out of the person, because only the opinions of key players count. It is here that the use of interviews in the local community distinguishes itself from the practice of the local press. Interviewing in the local community as a source of knowledge building is not about the spectacular, but rather the heart-felt and the discretely personally revealing.

5.22 Beyond the Written Word

Note that not all knowledge needs to be or should be written down. One of the errors of advocates of so-called knowledge management is that they believe they can get all useful knowledge into a computer. This is not only impossible but also undesirable. Much context-based knowledge, what is called "tacit" knowledge, is best kept out of a computer because it changes so rapidly and depends on a context that can hardly be categorised in a computer. For such knowledge, the Internet is an efficient means of locating and contacting those who have the knowledge that interests us rather than trying to cram their knowledge into a database. For the local community, this implies that the Internet needs to be used in conjunction with other local activities so as to put people in contact with each other or at least to offer the possibility of getting in contact with each other when it is desired or required. This applies to the work on interviews. It is not aimed at recording a person's memoirs, but rather in facilitating the contact and the exchange between people as well as building a minimum written memory. It is also about helping to identify those people who have experience that might be of interest to us. Following on from an interview with one of the local bakers, we received a great many e-mails from people who wanted to know more about the profession.

5.23 Structures for Sharing Knowledge

Sharing knowledge within the local community can take place at any time and in any place. But a number of contexts play a privileged role. Clubs and associations are good examples, as are the church, trade unions and political parties as well as the family. All are contexts in which individuals can share their experience and knowledge with others. There are, however, limitations to these structures when it comes to intensifying the exchange and development of knowledge – which after all is the underlying aim of a so-called "knowledge society". One of these limitations is the restriction to experience circulating between these structures. However, probably the most important limitation is that sharing experience is not considered a central activity of these groups. If it is perceived at all, it is seen as a secondary, additional advantage. Some work on a European level has pointed to associations as potential relays for important learning processes about citizenship and information and communication technologies use and more generally as a vehicle for lifelong learning. What is needed is a catalyst that works with existing structures to focalise more attention on sharing experience and developing knowledge. This is exactly what Saint-Blaise.Net is about. The aim of Saint-Blaise.Net is to encourage social, cultural and economic development within the village by a judicious use of the Internet. Having said that, the Internet, in a way, is only a pretext, even if it is also a very useful tool. The Internet is not a central preoccupation. The aim has nothing to do with the glorification of technology.

5.24 Why Share Knowledge?

Sharing knowledge consolidates and enriches the community not only by nurturing knowledge and thus facilitating decision making but also by strengthening the links and bonds between people. At the same time, it contributes to the well-being of each and every individual by reinforcing his or her sense of personal value as a human being (rather than purely as a wage-earner) and by strengthening his or her relationship to those around them. Sharing knowledge on a local level contributes lastingly to the sustainability of society.

5.25 And Knowledge from Elsewhere?

Do not misunderstand me. In talking of what might be called "home-grown" knowledge, I do not mean to suggest that knowledge "made elsewhere" should be neglected. On the contrary, it is extremely important. It is just that the balance in our society is disproportionately weighed in favour of knowledge from elsewhere. What I am stressing here is that knowledge acquired from outside the community needs to be integrated into the solid foundations of knowledge developed and shared within the community. What's more, knowledge acquired from abstract sources, for example by reading, needs to be likewise, integrated via our experience. Marc-Alain Ouaknin[1] expresses an attitude to reading that seems to me to characterise best what I mean. Rather than taking what is written at its face value, as dogma if you like, he pleads for an ongoing exploration that goes beyond the self-evident.

5.26 Building on Local Experience

One of the extremely delicate balances in society is that between constancy and change. Too much of one or the other can be unhealthy for all concerned. This is true at all levels of life from the single cell to the whole universe. It is particularly true of the local community. In talking of knowledge within a village like Saint-Blaise, we are at the heart of that problem

---

[1] Ouaknin Marc-Alain, Lire aux éclats-éloge de la caresse, Seuil, 1989.

of balance. People's lasting value to society does not necessarily lie in their current performance but rather in accumulated past experience and how this can be used today. The coherence of local society is not to be found in current trends or fads, but in time-honoured activities that structure time and space and give a meaning to life. At the same time, the village cannot exist isolated from the world around it. So-called globalisation is one aspect of the interconnectedness of the modern world. Unfortunately, as mentioned above, much of the discourse about globalisation ignores the local community and its rich past. It is this sense that one could say that globalisation leads us blindly to much misery in an unsustainable future. On the other hand, experiences flowing into the local community from the outside world can bring a refreshing, if not essential, breath of fresh air: new ideas, new ways of doing things, new ways of seeing the world, a wider, more enriching perspective as well as unexpected answers to otherwise incurable problems.

5.27 Integrating Knowledge from Elsewhere

Finding the right balance between the old and the new, between the tried and tested and the excitement of risk-taking, needs to be anchored in our approach to knowledge. I suspect although I do not have the proof yet, that opening up and exchanging experience within the local community will facilitate the judicious integration of knowledge and experience from elsewhere. Being solidly anchored in a feeling of the value of their own experience and at the same time having come to terms with diversity on a local level should make it possible for the local community to feel less threatened by innovation and more open to new experiences. At the same time, they should be better equipped to judge the appropriateness of what is available and freely choose what is best for them and their community.

6

Connected Memories in the Networked Digital Era: A Moving Paradigm

Federico Casalegno

6.1 Preamble

Where communal memory is concerned, we have at least three emerging paradigms that help us to understand the linked entities of evolving social memory and the diffusion of communication technologies.

In the first of the suggested paradigms, we may appraise the connected neighbourhood with the aid of a socio-anthropological "viewing lens"; in this case, interactive technologies may be called on to increase community cohesion, thanks to the possibility of better recording, storing, and sharing of local historical knowledge and social memories.

The second paradigm suggested is the "living memory"[1] vision, in which interactive technologies may provide members of a geographically based community with the instruments to capture, share, and explore the richness of their social memory and local culture. Civic nets and Internet cafés may be seen as part of this living memory paradigm.

The third paradigm, here designated as the "mobile casket", derives from the extensive use of mobile devices that have the twin functions of allowing users, first of all, to collect and store personal information (images, videos, audio, text) and, secondly, to share this information with family, friends, and colleagues.

---

[1] "Living Memory" was a European Commission – IST-sponsored project, funded in the late nineties and executed by a consortium of five international partners. For further information on this project, refer Stahis, K., Debruijn, O., Purcell, P. and Spence, R. "*Ambient Intelligence: Human–Agent Interactions*", Chapter 13 of this book, or visit http://web.media.mit.edu/~federico/living-memory, accessed 1 May 2005.

6.2 Introduction

The intricate relationship between information technologies, community, and social memory is critical to a fuller understanding of the evolution of our societies. Telecommunications and information technologies modify the process of accessing and storing data and knowledge, and consequently they also modify our relationship with both social and historical memory. Our various virtual communities continue to expand on a planetary scale, and hence the neologism "Global Village", while concurrently, their physical counterparts are progressively being contained within specific contexts and places, often referred to as a process of "localization". And then, there is the unpredictable evolution of today's communication systems, an evolution that has been defined by Albert Einstein as the third bomb of the 20th century, after the atomic bomb and the demographic bomb.

Therefore, as communication systems continue to develop, we face new scenarios with imprecise boundaries, that lead to endless new opportunities for establishing the relationship between social memory, community, and information systems.

How can the knowledge process dynamic, together with the storing and transmitting of information, sounds, and images through digital devices, affect the communal memory and the way we both conceive of and create communities?

We have at least three emerging paradigms that help us to understand the phenomena concerning the evolution of social memory linked with the recent diffusion of communication technologies.

The first paradigm views the connected neighbourhood with a socio-anthropological lens; from Web sites to blogging, interactive technologies tend to cohere community structures, thanks to the possibility of better recording and the better sharing of historical knowledge and social memories.

The second paradigm is represented by the "Living Memory"[2] project vision: interactive technologies providing members of a geographically-based community with the means to capture, share, and explore the richness of their social memory and local culture.

The third paradigm, the mobile casket, derives from the widespread use of mobile devices that have the double functions of allowing users to collect and store personal information (images, videos, texts) and personal data and subsequently share them with family, friends, and colleagues, thanks to recent developments in communication technologies.

---

[2] "Living Memory" project, European Commission IST sponsored project, see note 1.

6.3 The Socio-anthropological Lens: First Paradigm

A first level of analysis leads us to consider the dynamics that character-ize a community which is using interactive media to nourish and feed its socio-historical memory. From the origins of the World Wide Web to the present time, we have witnessed a simplification of the interactive informa-tion model. If, at the beginning of the nineties, only a restricted elite could enrich the electronically mediated community memory, we are witness now to a progressive social appropriation of the collective digital space. This is due to the lower cost of technologies, the ubiquitous access to networks, and the simplification of the current user–interface interaction models.

Recent technologies increasingly allow lay communities and associations to use these networks in order to sustain their collective memory, to inform, document, and share their social history and to keep their local traditions alive.

If we consider virtual communities and their Frequently Asked Questions (FAQs) section, we find an excellent example of how online communities self-organize to collect, classify, and categorize the areas of common knowl-edge, thereby systematically creating a collective community memory. For the new users who join such a community, it is easy to access this collective memory, the common knowledge thus progressively becoming part of the community itself.

6.3.1 Metive and the Oral Literature

Among the innumerable projects that we could mention as examples of how recent technologies can increase the process of creating a social collective memory, we can take the Metive [3] project.

This nonprofit organization, created during the seventies, wanted to pre-serve the rural traditions of the Poitou-Charentes regions in France. This association discovers and collects the social memories and the cultural pat-rimony that is orally transmitted by the Poitou-Charentes inhabitants. They have now integrated in their mission the use of the World Wide Web and related digital technologies to store all aspects of popular local culture: from language to music, from popular mottos to legends, from dances to songs, from oral arts to local customs to manual know-how. This precious evi-dence of "oral literature" is available on the Internet, where people can read texts and listen to original audio excerpts, access this very rich communal memory and participate and enrich it, uploading their individual memories and knowledge.

---

[3] http://www.metive.org, retrieved 8 May 2005.

This process leads to two related and extremely important phenomena. The first is that they capture and store local memory of the region simply and accurately. Tradition, habits, songs, and local folklore are stored and made accessible to the local and worldwide population over the Web.

The second is that the local population, being very active in the collecting, storing, and nourishing of this oral memory, and seeing their social collective memory organized online, they have acquired an enhanced consciousness of the richness of their local oral tradition, thereby increasing the community awareness of this.

The combined results of these two phenomena stimulate social cohesion and catalyze the emergence of social ties.

6.3.2 Christo's Gates

We can also observe ephemeral events and "just-in-time" communities that emerge on the top of a singular event or particular facts, using blogs to publish, capture, and share memories. The artist Christo, in February 2005, created a temporary art event in the New York Central Park, decorating the park with more than 7000 orange gates as showed in Figure 6.1. The artwork was on display for 2 weeks, and visitors came across the park walking through the gates, improvizing picnics, happenings, parties, and talks. Digital cameras, video cameras, mobile phones, etc., were used to collect, document, and store both the artwork and the atmosphere that emerged during those 2 weeks. Now that the gates are gone and the Park has regained its normal appearance, there is an online blog[4] that remembers the event in its entirety. What is important is not only the simple fact of remembering and reproducing the images and sound of the event with digital media, but more importantly the fact that everyone has the opportunity to participate in the creation of this collective memory: crowd, snow, sun, cold, conversations, memories, kid's scratches, workers. The location in a digital place on the Web allows for the sharing of conversations, emotions, and images, where the memories of all the participants are connected and shared. A group of individuals that experiences the same event individually or in small groups becomes a postexperiential community, thanks to the creation of this collective memory.

Similar projects, more or less complex than Christo's gates, are very frequent on the World Wide Web. One of the most powerful examples of just how people participate in building the collective memory, in another scale, is represented by the Wikipedia (or Cellphedia [5] with mobile devices), where people are collectively engaged in building the collective memory.

[4] http://www.gatesmemory.org, retrieved 8 May 2005.
[5] http://www.cellphedia.com, retrieved 30 May 2005.

Figure 6.1 Pictures from the Christo's gates event, February 2005, Central Park, New York City.

In general then, it may be said that people are progressively more able to participate in the construction and the sharing of collective memories and, as Serge Moscovici (2001) claims, everybody participates in the making of small myths that circulate within our communities, especially the myths that everybody talks about and everybody is aware of. In doing so, we all create and participate in the circulation of gossip and the local urban legends.

As a matter of fact which extends beyond the possibility of acting, every human being needs to rationalize their actions through speech: our actions would not have any value if we did not transform them into something that can be expressed through speech. Narrating our actions to others is at the

115

same time telling ourselves and in this sense we are all untiring and insatiable *mythmakers*. Thus cyberspace becomes a perfect environment that allows users to express their potential of being *mythmaker*. Blogs, photoblogs, newsgroups, discussion forums, or mailing lists are based upon the contribution that every participant makes by sending information and evidence and enriching this communal space with personal content and experiences.

6.4 The Living Memory Vision: Second Paradigm

With the European Living Memory project approach, a particular vision for the social construction of the collective memory with digital media emerges. This project, completed in at the beginning of the decade by a team of five European national partners (academic and industrial), eventually resulted in the design and the creation of an innovative communication environment. The overall goal was to create a networked communication environment to support the collection of local history, the diffusion of local news and the sharing of personal experiences and memory through multiple media that were integrated in the physical community.

The challenge of the project was not only to design a communication system to favor the storing of historical and formal memory of a defined suburban community in Scotland, but a further challenge consisted in creating a communication environment to help the members of the local, physical community to share information concerning their daily, ordinary lives. It is not just a matter of projecting a merely technological means that can collect and provide formal information about a group, but it is more importantly a matter of conceiving a real communication environment to collect, keep, and spread daily communications and informal exchanges between the inhabitants of a community (Mc William, 2001).

The early design solution for the project was based on three elements that together created a communicational environment to help a local community share its everyday memory.

6.4.1 The Design Solution

1. *Collective memories.* The first element is made up of different large "screens" that continuously provide information about the activities of the community, and the information is provided by the inhabitants themselves. These screens can be installed in squares or shopping malls, bus stops, or public squares.

2. *Strategic memories.* The second means is an electronic personal "token" that the community members carry with them. When they read an interesting piece of information on the large screen, they can "catch" it by simply sliding the token in the device under the screen. The token does

credit: Philips Design

Figure 6.2 Screen with personal electronic token.

not contain the piece of information: its function is that of letting us access the data, it is a strategy of accessing memory just like a bookmark of a browser.

3. *Personal memories.* The "interactive table" is the third element and it is made up of a tactile screen that enables the users to access and send information, directly interact with the contents of such piece of local information and selectively communicate this information to other users. These "interactive tables" are connected with the network and can be placed in public places like bars, libraries, schools, and shopping malls.

Local users, who create the content and interact with it, create the collective memory of the community: in order to help the search for information within this flow of memories and contents, a number of software "intelligent agents" with various roles were developed. This software agency helps us in searching and finding the information we are interested in. Members of a local community can use these interactive multimedia stations to access and post specific information in the Living Memory system and thus become a part of the constant shaping of their communal living memory.

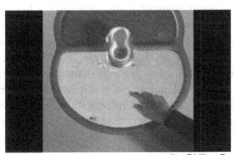

credit: Philips Design

Figure 6.3 View interactive table with token holder.

117

These three nodes of memory – large screens, personal electronic token, and the interactive table – combined together create a very special communicational environment. However, more than the design solution, which evolves along with the technological development, we would like to emphasize the particular vision on the relationship between memory, new media, and community that emerges with the Living Memory project.

6.4.2 Territory as Interface

Within the communication environment provided by Living Memory, the physical dimension of the neighbourhood and the digital information infrastructure are superimposed on and are sustained by each other. The growing volume of available digital information scattered over the territory emphasizes the importance of physical places; the environment becomes a real memory interface, a connective tissue. It is not a matter of substituting the "real" dimension of the territory with "virtual" dimension made possible with the digitization of the information, but to integrate this information in the urban physical territory. The cyberspace superimposes the space: in fact, real and digital do not reject each other but they create a wider topography of the places, which William J. Mitchell discusses in *E-Topia* (Mitchell, 1999). There is a superimposition of physical space and interconnected electronic spaces that support each other in a mutual relationship.

Following the invention of the Internet and the subsequent wide diffusion of the World Wide Web, we have promoted a new vision of the interaction paradigms defining collocated communities as "connected" rather than "virtual", as the "EC i3": intelligent information interfaces[6] projects approach showed.

6.4.3 Local Content Shared in Space and Time/Not "Anywhere and Anytime"

Another important element is the relationship between information and space–time. The emerging communication paradigm (as represented in the Living Memory project) is radically different compared with the Internet one, where you may access information "anywhere and anytime". Here, the information is not linked with a well-defined *place*, you can access the same content from any Internet access point worldwide, and it is not tied to a particular *time*, even if for some information the temporal element is crucial.

In the paradigm proposed by the Living Memory project, the information is local and highly contextualized. The neighbourhood's inhabitants feed

---

[6] http://www.i3net.org/, accessed 8 May 2005.

the collective memory with local content: the information is shared and contextualized, and directly tied to the local residents' life.

6.4.4 Fractal Village/Not the Global Village

Another innovative aspect raised by the Living Memory vision concerns the shift from the Global Village toward the Fractal Village paradigm.

When the Living Memory project team began to design the communication environment, the Global Village paradigm was dominant and we were facing the global spread of the World Wide Web and a progressive democratization of the opportunity of accessing the Internet; simultaneously we were experiencing the feeling that the spreading of communication systems could finally free us from space-time constrictions, enabling us to interact with the most distant people and cultures. With the Fractal Village, on the contrary, we pointed out the importance of the local-based community and the importance that every single community's member has for the enrichment of the local memory and culture. In fact, as in a fractal image, every single part of the image contains the whole image and vice versa: within a local community, every single person incorporates the whole historical, cultural, emotional, and social elements of the community to which they belong. And the entire community is expressed through each individual. Moreover, in order to let this happen, the communication system has to allow all the community members to take part in both the creation and the enrichment of the communal memory, nourishing it with the information that the inhabitants themselves believe to be relevant and useful for both themselves and for the community itself.

The Global Village, together with the associated virtual communities, made up of both symbolic and abstract entities, and both obscure and desired entities, all having vague boundary outlines. The vision proposed by the Living Memory project is based upon the fact that new communication technologies both have the capacity to help and are needed to help users be part of the Global Village, allowing such people to create bonds and additionally create a sense of community between people who shared common interests. However, they also have to allow members of a local community to get to know each other better and to concretely participate in their own social and communal life.

6.4.5 Universal Access

Communal memory can be collective only if each community member participates in its creation. In this sense, user interfaces, interactions with the relevant information and associated devices, and access points promote the access of every member of the local community.

119

Furthermore, in order not to upset existing social dynamics and social rhythms, we thought about a system that could provide access from public places and to group media; instead of designing a system that could isolate users in their homes, we worked on a system that could encourage people to use media in public places.

6.4.6 *Digital Grandmother*

In the mid-nineties, Nicholas Negroponte published *Being Digital* showing how, in the foreseeable future, a *digital grandmother* will help the user to better navigate and retrieve the appropriate information from cyberspace (Negroponte, 1995). Personal customizable intelligent software will help people to have the right information at the right time, and to provide assistance to appropriately interact with the enormous amounts of data and information. Moreover, because these software agents know the users, they will help them in their everyday tasks and activities. Along with this paradigm of the *digital grand-mother*, the Living Memory project team defined another paradigm where the *real grand-mother* could "seed" the system's information from her personal experience.

For example, if students learn at school about the Second World War, they have institutional media that relate institutional information with fact and events, transmitting an historical memory. The *digital grand-mother* software agent, in this case, helps them access the relevant information on the Web, automatically records TV shows, makes a press review, and informs them if there is a radio talk that could be interesting for the matter. With the Living Memory, on the contrary, we promoted a system where the *real grand-mother* can feed the system with personally experienced memories. She can add her direct experience to school texts and institutional media telling, for instance, the story of her life during this period: she augments the official information from her repository of personal memories. Institutional and informal memories resonate, creating a better sense of community belonging and social cohesion.

6.4.7 Informal and Tactical Communication vs Formal and Institutional Information

Within both the local and territorial communities, people exchange a great number of messages, personal announcements, and informal communications that mirror the needs and facts of everyday life. These messages clearly testify as to how the members of a neighbourhood need to communicate with each other and how they need to share a common memory. The communication environment projected within the Living Memory project supports these communication trends that have always been present in local communities without upsetting the rhythms and the already existing social configurations. New media environments, such as Living Memory, support the informal and interstitial memory, that is the one made up of ordinary and daily conversations that let the inhabitants of a specific and located place establish social relationships with each other, get to know each other, and exchange favors, services, and useful information.

These visions, embedded in the Living Memory project, were essential to the design solution and the vision that drove the Living Memory concept. Beyond this specific project, these six points help us to understand the relationship between communal memory, community, and new media. Nowadays, with the diffusion of wireless mobile devices, memory, and community, the media ecosystem becomes more complex. A community, in order to exist and sustain itself, must have the opportunity to share a communal memory; this happens when the community as a whole can access and nourish this memory in a constantly enriching, modifying, and appropriating process. As a matter of fact, it is not enough just to build piece after piece of a fact in order to construct a social memory. However, it is necessary that this reconstruction is based upon commonly held memories that are to be found more in our individual personality than in that of the other community members, as these memories are continuously passed from one to the other (Halbwachs, 1986). The aim of the Living Memory team was that of designing a system that allowed this constant flow of information supporting the storing and the exchange of commonly held and shared memories.

6.5 The Mobile Casket: Third Paradigm[7]

At the dawning of the development of new communication technologies, with the paradigm of the Global Village, the virtual communities have been blurring the geophysical and territorial dimension of the community. The social space of these communities in the *Global Village*, in opposition to the idea of a place, was "anywhere": anywhere, generic and neutral, abstract from the physical territory. Now, on the contrary, we can better understand the synergic superimpositions of digital information ambience on both the surrounding physical and social territories, and how these complementary dimensions create an hybrid, richer, and original space. A "third space" therefore emerges, created by the synergy between the physical and the electronic space, that has peculiar characteristics: it is fluid, dynamic, intangible, but it can be "inhabited". The emerging of this "third space" represents a new paradigm: the social space is no more "anywhere" as for the

---

[7] For this point 6.5 see Casalegno, F., Susani, M., Atlas of Aural knowledge book under review for publication. For further information see: http://www.auralknowledge.net, retrieved 30 May 2005.

virtual communities of the Global Village, but it is "here and now". There are multiple explanations that have made the emerging of this "third space" possible. The first reason is the wide diffusion of the *wireless mobile media*, and the second considers the differences between *wireless media* (that are personal and individual, mobile and integrated) and personal computers. Briefly, *wireless media* are

- *personal and individual.* Objects that we always carry with us and that we customize to our own pleasure, integrated interactions in our daily life that are dense with emotional association.
- *mobiles.* Objects that, unlike personal computers, are always with us and operate in every social, professional, or familial situation.
- *integrated.* With the wireless media, we can send/receive information and add personal comments at the same time: in fact, the same communication device allows both to access information, stored or networked, and to manage interpersonal communication with our friends, family, and colleagues. The integration between communication and the access to information create a hybrid narrative.

With the diffusion of *wireless mobile media*, we face the emergence of a new relationship between media, community, and memory. *Wireless mobile media* become an extraordinary digital memory repository where users keep the electronic memory of our personal experiences, pictures of our family and friends, videos from trips, music, and vocal and text messages. *Mobile wireless media* are personal communication instruments that not only allow us to exchange information with our community, but through which we can also communicate personal experiences, memories, and events, and all this personal history is digitally recorded, stored, and capable of being shared with our community.

Blogs show how users like to share their personal memory: they publish information that is personal, intimate, and that comes from their lived experience. Blogs, intimate personal shared journals, become an individual published memory that can be enriched even from an enlarged group of friends or from unknown visitors.

6.6 Conclusion

We have focused on three emerging paradigms concerning the relationship between the diffusion of communication technologies, social memory, and society. This relationship is extremely complex and constantly changing, and there are many other aspects that we have neglected. The embedded connected memory, where human tissue meets electronics, and where people, spaces, and objects are interconnected through electronics and microchips has an increasing significance and topicality. Moreover, objects can

The scale of the individual and the sphere of their personal surround: an interaction between the individual and the individual's immediate ambience.

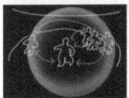

The scale of the community, living in physical proximity: interactions between friends, neighbours, and the environment in which they live.

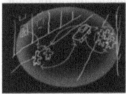

The scale of the city: interactions between physical sites, (buildings, neighbourhoods) and the people who visit or reside in them.

The scale of the territory: all of the above scales combine to form an "interactive tissue" where different scales of interaction are combined.

Figure 6.4   Ascending spatial scales of human activity.

be expected to have an embedded networked, interactive memory, which will have the capacity to improve the interconnection between people and places, defining interesting new paradigms for collective human memory.

There is certainly a range of interesting new issues concerning the potential relationship between communication technologies, memory, people, and society, from a biological, cultural, and technological point of view.

With the diffusion of information technologies, we may expect radical change in the way people and communities will communicate, both in the manner they access information and knowledge and in the way they may be actively involved in the creation of a common memory. We propose[8] four ascending scales in which we may characterize this change.

If at the dawning of the Internet we were impressed by the opportunity of exchanging messages with people at the other side of the world, nowadays

---

[8] Drawings and texts are from a research project on "aural knowledge": Casalegno, F., Susani, M., Atlas of Aural knowledge book under review for publication. For further information see: http://www.auralknowledge.net, retrieved 30 May 2005.

we are increasing the level of communication with our local, social, and cultural environments. In the interactions taking place in cyberspace, the impressive aspect that facilitates travel at hyperbolic speed freeing us from space bonds and be part of distant, virtual communities is still present. But it is because *man is human*, that he uses communication networks in the hope of leading a better everyday life in the "physical" world rather than in virtual architectures. After all, cyberspace is not a distant galaxy, but a reflection of what happens on the earth; it is the result of a complex convergence of human feelings and associated converging technologies.

The three paradigms we have presented give us a scenario as to how new technologies may focus on local aspects of human life, reinforcing the social cohesion at a local level, promoting the creation of a communal memory, favoring the storing of historical and "official" information, and, at the same time, sharing of ordinary and informal information.

The potential that new technologies offer, if well utilized, toward the creation of a communal memory to improve social cohesion is enormous. Even if the risk for the development of an ominous *big brother* is a real issue, new information technologies not only increase the possibility to store data and to efficiently organize information and knowledge, but also give the opportunity to community members to be actively involved in sustaining a communal memory and be part of the local culture. As Roger Bastide reminds us, we do not have a communal memory when the identities of "me" and the "other" are simply immersed in the same collective knowledge, but we, on the other hand, have a communal memory when our personal souvenirs are inextricably articulated with the souvenirs of the others', when our mental images resonate with the images of the other community members (Bastide, 1970). Interactive media have the potential to reinforce this process of communal souvenir building.

The opportunity of nourishing social and communal memories allows us to create, or so it is our hope, what Edgar Morin defines as a *poetical vision of our existence*: he reminds us that every human being feels the need to retain a cultural heritage so that it can conceivably conquer the present, that is to lead not only toward a useful and functional existence but also toward poetical existence (Morin, 2001). The different forms of communal empathy, from love to celebrations, from parties to festivities, are paths that lead man toward this idea of a poetical state of living.

"New technologies" are man-made creations. Prolonged use and experience of them endow them with a value and a sense of societal purpose; it is the task of mankind to use them to make this *poetical vision* come true.

6.7 References

Bastide, R. (1970) Mémoire collective et sociologie du bricolage. *L'année sociologique*, (21), 94.

Halbwachs, M. (1986) *La mémoire collective.* Presse Universitaire de France, Paris.

Mc William, I. (2001) The story of the lost cat. In Casalegno, F. (ed.), *Memoria quotidiana,* Le Vespe Editore, Milan, pp. 17–19 (French edition *Mémoire quotidienne,* 2005, Presse de l'Université de Laval, Montreal, Québec, Canada).

Mitchell, W.J. (1999) *E-Topia.* MIT Press, Cambridge, MA.

Morin, E. (2001) Memorie vissute per un'esistenza poetica. In Casalegno, F. (ed.), *Memoria quotidiana,* Le Vespe Editore, Milan, pp. 141–154 (French edition *Mémoire quotidienne,* 2005, Presse de l'Université de Laval, Montreal, Québec, Canada).

Moscovici, S. (2001) Memorie, rituali e ciber-rappresentazioni. In Casalegno, F. (ed.), *Memoria quotidiana,* Le Vespe Editore, Milan, pp. 79–92 (French edition *Mémoire quotidienne,* 2005, Presse de l'Université de Laval, Montreal, Québec, Canada).

Negroponte, N. (1995) *Being Digital.* Alfred A. Knopf, USA.

Community and Communication: A Rounded Perspective[1]

Jennifer Kayahara

7.1 The Intersection of Internet and Community

One of the big questions in the early days of Internet research was how the Internet would intersect with community. Scholars and policy makers were curious (and sometimes doubtful) about whether people would successfully carry community with them into the online realm, and watched with interest as users sought to make connections with people who shared an intellectual or emotional resonance but lacked physical co-presence (e.g. Rheingold, 1993; Turkle, 1995; Reid, 1999).

Research interests have since broadened, but these preliminary concerns continue to shape much of the ongoing work, largely because the community concern has not yet been satisfactorily addressed. While some scholars believe that online communities can and do exist (e.g. Rheingold, 1993; Turkle, 1995; Wellman and Gulia, 1999), others express more caution (e.g. Etzioni and Etzioni, 1999; Slevin, 2000; Driskell and Lyon, 2002) or outright skepticism (e.g. Freie, 1998; Nie, 2001). Despite a growing quantity of data and a great deal of thoughtful analysis, there remains little concurrence as to whether it is possible to form a genuine community online. This suggests that perhaps the main barrier to agreement is not a lack of data or analysis, but rather a misspecification of the question.

7.2 Community: Past and Present

The subject of community has a long history in social science literature. In daily life, the term "community" conjures all sorts of positive impressions of

---

[1] Author's Note: I am grateful for the advice and aid of Lorne Dawson, Kieran Bonner, John Goyder, Barry Wellman, and Brent Berry in preparing this chapter. Thank you as well to Patrick Purcell for suggesting the title.

tradition and loyalty, intimacy and belonging, safety and support. However, the elusive nature of community and its many uses rapidly become apparent whenever an attempt is made to systematically study it. Community, Cohen has observed, "is one of those words – like 'culture', 'myth', 'ritual', 'symbol' – bandied around in ordinary, everyday speech, apparently readily intelligible to speaker and listener, which, when imported into the discourse of social science, however, causes immense difficulty" (Cohen, 1985, p. 11).

Part of the difficulty in determining precisely what community means comes from the substantial number of conceptualizations that scholars have produced. In 1955, Hillery identified 94 separate definitions – a compilation he explicitly stated was by no means complete – and the number has since grown. Another part of the difficulty also arises from the way the meaning of community has branched out over time, expanding beyond the original neighbourhood focus to encompass geographical conceptualizations based on shared location (e.g. Wirth, 1938); ideational conceptualizations based on shared interests, ideas, or values (e.g. Anderson, 1983); and relational communities based on shared bonds between members (e.g. Wellman, 1999b).[2]

In addition to creating confusion, this expansion or fracturing in ideas about what a community can include has led to debate about what rightfully qualifies as a community.[3] The depth of the disagreements has resulted in questions of whether community remains a useful construct for scholars (Stacey, 1969, p. 134); however, this seems like a rather extreme response to a common problem in studies of the social world. A more positive approach is to choose a conceptualization grounded in the knowledge of the possibilities.

With so many conceptualizations and definitions, it is impossible to do justice to all of them. In the following pages, I will describe a few of the most prominent and influential classic and contemporary conceptualizations in order to highlight one of the major divisions in contemporary community studies: the divide between neighbourhood communities and networked communities. In doing so, I will attempt to capture both the historical sweep and the array of contemporary options. I will also describe one particular case which demonstrates some of the implications of choosing one school of thought or the other: the debate over the existence of Internet-based communities.

7.3 Classic Conceptualizations of Community

In order to understand the roots of the disagreement about the nature of community, we must go back to the early work of scholars: Tönnies,

---

[2] This is only one way of organizing conceptualizations of community. For examples of alternate typologies, see Komito (1998) and Brint (2001).

[3] For examples of discussions of this debate, see Fowler (1991), Freie (1998), and Brint (2001).

Durkheim, Marx, and Wirth. Their conceptualizations have shaped many of the popular notions about community, and have informed scores of contemporary theories.

Many of the most influential early scientific theories of community arose in the 19th century, a time of rapid change. Western Europe was undergoing various phases of the Industrial Revolution and its accompanying effects, including a sharp increase in the rate of urbanization. Many observers of the time were worried about what effect these great shifts would have on societal cohesion, social order and the known ways of life, and worried that European society would be unable to adjust to the changes.

7.3.1 The Clash of Past and Future: Ferdinand Tönnies

The most influential classic community scholar may be Ferdinand Tönnies. Tönnies was very much concerned with the question of how social order is maintained (Turner and Dolch, 1996, p. 21). Tönnies believed that all social relations are rooted in human will (Tönnies, 1887, trans. 1957, p. 33). He argued that in nonurban societies the dominant form of will is natural will, which leads to what he referred to as *Gemeinschaft* (community) relations (Tönnies, 1887, trans. 1957, p. 42). *Gemeinschaft* relations are characterized by intimacy, sympathy, trust, and common values (Tönnies, 1887, trans. 1957, p. 250). These are relations whose participants share benefits and misfortunes in common, if not necessarily in the same proportion (Parsons, 1973, p. 143). *Gemeinschaft* relations reflect what many people think of when they think of community.

Tönnies identified three types of *Gemeinschaft*: *Gemeinschaft* by blood, *Gemeinschaft* of locality, and *Gemeinschaft* of mind (Tönnies, 1887, trans. 1957, p. 42). The first, *Gemeinschaft* by blood, refers to kinship relations. This includes both the mother–child relationship, based on instinct, physical dependency, and fond memories, and the sibling relationship, based almost entirely on force of memory and therefore, in Tönnies' opinion, the most human relationship possible between people (pp. 37–39). *Gemeinschaft* by blood is primarily a relational orientation to community as it is not dependent on physical proximity, but can thrive on memory and feeling. Tönnies, however, believed that the average person prefers to be surrounded by kin and is happier so (p. 43), which lends a geographical focus to this conceptualization.

The second, *Gemeinschaft* of locality, refers to the community of common property and the local neighbourhood, and in that sense is extremely geographical. As with all forms of *Gemeinschaft*, however, the quality of the relationship also plays a role. According to Tönnies, living together in rural villages forces inhabitants to deal with each other, cooperate in the management of their holdings, and provides them with intimate knowledge of one another (Tönnies, 1887, trans. 1957, p. 42). It is *Gemeinschaft* of locality that people often have in mind when they proclaim that community is in danger of disappearing, and in that sense, it is arguably the most influential element of Tönnies' conceptualization. There has also been some corresponding criticism, of which one key argument is that the type of community epitomized by *Gemeinschaft* of locality never really existed. As Brint has observed, it is "an oft-repeated message of the community studies literature that communities are not very community-like" (Brint, 2001, p. 6). He notes that communities are "rife with interest, power, and division" and "people in even the most enclave like communities do not necessarily associate with one another more frequently than they do with people outside the community" (Brint, 2001, p. 6). While it could be argued that having the option to seek relationships outside of one's local community is a product of modernity (Berger, 1988, p. 51), social network analysts such as Wellman and Wetherell (1996) argue once you eliminate the assumption that the bulk of interaction took place within a small, localized group, the evidence shows that people were not as restricted in their interactions to their local circle as is widely believed.

Finally, the third type, *Gemeinschaft* of mind, is unrelated to both blood and place. It is a relationship between friends, based upon similarity of work and intellectual attitude (Tönnies, 1887, trans. 1957, p. 43). Like *Gemeinschaft* by blood, *Gemeinschaft* of the mind does not require physical proximity, though Tönnies argues that it most easily comes into existence through frequent meetings and that frequent meetings are most easily accomplished if people live in near to each other (Tönnies, 1887, trans. 1957, p. 43). This is very much a relational conceptualization of community – perhaps the most completely relational conceptualization of the three since even *Gemeinschaft* by blood presumes a fair degree of face-to-face interaction and thus physical proximity at some point. The defining characteristic of *Gemeinschaft* of mind is the shared bond, the quality of the relationship.

Gemeinschaft is the typology that most closely captures what many people think of when they hear community; Tönnies, however, believed that relationships in a modern, urban society no longer took the form of *Gemeinschaft* relationships. For this environment, he developed a second typology: *Gesellschaft*. Where *Gemeinschaft* is driven by natural will, *Gesellschaft* is driven by rational will. Tönnies believed that *Gesellschaft* relations are the dominant form of relations in the industrialized world. *Gesellschaft* relations are characterized by artificiality, isolation, self-interest, and tension (Tönnies, 1887, trans. 1957, pp. 64–65). It is the coexistence of people independent of each other (Tönnies, 1887, trans. 1957, p. 34).

Steven Brint has criticized Tönnies for developing his typologies with an eye to creating the greatest possible contrast between communal and associative relationships rather than identifying the defining characteristics of each type of relationship (Brint, 2001, p. 2). While this is in keeping with a long sociological tradition of relying on extreme typologies in order to define a range into which other social forms can be fitted (Loomis and McKinney, 1957, p. 12), Brint argues that it resulted in a situation in which essential elements were neglected (Brint, 2001, p. 3). This suggests that any attempt to apply Tönnies' typologies to any actual social organization may result in problems. Tönnies' ideas clearly resonated with a great many people, providing compelling descriptions and terms that seem to describe some an inherent difference in the quality of relationships that many people perceive. However, the use of Tönnies' typologies as the yardstick of community, rather than as a way of organizing the landscape, creates the risk of an impossible standard.

7.3.2 The Question of Social Order: Emile Durkheim

Like Tönnies, Emile Durkheim was strongly concerned with the problem of maintaining social order in the midst of rapid change. Durkheim was also more explicitly interested in the issue of morality, which he described as "the daily bread without which societies cannot exist" (Durkheim, 1893, trans. 1969, p. 51). These concerns formed the foundation of Durkheim's conceptualization of community. For Durkheim, community was important because it equipped humans with social support and moral sentiments (Brint, 2001, p. 3). In keeping with the French philosophe tradition, he believed that the best way to approach the issue of morality was to study the nature of the collective conscience (Turner and Dolch, 1996, p. 23). To facilitate this study, he developed an ecological model that described what he perceived as a transition from a traditional society based on mechanical solidarity to a modern society based on organic solidarity (Durkheim, 1893, trans. 1969). These typologies were not intended to describe a range, but rather an "irreversible historical trend" (McKinney and Loomis, 1958, p. 560) that Durkheim linked to the introduction of industrialization and the subsequent increase in labor specialization (Durkheim, 1893, trans. 1969, pp. 3–5). Unlike Tönnies, who set community up in opposition to modernity, Durkheim believed what existed in modernity was also a form of community. For him, there was no inherent conflict between modernity and community; only changes to the nature of community and thus to the nature of the collective conscience that was the core of his concerns.

Durkheim defined the collective conscience as the determinate system formed from the "totality of beliefs and sentiments common to average citizens of the same society" (Durkheim, 1893, trans. 1969, p. 79). He believed that the collective conscience exists external to any given individual and is passed from generation to generation, binding them together (Durkheim, 1893, trans. 1969, p. 80). Durkheim argued that the collective conscience changed according to the nature of society. Many of the differences, he suggested, could be observed in changes in the societal reactions to crime and criminals.

The first type of collective conscience described by Durkheim is the type he associated with societies based on what he referred to as mechanical solidarity. In these societies, people are bound together by similarities in conscience and lifestyle (Durkheim, 1893, trans. 1969, p. 130). A mechanically solidary society is characterized by mental and moral homogeneity. The collective conscience of this society is marked by exteriality, in that it exists external to the members, and constraining, in that members cannot morally refuse to be part of it (Loomis and McKinney, 1957, p. 13). At the other end is organic solidarity, which is found in societies whose members are bound together by interdependence and the complementary nature of their differences (Durkheim, 1893, trans. 1969, p. 131). An organically solidary society is characterized by role specialization and a division of labor. The differences between the two can be highlighted by their different approaches to law and crime. In a mechanically solidary society, crime is an offence against common moral sentiments and thus of a sacrilegious nature (McKinney and Loomis, 1958, p. 560). In an organically solidary society, crime becomes an offence against personal rights and the concern of the justice system becomes restitution rather than repression (McKinney and Loomis, 1958, p. 560).

There are some obvious similarities between the work done by Tönnies and Durkheim on the changing nature of society and community. Both men emphasized society's need for cohesion, whether through solidarity as proposed by Durkheim, or affirmative social relations as argued by Tönnies (Cahnman, 1995, p. 87). Both theorists also clearly regretted the gradual loss of the older forms of social life, although neither saw any use in arguing for that which could not be recovered (Cahnman, 1995, p. 87). However, despite these parallels and despite the superficial resemblances in their typologies and terminology, there is a fundamental difference between them. According to Tönnies, *Gemeinschaft* and *Gesellschaft* are both present in every society though the proportions may vary; relationships with friends and kin will always have a *Gemeinschaft* orientation. Durkheim's model describes a transition from one type of society to the other. Overlap between the two types is temporary, leaving mechanical solidarity nonexistent in contemporary society. This has important implications for their respective conceptualizations of community. What Tönnies brings is a vision of community that is oriented toward relationships bound by blood and tradition, relationships that are not possible to achieve on a large scale in modern society (assuming that the *Gesellschaft* mentality has not, in some sense, eliminated altogether the possibility of forming such relationships).

7.3.3 Karl Marx and Political Economy

Marx viewed humans as innately social (*Gemeinswesen*) and asserted that society serves as a means of humanizing people with true community as the final goal (Mahowald, 1973, p. 475). The difference between the final community (*Gemeinschaft*) and its previous incarnations (*Gesellschaft*) is the possibility of personal freedom. According to Marx, "[i]n the previous substitutes for the community, in the State, etc., personal freedom has existed only for individuals who developed within the ruling class and only insofar as they were individuals of this class" (Marx, 1847, trans. 1978, p. 197). The eventual goal is for *Gemeinschaft* and *Gesellschaft* to be one (Mahowald, 1973, p. 482), at which point real freedom will come from association with others (Marx, 1847, trans. 1978, p. 197).

Marx's work served as the inspiration for another conceptualization of community. Researchers working from the political economy perspective in the middle of the 20th century found a high level of dehumanization and dissatisfaction among people living in a contemporary industrial society (Hale, 1995, pp. 66–67). Drawing on Marx's work, they attributed the dissatisfaction to the dehumanizing effects of capitalism rather than a change in village size. Poverty, inflation, unemployment, and exploitation were all seen as factors that corroded community and increased individual insecurity and dissatisfaction (Hale, 1995, p. 67). According to proponents of this perspective, satisfying social relations arise when economic security and control is assured (Hale, 1995, p. 67). The danger to community is not urbanization, but exploitative labor practices that pit people against each other in all environments.

7.3.4 Urban Communities: The Chicago School

Community studies became popular at the University of Chicago during the 1920s and 1930s. While this places it a little later than the other conceptualizations of this section, the concerns of the scholars of this period were similar to their 19th-century counterparts – how was urbanization affecting the community? The answer from the Chicago school was a little different than that arrived at by Tönnies. When they went out into the city, they discovered that communities had not been left behind in the rural villages during the great rush to the cities; rather, rural migrants had brought it with them.

The Chicago researchers selected an ecological approach to community as a path to explaining the logic of cities (Bernard, 1973, p. 35). They believed that people sort themselves into physically distinct "natural" communities and sought to examine how people fit into their communities (Effrat, 1974, p. 5). Proponents of the ecological approach conceptualize communities as "clearly discernable, spatially delimited entities with well-defined

133

boundaries" (Goldenberg and Haines, 1992, p. 302) and treat them as relatively autonomous social units (Effrat, 1974, p. 5). The latter point led to criticisms that assumptions about the autonomy of local communities resulted in researchers ignoring the influence of external but still influential agencies such as the government (Simpson, 1974, p. 314) and the importance of external ties (Wellman, 1979, p. 1202).

One of the best-known community researchers in the Chicago School is Louis Wirth. Wirth developed a theory of community in which he argued that cities possess specific characteristics that lead to the patterns of culture identified by theorists such as Tönnies (Hale, 1995, p. 65). Wirth believed that density alone could not be used to distinguish the urban from the rural; he argued "unless density is correlated with significant social characteristics it can furnish only an arbitrary basis for differentiating urban from rural communities" (Wirth, 1938, p. 5). Differentiating by size posed a similar problem. In the end, Wirth selected three factors that he believed worked together to differentiate urban communities from rural communities: size, density, and heterogeneity (Wirth, 1938, p. 8). Size was important because a larger population made it more difficult for everyone to meet face to face, forcing people to rely on delegates to make their will known. It also made it very difficult to get to know other people well (Wirth, 1938, pp. 13–14). Density, according to Wirth, led to "a spirit of competition, aggrandizement, and mutual exploitation. To counteract irresponsibility and potential disorder, formal controls tend to be resorted to" (Wirth, 1938, p. 15). Finally, heterogeneity led to distrust because people could no longer be sure that those around them held values common to their own (Hale, 1995, p. 66). Together, these three factors led to a loss of community values (Hale, 1995, p. 66).

The Chicago School is a key example of the neighbourhood approach to community, which is based on the idea that community is a product of physical proximity; members of the Chicago School focused on physical neighbourhoods to the exclusion of all else. While some attention was paid to the nature of relationships between people within a neighbourhood, the physical structure of the neighbourhood was assigned a strongly deterministic role and in this sense, it stands in almost direct opposition to network conceptions of community.

7.4 Contemporary Views of Community

With the 20th century came the diffusion of a number of communication and transportation technologies that carried with them the seeds of change. While letter mail and travel are both old and people have never been entirely limited to their neighbours for company (Wellman and Leighton, 1979; Wellman and Wetherell, 1996), the introduction of technologies such as trains, telegraphs, cars, and telephones made it much easier for people to look beyond their neighbours for information, companionship, and support. For some scholars, these changes presented a grave threat to community, or even marked its end. For others, this was merely a natural shift from dense, local communities to sparse, ramified communities and the individualism that goes with them.

The contemporary visions of neighbourhood communities bear many similarities in style to Tönnies' *Gemeinschaft* of locality. They are visions of communities based on physical proximity, places where the residents watch out for each other and dense interactions foster common values. These conceptualizations tend to include a richness of experience based on continuity of experiences and ties. Many proponents of these conceptualizations express concern that information and communication technologies are tearing communities down, and worry about what will be lost from human experience as a result.

7.4.1 Brave New World: The Communitarian Perspective

One of the most coherent examples of a contemporary, neighbourhood-based conceptualization of community is communitarianism. Communitarianism is based on the premise that communities are necessary to uphold moral standards (Etzioni, 1993, p. 32). Sociologist Amitai Etzioni developed the idea of communitarianism, driven by concern that too many people in contemporary society sought rights without responsibilities (Etzioni, 1993, p. 5). He believes that building stronger communities was the way to counter this situation. Etzioni is not a reactionary; he believes that the traditional family deserved to be critically rethought (Etzioni, 1993, p. 12) and that challenging the old values was a good idea. The problem, in his mind, is that once the old ways were eliminated, the process of change stopped and nothing new was developed to replace the old institutions (Etzioni, 1993, p. 24). Communitarianism is an attempt to fill that perceived void.

A communitarian community is one characterized by common values, consistent membership, regular social interaction, and the ability to exercise control over its members (Etzioni, 1993, pp. 381–382). It consists of a densely woven network of affective relationships among a group of individuals, not merely a series of one-on-one relationships. It also requires a commitment to "a set of shared values, mores, meanings, and a shared historical identity – in short, a culture" (Etzioni and Etzioni, 1999, p. 241). In order to build this community, Etzioni advocates that people be encouraged to organize events that will facilitate the building of community bonds (Etzioni, 1993, p. 126). Other suggestions for how to facilitate the creation of communitarian communities include embracing civic activism, making divorce more difficult, and encouraging parents to spend more time with their children and less time on their careers once basic material needs have been met (Freie, 1998, p. 159).

7.4.2 Reciprocity: Social Capital and Community

A second contemporary conceptualization of community was developed by Robert Putnam, around his ideas about social capital. Putnam defines social capital as "features of social organization such as networks, norms, and social trust that facilitate coordination and cooperation for mutual benefit" (Putnam, 1995a, p. 67). According to Putnam, social capital helps citizens resolve collective problems, increases the level of trust in communities, increases people's awareness of their interdependence, improves the flow of information between people, and offers individual psychological and biological benefits (Putnam, 2000, pp. 288–289). He argues that levels of social capital have declined in the USA in recent years, as evidenced by declining levels of membership in unions, PTAs, and civic associations (Putnam, 1995a, p. 69). The best way to restore it, he writes, is to build structures and introduce policies that will encourage renewed civic engagement (Putnam, 2000, p. 403). Although social capital is not by itself a conceptualization of community, Putnam's ideas about how social capital is created are very strongly related to both community and place. The evidence he gives for the decline of social capital is the decline in local organizations built on face-to-face contact and the decline in family activities (Putnam, 2000).

Putnam's attempts to link social capital to community have also met with some criticism. Alejandro Portes has criticized Putnam for applying social capital to the community level, arguing that it was originally conceptualized an individual variable, and that shifting it to the community level, turns social capital into both cause and effect (Portes, 1998, p. 19). He also argues that Putnam assigns desirable outcomes to social capital and then infers its existence from those same outcomes, thus rendering the concept meaningless – a practice Portes attributes to Putnam's desire to find one overriding cause for all of his observed differences (p. 20).

Putnam has also been criticized by Mouritsen for making trust a central element in his definition of social capital without giving due attention to the role cultural traditions play in establishing types of trust or adequately explaining how trust can accomplish the tasks he sets for it (Mouritsen, 2003, p. 660). As Mouritsen notes, trust can often go along with intolerance as well as tolerance, especially when generalized to the group (community) level, which can just as easily lead to trust of certain categories of people within the community (frequently those most closely resembling us in some socially significant way) rather than trust of members of the community as a whole.

Although subject to criticism, Putnam's approach also has a certain appeal. His ideas about community conform fairly well to popular conceptions of community, and at first glance, social capital seems to offer promise as a way of empirically measuring community. However, to do so, one would have to ignore the problems associated with treating social capital as a community-level variable. It also offers little information about the boundaries of a community except in rural areas with geographically enforced boundaries because any attempt to measure social capital (as opposed to participation in associations) will lead away from the community into the extra-community networks of each individual. If social capital is treated as an individual level variable, measuring its strength and presence could be a very effective way of determining the boundaries of any given individual's personal community, but it would lack the totality of experience for a group.

7.4.3 Boundaries: The Social Constructionist Approach

Social constructionism is grounded in the work of William Isaac Thomas, Alfred Schutz, and Schutz's students Peter Berger and Thomas Luckmann. It is predicated on the belief that humans actively create their social world and are in turn shaped by that world. Social constructionists who turn their attention to community seem inclined to avoid definitions in favor of examining how people live communities.

Anthony Cohen, a community scholar in the social constructionist tradition, suggests that we need to consider two main elements of communities: consciousness and boundaries (Cohen, 1985, pp. 12–13). In order for a community to exist, its members must be aware of both its existence and their membership; this is implicit in the idea that the community is constructed. Cohen argues that the "consciousness of community is, then, encapsulated in perception of its boundaries, boundaries which are themselves largely constituted by people in interaction" (p. 130). Cohen contends that community implies two things: "that the members of a group of people (a) have something in common with each other, which (b) distinguishes them in a significant way from the members of other putative groups" (p. 12). Thus symbolic boundaries are important to separate members from nonmembers and serve to "encapsulate the identity of a community" (p. 12).

The problem with the social constructionist approach is that it may leave community too broad a term to be useful. There is no demand for responsibility or commitment or even basic interaction. Under this definition, the viewers of a television show may judge themselves a community and meet the criteria above although they have never spoken to a single other viewer. Although acknowledgment of membership and the presence of a sense of community are important to the phenomenological experience of community, it seems like the bar should be set a little higher in identifying what constitutes a community.

7.4.4 Community Liberated: Network Communities

Network communities are one of the main alternative conceptualizations to neighbourhood communities. Network analysts conceptualize communities in terms of the quality of the relationships rather than the physical proximity of the members. "Communities", says one prominent network analyst, "are about social relationships while neighbourhoods are about boundaries" (Wellman, 1999a, p. xii). Network analysts tend, on the whole, to be less concerned with the decline of community than scholars who take a more traditional stance. Neighbourhood ties may be declining, but network ties are not.[4]

Although all communities can be described in community terms not all networks are communities, as evidenced by the fact that not all network analysts consider themselves community scholars. However, distinguishing between a noncommunity network and a community network can prove difficult. Since many network studies take the form of ego-studies, focusing on collecting information about all of the major ties of the research participants involved, it might make sense to refer to personal communities. In this conceptualization, each person has his or her own community which provides many of the same experiences as a traditional, geographically bounded community while allowing for greater choice and flexibility. This is a very different vision of community from the traditional, geographically bounded version, but may more accurately capture some of the interaction elements of communities. Network analysis reveals that support tends to be very specialized; members select which resources they provide to any given member based on their relationship to a particular member (Wellman, 1999a, p. xiii). It is not absolutely certain whether this is a product of the disintegration of communities or a long-standing feature, but if it is the latter it is one not often addressed in broad theories of neighbourhood community.

Network communities may be distinguished from noncommunity networks by imposing additional criteria. Wellman, for example, defines communities as "networks of interpersonal ties that provide sociability, support, information, a sense of belonging and social identity" (Wellman, 2001, p. 228). A geographical boundary could also, in theory, be imposed. However, network research has revealed that while local ties are strong and important, they constitute a minority of most urbanites' ties (Goldenberg and Haines, 1992, p. 308). This suggests that imposing a location requirement is likely to result in researchers overlooking a vast portion of an individual's ties, and vast portion of the social benefits often associated with community membership.

Several criticisms have been leveled at network conceptions of community. Calhoun argues that network analysis offers a vague and weak

---

[4] The average number of weak ties may actually have increased with the introduction of the Internet; see Chapter 8 in this book for data.

conception of community that neglects a vital component: a sense of community derived from its relational structure (Calhoun, 1998, p. 374). Frankfort-Nachimas and Palen agree, suggesting that defining community only in terms of interpersonal or friendship ties neglects other dimensions common to definitions of community (Frankfort-Nachimas and Palen, 1993, p. 2). Elements such as ritual, institutions, and group gatherings are generally absent from the requirements of network analysts, and even those who favor the relational approach to sociology could reasonably consider this an important absence.

7.5 The Clash of Conceptualizations: The Internet-Based Community

Online communities, or the online aggregates commonly referred to as communities, provide an interesting intersection of the arguments for the various types of communities. The long-standing debate over whether community can exist online has provided both sides with a chance to make a case that helps to reveal the implications of their respective positions while clarifying some of their assumptions about what a community should be and enabling a direct comparison of their positions in a somewhat easier fashion than usual. The difficulty with using the online community debate as a means of exploring the different views of community is that the people most engaged in the debate are often different from those who generated the comprehensive theories previously discussed. However, in many cases the same lines of thought persist, making it possible to ascertain how the original theorists might contribute to this debate.

7.5.1 The Case For Community Online

In making their case for or against the existence of online communities, each side tends to emphasize those elements which believe are key to the experience of community. In the case of opponents to the idea that online communities can exist, the emphasized elements also tend to be those which they believe cannot be found online. The qualities of community emphasized by people who believe that online communities exist are sociability, support, and sense of belonging.

7.5.1.1 Sociability

Companionship is a basic human need, and sociability therefore a common human trait. Simmel defines sociability as "the art or play form of association, related to the content and purposes of association in the same way as art is related to reality" (Simmel, 1949, p. 255). For Simmel, sociability is association for its own sake, a realm to be kept free of the personal and the material:

> Riches and social position, learning and fame, exceptional capacities and merits of the individual have no role in sociability or, at most, as a slight nuance of that immateriality with which alone reality dares penetrate into the artificial structure of sociability. (p. 256)

Sociability as it is used in the community literature typically bears some resemblance to Simmel's definition, comprising association for the pleasure of association, although other researchers are not always as strict as Simmel about separating the material self from sociability.

In the classical literature, sociability appears most prominently in Tönnies' work, communitarianism, and network analysis. Where support was the key to *Gemeinschaft* of blood and *Gemeinschaft* of location, sociability is the key to *Gemeinschaft* of mind. *Gemeinschaft* of mind, according to Tönnies, is based on a relationship between friends and a level of intellectual accord (Tönnies, 1887, trans. 1957, p. 43). This concept implies an element of sociability that can be seen particularly clearly in Tönnies' argument that frequent meetings are important to establishing and maintaining a *Gemeinschaft* of mind relationship. Here again we see an interesting reversal; in Tönnies' view, sociability is both cause and consequence of a *Gemeinschaft* relationship and not merely a side effect that can be used to identify a positive relationship. Among more recent community theorists, sociability is most evident in communitarianism and network analysis. Communitarian theorist Etzioni (1993, p. 126) argues that events are important for building community bonds. If a community is to have the force of moral persuasion, as communitarians believe it should (Etzioni, 1993, p. 32), then members need to know each other. Sociability is an effective way of building social bonds.

Contemporary Internet researchers tend to use sociability less strictly than Simmel, although the idea of association for its own sake still arises. Sociability in the contemporary sense tends to focus on companionship, conversation, and communication of shared interests. Like support, sociability is arguably necessary to but certainly not sufficient for the formation of communities. Most communities involve some sociability, but sociability can occur outside of a community context. The importance of sociability in communities tends to be most strongly emphasized among researchers who apply a social constructionist framework to their study of online communities. This is perhaps not surprising given their emphasis on the importance of interaction in creating a community; the mere act of socializing – even according to Simmel's limiting definition of sociability – will help to create the shared definition of the situation that leads to the gradual building of a community.

One study that focuses on the role of sociability in creating an online community is Shelley Correll's (1995) account of a BBS-based electronic bar targeted at lesbians. A BBS is a system that enables real-time, text-based interaction among participants. Participants connected at the same time can type to each other and receive immediate written responses visible to all other participants. The BBS studied by Correll consisted of a single, bar-themed room in which all of the participants gathered to converse. Correl describes the BBS as primarily oriented toward sociability: it had no common topic of interest beyond common gender and sexual orientation, and the bar metaphor that served as a conversation starter when people ran out of things to say. There was no goal and no plan; people merely gathered to converse. While those concerned with the quality of relationships might argue that electronic sociability is not the same as face-to-face sociability, no one is likely to argue that sociability does not exist online.

7.5.1.2 Support

Support in this context refers to any form of social interaction that involves providing some form of assistance. The assistance can be either tangible or intangible. Tangible support typically takes the form of physical goods, while intangible support can include information, psychological, and emotional support. Types and levels of support can range from a cup of sugar borrowed from a neighbour to a down payment for a house borrowed from parents, from a quick note on how to fix a computer problem to extensive legal advice.

In the classical literature, support appears in the work of both Tönnies and Durkheim. Durkheim subscribes to the idea that social support is one of the key functions of a community (Brint, 2001, p. 3). In Tönnies, support is suggested by the description of *Gemeinschaft* relationships. *Gemeinschaft* by blood is based in part on either a one-way (between parent and child) or reciprocal (between spouses) exchange of support. Interestingly, Tönnies argues that *Gemeinschaft* by blood is a consequence rather than cause of support (Tönnies, 1887, trans. 1957, pp. 37–39). In *Gemeinschaft* of locality, support is implied by Tönnies' comment that one of the characteristics of a rural village is that inhabitants are forced to deal with each other and cooperate in the management of their holdings (Tönnies, 1887, trans. 1957, p. 42). In *Gemeinschaft* of mind, support is suggested by references to friendship and cooperation (p. 43). Support in this case might include support for ideas, beliefs, and perhaps identity, Tönnies' "mutual furtherance and affirmation" (p. 44).

The importance of support becomes somewhat more prominent among community theories once we move later into the 20th century. The two perspectives that make greatest use of support are the social capital/traditionalist perspective and network analysis. Putnam, on the social capital side, argues that "networks of civic engagement foster sturdy norms of generalized reciprocity and encourage the emergence of social trust" (Putnam, 1995a, p. 67). According to him, the expectation of receiving support is one of the advantages of belonging to a community with a high level of social capital. Other network analysts focus less on generalized reciprocity, and more on the individual levels of support available from one's personal community.

Two interesting studies of online support were conducted by Cummings et al. (2002) and Dunham et al. (1998). Cummings et al. (2002) surveyed a random sample of people participating in an online discussion for individuals who had suffered hearing loss to evaluate the levels of support the individuals received online and offline. They discovered that most people do report receiving emotional and information benefits from their online groups (Cummings et al., 2002, p. 86). The beneficial side is particularly apparent among people who have no offline support.

The second study focused on computer-mediated social support for young, single mothers. Dunham et al. set out to discover "whether a properly designed CMSS (computer-mediated social support) network for single, young mothers might provide long-term multidimensional social support for the constantly changing needs of these young families" (Dunham et al., 1998, p. 2). For this project, they recruited 50 young, single mothers with at least one child less than 1 year of age, all living in Halifax. Each user was given a user-id and password connecting them to the BBS, giving them access to a public message board called Moms and Kids where messages remained for 60 days, private e-mail, and access to multiuser text-based teleconferencing. What Dunham et al. found was that the majority of the messages on the system dealt with aspects of the mothers' mental health, including boredom, social isolation, and social alienation. Of the replies, 56% were emotional in nature, 37% were informational, and 3% were tangible (Dunham et al., 1998, p. 8). The last figure is of particular interest because it indicates that support in online social environments can extend beyond the virtual realm; under some circumstances, people are willing to tangibly aid individuals who they have never met face to face. The other important point arising from this case study is the evidence that both receiving and sending support lowered mothers' stress levels, with the former having a stronger effect than the latter (Dunham et al., 1998, p. 10). This indicates that emotional support, which is exchanged relatively easily online, can be effective even when not offered face to face.

7.5.1.3 Sense of Belonging

Sense of belonging refers to the intangible sense that one is a part of something larger than oneself. In a community context, it often suggests acceptance, and the perception that one has a home, whether literal or metaphorical.

Returning to Tönnies, it is interesting to note that an element of both *Gemeinschaft* by blood and *Gemeinschaft* of location is that people belong to each by virtue of birth rather than any personal qualities. Although kin can reject kin and communities can ostracize their members, both events are relatively rare because interdependency tends to force people to learn to tolerate each other. This is an aspect of belonging that tends to be lacking in online communities; most communities have a leader or moderator who decides who can belong and who cannot. While the rules governing which individuals are granted membership may be fair, they still allow for the possibility of being ejected from the community. An intriguing question, then, is whether sense of community develops more easily in a forum where exclusion is more difficult? If so, would a (perceived to be) fair administrator who made ejection from the community a predictable matter be sufficient to overcome the limitations on sense of community imposed by the possibility of rejection and ejection?

Among the more contemporary schools of thought, communitarianism, social constructionism, and network analysis all make use of the idea of sense of belonging. For communitarians, encouraging the growth of a sense of belonging is the key to fulfilling the other requirements of community: common values, consistent membership, regular social interaction, and the ability to exercise control over its members (Etzioni, 1993, pp. 381–382), while some network analysts suggest that it is one of the criteria by which a community is distinguished (Wellman, 2001). For social constructionists, sense of belonging is almost identical to the community consciousness that Cohen described as crucial to the creation of communities (Cohen, 1985, p. 12). The feeling of belonging to the community makes the experience real. It also appears in the work of theorists such as Freie, who contends, "People within a community actively participate and cooperate with others to create their own self-worth, a sense of caring about others, and a feeling for the spirit of connectedness" (Freie, 1998, p. 23). The spirit of connectedness is one of the main factors distinguishing community from other types of collective gatherings.

There have not been many studies done on sense of belonging in online communities, although Anita Blanchard includes it in her own evaluation of the criteria for determining whether an online aggregate is a community (Blanchard, 2003). It is possible that sense of belonging may be related to another element commonly associated with communities: social control. This is suggested by Anne Hornsby's application of Durkheim to the Internet, and her suggestion that people are willing to submit to some control in order to gain the benefits of community membership. It could be argued that online willingness to submit to social control could indicate that users are receiving benefits, one of which might be a sense of belonging.

7.5.2 The Case Against Communities Online

The major community elements raised most often by people who do not believe that online communities exist can be broken down into two groups. The first set of elements is concerned with the quality of the relationships within a community, while the second set is concerned with community as a platform for political action. The arguments around elements of the former type tend to focus on things like broad relationships, high levels of commitment and continuity, the strength of tradition, consequences for one's actions, and responsibility to one's fellow citizens. Underlying these concerns is the belief that these things require ongoing face-to-face interaction and geographical boundaries. Arguments around the latter set of elements tend to focus on the link between community and democracy, the role community plays in teaching people the skills necessary to take political action, and of the need for the bonds that only face-to-face interaction can form if people are to effectively defend themselves and take action on their own behalf. Some individuals invoke arguments from both sets, while others refer to only one set. In general, those who references arguments from only one set tend to focus on the quality of relationships rather than political action.

7.5.2.1 The Quality of Online Relationships

The quality of relationships is a set of interrelated arguments that mostly relate back to the idea of *Gemeinschaft* relationships. Whether there is a qualitative difference between an online and face-to-face relationship and what effect such differences as exist might have is still an area under research. However, what limited research has been done and what we know about technological limitations has led both sides to make arguments about how the quality of online relationships affects community building.

The first element is the relative narrowness or broadness of online relationships. Communities have long been associated with broad, complex relationships – an idea that can be traced back to Tönnies – and many people believe that these relationships cannot be formed online. Calhoun, for example, argues, "where electronically mediated groups and networks are not supplements to those with strong face-to-face dimensions, they typically reach a category of people who share a common interest" (Calhoun, 1998, p. 392). Similarly, Crang suggests that "the net allows fluidity of identity and differentiated performances to different audiences", leading to "mutual support through peer groups in specific and narrow fields" (Crang, 2000, p. 308). Both argue that because much of online interaction is focused around narrow interests, relationships will also necessarily be narrow. However, while this may be an argument against classifying narrow interest groups as communities, it only holds as a general argument against the possibility of all online communities if it can be demonstrated that all online interaction is built around narrow interaction. This is an empirical question, and the early evidence suggests that this is not the case (e.g. Reid, 1999; Rheingold, 2000; Kendall, 2002).

A slightly different approach to this is taken by Freie, who contends, "Primary relationships are developed through primary experiences, active engagements with our world, with others in our environment, and with our physical environment itself" (Freie, 1998, p. 150). That is, online relationships are inherently inferior because they occur online and thus lack physicality. Along comparable lines, Calhoun notes that people leave their computers at home to go to cyber cafés, and suggests that this indicates that there are "dimensions of publicness and sociability reproduced poorly if at all in computer-mediated communication" (Calhoun, 1998, p. 373). Even Wellman and Gulia, supporters of the concept of online communities, have observed that the lack of situational cues in most online relationships may discourage the formation of strong ties (Wellman and Gulia, 1999, p. 176), although Wellman points to Granovetter's (1973) work demonstrating that weak ties are better than strong ties for locating new resources and are thus still useful, although perhaps not for sociability. Kraut et al. (1998, p. 1027) had similar findings in their studies, noting that most of the online connections they found among HomeNet project participants were weak ties.

Walther argues the contrary position, stating that people manage to find ways to adapt to the narrower bandwidth associated with computer-mediated communication (Walther, 1996, p. 9). He suggests that the narrower bandwidth means that relationships may take longer to develop because of the increased length of time required to share information (Walther, 1992). Chan and Cheng (2004) found further empirical support for this position. In a survey of Hong Kong residents, they discovered that people rate their offline friendships as stronger than their online friendships during the early stages of development, but that the differences tended to disappear after a year. Their findings suggest that online and offline friendships develop along a different course, but friendships that last eventually converge. There are some other tantalizing hints about differences between online and offline friendships, such as Lori Kendall's suggestion that online friendships may weather change better than offline friendships, but at the cost of depth or importance (Kendall, 2002, p. 142). For the moment, there seems to be enough evidence that strong ties can form online to make dismissing all online relationships unwise. More study is needed, however, before definite conclusions can be drawn about the nature of online relationships, and about the validity of this concern.

The second issue raised with the quality of online relationships is the ease with which anonymity is maintained online, and the consequent low commitment levels to online relationships. In a neighbourhood community, anonymity is nearly impossible because everyone is recognizable on sight, and commitment is high because people are invested both materially and psychologically in the community. Online, it becomes much easier to hide one's identity, which arguably reduces commitment to the group by allowing people to leave and join groups easily without being followed by an undesired reputation. In this context, anonymity is considered problematic because it frees people from having to take responsibility for their actions, and makes it difficult to enforce sanctions. In face-to-face communities, people are recognized by appearance, which is difficult to effectively disguise. Online, a user who has been banned can return with a new name (Etzioni and Etzioni, 1999, p. 243). This ease of return may weaken the group participants' faith in the ability of group leaders to sustain civil interaction because of the ease with which uncivil people can escape punishment. Anonymity can also allow people to act in ways they would not normally act, leading them to assume roles and personalities that by their own statements do not resemble their offline roles and personalities (e.g. Donath, 1999; O'Brien, 1999). This can include everything from gender swapping and disguising their age to a complete personality change, with shy people becoming outgoing and polite people becoming brash. This raises questions as to whether a community is a community when the people within it are not behaving in a fashion consistent with their own self-perception.

The other factor, low commitment levels, refers to the ease with which people can withdraw from online groups. Online interaction is elective – and often interest-based – and membership in online groups is rarely necessary for life or career (Crang, 2000, p. 308). Thus, the nature of the medium could conceivably encourage low ongoing commitment to the group. It is relatively easy, in a practical if not always emotional sense, to abandon an online relationship if things start to go badly. Nickname changes, e-mail filters, and blocking features on instant messaging platforms make it much easier to avoid virtual ex-friends than it is to avoid real life ex-friends, who tend to have an inconvenient habit of showing up at work or social functions or the corner store. In addition to making it more difficult to enforce social norms outside of the immediate community, this ease of departure also decreases the need for people to tolerate those they do not care for, and deprives people of the opportunity to learn conflict resolution. What must be added to this concern, however, is the point that technological affordances are still under human control. The technology makes it possible for people to withdraw from online groups, but it is also possible for them to choose to stay and work out differences. Technology does not demand that they run away, and emotional commitments – or time invested – might lead them to ignore the technological affordance and stay.

The third major objection is the difficulty in enforcing social control through online communities. Social control includes both the creation and enforcement of norms. Social control is considered ineffective online for several reasons. The first is that the anonymity facilitated by the Internet makes social control difficult because people can easily escape the consequences of their actions. Ryan, for example, argues:

> For the most part, disembodied communication where nobody has to say who they actually are results in exchanges that are vapid and repetitive just because they lack the constraints of real life. Free speech is free when it is responsible – not in the sense of being dreary and commonplace, but in the sense of the utterer having to live with the consequences of their utterances. (Ryan, 1997, pp. 1170–1171)

In this case, the argument is that people do not have to live with the consequences of their actions because they can easily escape repercussions by reappearing under a different identity. While there are sometimes ways around this limitation if the moderators of a community are dedicated and technically proficient, the legalistic aspect of top-down governance stands in stark contrast to the self-administered sanctions typical of many small towns and other neighbourhood communities.

There are a number of studies that address boundaries and social control within online communities (e.g. Reid, 1999; Birchmeier et al., 2005), leaving little doubt that online groups do make attempts to exercise control over the behavior of their members. The other factor that some point to as a limitation of online groups is the effect of the limited scope of most online groups on their ability to exercise social control. Real life communities, in addition to being difficult to leave, also tend to be all-encompassing. Real life friends, family, and neighbours are hard to avoid, possessing multiple ways of contacting each other, and the ability to communicate with each other. Almost every aspect of life is touched by other people, and is therefore under their scrutiny. An online group might establish norms, either through fiat or negotiation (Gotved, 2002, p. 410), and it may enforce those norms among participants through a variety of means including the electronic versions of the traditional methods of criticism and ostracism (Kollock, 1999; Reid, 1999; Coates, 2001), but an online-only group cannot reach beyond the screen.

The idea of community as an agent of social control appears frequently in the classical community literature, which is not surprising given the interest of some early theorists in the maintenance of social order and their beliefs about the role of community in structuring behavior. Durkheim in particular focused on social control, noting that the conscience collective in mechanical societies led them to treat crime as an offence against common moral sentiments (Durkheim, 1893, trans. 1969). Looking at the Chicago School, the role of community in exercising social control can be seen implicitly in Wirth's description of the decrease in social control that results from increases in the size, density, and heterogeneity of the population, all of which lead to a break down in community (Wirth, 1938). Similarly, the traditionalists value social control as an element of past communities while the communitarians argue that social control is vital for communities to evolve and thrive (Etzioni and Etzioni, 1999, p. 243).

7.5.2.2 Community as a Political Force

Many scholars in the neighbourhood community tradition value community for its role in both preparing people for collective (often political) action and actually engaging them in the process of attempting to bring about change. Calhoun, for example, argues, "Strong communities provide people with bases for their participation in broader political discourse. They provide them with informal channels of information, chances to try out their ideas and identities before they enter into the public sphere" (Calhoun, 1998, p. 385). The image conjured by this conceptualization of community is that of the town hall where local citizens gathered to discuss issues, hash out problems, and select a course of action. This is a role critics argue that online groups simply cannot fulfill. Ryan writes, "The Internet is good at reassuring people that they are not alone, and not much good at creating political community out of the fragmented people that we have become" (Ryan, 1997, p. 1170). This conceptualization of community is closely related to the idea of social capital as defined by Putnam: "features of social life – networks, norms, and trust –that enable participants to act together more effectively to pursue shared objectives" (Putnam, 1995b, pp. 664–665). Although the two ideas are similar, it remains useful to differentiate between them; not every person who discusses collective action does so in social capital terms and some of the elements of Putnam's definition such as norms are addressed separately.

Despite the belief by some recent community theorists that political activity and collective action are important elements of communities, it is important to note that most contemporary communities – including most geographical communities –do not engage in collective actions of this type. Thus, to make this a requirement of community is to eliminate as communities a large number of gatherings that would otherwise be fairly noncontentiously labeled communities. Rather than limiting the term community to those groups that are politically oriented, a more fruitful approach might be to consider whether communities engage in joint action of the type described by Blumer. Joint action in this context refers to actions for which people engage in an interpretive process where they make indications to each other as well as to themselves, informing others of what their intentions are so that the actions have meaning for everyone involved (Blumer, 1986, p. 16). Joint action can occur on a smaller scale than the collective action described by scholars such as Calhoun, encompassing activities such as weddings, funerals, and other community rituals. The importance of rituals should not be underestimated. As Freie notes:

> Rituals transmit knowledge about the community, confirm and maintain the existence of social groups and cultural forms, justify the power of the dominant groups as well as provide legitimacy to those groups that are weaker, moderate conflict, and affirm the legitimacy of the community. (Freie, 1998, pp. 21–22)

Focusing on evidence of joint action captures group-level cooperation without setting the barrier so high as to be insurmountable. It also captures the integration element of community proposed by Durkheim and further emphasized by scholars such as Hornsby (2001, p. 100) and Freie (1998). There is some indication that joint action can occur online, such as the online, in-game wedding of Achilles and Winterlight for which their guests showed up with virtual gifts (Turkle, 1995). A more relevant example to the real world might be the various fandoms that raise money for charity (and raise the profile of their favorite television show) through auctions and blood drives.

A second and related argument made by people who view community as a political force is that online groups lack the diverse membership of real life communities, and thus fail to teach their members tolerance, where tolerance is viewed as an important element of democracy. Freie, for instance, argues that Netizen communities tend to be composed of members so similar that little difference of opinion actually occurs (Freie, 1998, p. 147). Similarly, Calhoun states that the interest groups that dominate the Internet are categories of people linked by a single concern rather than networks that bind people across lines of significant social differences (Calhoun, 1998, p. 385). He goes further, arguing that the tolerance fostered by sustained interaction with different people is necessary not only for the collective action, but for the continuance of democracy:

> Democracy must depend also on the kind of public life which historically has flourished in cities, not as the direct extension of communal bonds, but as the outgrowth of social practices which continually brought different sorts of people into contact with each other and which gave them adequate bases for understanding each other and managing boundary-crossing relations. (Calhoun, 1998, p. 391)

Hern and Chaulk agree with him, suggesting that "Genuinely participatory and direct democracies require the kinds of human-scaled social relationships that only face-to-face living, commitment and unshakeable love of place can support" (Hern and Chaulk, 2000, p. 115). This belief offers an interesting insight into one of the reasons that some opponents of online communities believe their battle is worth fighting; if community is necessary to democracy and online communities are perceived as communities while failing to fulfill the democratic functions of community, then continued belief that online communities are communities could potentially lead to the further decay of offline communities, thus endangering democracy.

A third concern is that offline communities teach their members skills that online communities cannot, such as the leadership and organizational skills needed for political activity. The opponents of the idea of online communities further argue that offline communities increase citizen access to information and thus increase government accountability, and build trust between citizens, permitting them to more effectively resist nondemocratic regimes (Ryan, 1997, p. 1169; Kavanaugh and Patterson, 2001, p. 498; Paxton, 2002, p. 254). It could be argued that organizational skills can be improved online, and citizen access to information is almost certainly improved online, but it is also worth considering Calhoun's contention that the "corporate structure behind computers and the Internet is impressively centralized" (Calhoun, 1998, p. 382) and that the Internet greatly enhances existing power structures. He claims, "The more a particular possible use of the Internet depends on social organization and the mobilization of significant resources, the more it will tend to be controlled by those who are already organized and well-off" (p. 383). Offline, everyone who is concerned with an issue is free to speak up; online, there may be more limitations to citizen organization than are apparent at first glance. Related to this is another argument: the issue of privacy. Rather ironically, considering the earlier comments about anonymity interfering with community building, online communication is not all that private or even necessarily confidential. While the average community leader might not be able to ascertain the identity of a miscreant, those with sufficient power and knowledge – and sometimes the aid of the courts – can learn the real life details of almost any poster. As Kling has noted, "electronic cafes do not offer the protection for privacy that could be found in some of the traditional face-to-face public spaces: low persistence, low permeability, and relatively high control of messages" (Kling, 1996, p. 51). Once something is posted online, it is impossible to take back, and knowledgeable individuals who are truly motivated can find out who said it. In such an environment, people might be hesitant to agitate too much or argue too radically for fear of the consequences.

Accompanying the community and democracy argument is an underlying fear that online communities will displace offline communities, and thus potentially damage democracy. This fear tends to be accompanied by an assumption that time spent on the Internet replaces time spent socializing or otherwise participating in the offline community. The truth of this assumption is difficult to determine, both because of contradictory studies and because of the difficulties inherent in measuring quantity and type of Internet use. In his review of four studies on Internet use and social relations, Nie (2001) concludes that Internet use does cut into time spent with family and friends. There are also indications that Internet use may adversely affect offline relationships. An Israeli study of the effects of adolescent Internet use on family relationships found a negative relationship between frequency of Internet use by adolescents and their perceptions of the quality of their family relationships (Mesch, 2003). The researchers concluded in this case that time inelasticity was not the cause of the increasingly negative perceptions of family relationships and instead suggested that the Internet might lead to intergenerational conflicts on issues such as privacy and gaps in expectations (p. 1049). On the opposite side, Anderson and Tracey (2001) conducted a longitudinal study of 1000 UK households and found that users reported displacement in a wide range of activities, with no particular activity singled out by more than a small number of users. Furthermore, the users stated that any displacement was minimal. In a time diary study of UK users, Gershuny (2003) found a positive relationship between Internet use and sociability rather than the negative relationship that he was expecting, indicating that, at the very least, Internet use does not prevent people from going out.

The other issue that might be addressed with regard to this perspective is the question of whether it is worthwhile defining something by its purpose. While the quality of relationship arguments focus primarily on what community is –relationships of a certain broad and supportive quality – the political action approach is almost entirely consumed with what community does (or what proponents believe it should do), and gives only brief mention to what it actually is in terms of the types of relationships that are formed. Declaring community inseparable from its supposed (and highly contestable) function is to set up a situation in which community is defined circularly and loses meaning altogether.

7.5.2.3 Evaluating the Differences

Almost everyone on both sides of the online community debate agrees that community involves knowing the whole person and almost everyone agrees that trust, intimacy, and reciprocity are important elements of a community. The main difference between the two groups seems to be what they focus on; which traits they elevate and deem absolutely essential. Those who believe in the existence of online communities tend to emphasize the individual aspect of community. Support and sociability are both characteristics that lead to benefits for the individual. In this case, what benefits the individual – and leads to content and productive citizens – also benefits the group, but the focus is on individuals.

Those who oppose the existence of online communities tend to be more focused on the group. Social control and collective/joint action are both group-level concepts. People on this side of the debate seem to be more concerned with the role of community in encouraging people to look beyond self-interest and cooperate, not just by offering support to a friend they happen to like, but by working with the group and for the group. There is an element of arbitrariness to this distinction – support and sociability can both involve multiple people and collective action waged on behalf of one person – but I believe it also reflects a real philosophical difference about what community is and should be.

7.6 Conclusion

The reasons for the importance of the issue of online communities vary according to the group being addressed, and according to individual interests and concerns. Since it is not possible to address each individual concern, the focus will be on the importance of this question to four groups: scholars, policymakers, Internet users, and the general public. In some cases, there is overlap between these groups: scholars advise policymakers, both may be Internet users, and everyone is part of the general public. There is also sometimes overlap between their concerns, especially when those concerns are related to the welfare of Internet users or the public. Nonetheless, each group also has its own reasons for caring about the possible existence of online communities when acting in that role.

Community has long played a central role in the social sciences, which means that scholars who use the concept have a vested interest in determining how it is used and how it or if it evolves. Social scientists – and policymakers pragmatically – are also interested in how social relations change over time, and there are suggestions that the Internet is leading to changes in how we interact. Slevin, for example, argues that the Internet – particularly Internet communities – has led to a blending of the intimate and the distant, and suggests a need to consider "why we are increasingly prepared to subject ourselves to these mixed feelings of intimacy and estrangement in our day-to-day lives" (Slevin, 2000, p. 92). For him, the desire to form social relationships online could be indicative of the absence of some crucial element in modern social life. Giddens, on the other hand, argues that this combination of local and global is a natural consequence of heightened modernity. He states that modernity is characterized by a "disembedding" of social relations: the separation of relations from local context (Giddens, 1990, p. 21). He traces this to the growth in technology and the effect it has had on modern life, leading local activities to be influenced by remote events and agencies, and making them globally consequential (Giddens, 1996, pp. 9–10). It makes little sense in this view to limit our social relations to those we see on a daily basis when our actions can affect everyone, and indeed, it might be beneficial in an age of growing interdependence to encourage people to feel connected to people in other places and other cultures.

Not everyone is as neutral about the outcome of the evaluation of online communities as most scholars. Community is strongly linked in many minds to the idea of "nation-ness" (Slevin, 2000, p. 93). Through the link to nationalism, community becomes linked to patriotism. Community, as Putnam (2000) is quick to remind us, fosters the civic participation that underpins the state. To be a pillar of the community is to be a patriot, and so it is not surprising that community membership is viewed as a positive attribute. To achieve membership in a community, particularly a voluntary community such as an online community (from which one can be rejected), could be regarded by those who believe in the community–patriot link as affirmation of their basic value as human beings.

The interest in online community of some Internet users stems from a desire to receive validation of their own value and evidence that they are loyal citizens, and therefore might be eager to have the official stamp of community applied to their group. For participants in online groups, this label has the added benefit of justifying to a skeptical society the time they spend online, and validating their experience of feeling connected to other people online. If community is good and community can exist online, then time spent online can also be good and worthwhile. The alternative in many cases is to have their activities labeled either frivolous or deviant. The latter label can lead to attempts to separate people, particularly minors, from participation in online groups. Some young participants in the popular online multiplayer game Everquest have been sent to a psychiatrist for counseling by parents concerned about how much time their children devoted to a game (Chee and Smith, 2003), particularly one popularly known among some of its players (and its players' friends) as Evercrack.

For policymakers and Internet users, one reason to be interested in online communities, put forth by Bakardjieva, is that the community label affirms the value of the participatory model of Internet activity and thus provides a legitimate alternative to the consumption model that threatens to dominate the Internet (Bakardjieva, 2003, p. 294). Kling agrees that the market model threatens to dominate, noting that "Increasingly, the moral frameworks for organizing social relationships in cyberspace have been best articulated by libertarians (with a focus on privacy) and market enthusiasts who argue that market forces alone should shape the nature of on-line services" (Kling, 1996, p. 51). Virtual community is another model. Bakardjieva contends, "in all forms of virtual togetherness, unlike in the consumption mode, users *produce* something of value to others – content, space, relationship and/or culture" (Bakardjieva, 2003, p. 294). She argues that if the value of online gatherings is denied by denying that they are communities, we may lose this alternative. This seems like a bit of an oversimplification – it is possible for group activities to be considered valuable even if they do not take place in a community context – but the point is not entirely without merit given the resonance that community has in modern society, and the possibility that denouncing online communities as false may strip them of all meaningfulness in the eyes of the public.

Finally, for the general public, interest in online communities may arise from the belief that the term community refers to something important. These conceptions of community tend to focus on community as a collective endeavor that leads to a world that is "more united, more connected, more sharing, and more meaningful for human action than the one we live in now" (Freie, 1998, p. 23), a means of promoting moral standards and more responsible behavior (Etzioni, 1993), and a way of sustaining democracy (Calhoun, 1998). Those who hold these beliefs tend to argue that online gatherings cannot meet these needs, and thus are not real communities. They believe the debate is important because they fear that people, convinced that community has safely been transplanted online, will turn away from real life communities and the labor required to maintain them, and instead seek emotional satisfaction from virtual relationships (Freie, 1998, p. 35). Associated with this is the fear that online community is a way of escaping from dealing with the problems of the real world (Freie, 1998, p. 33). Although the digital divide has decreased, wealthier and better educated people remain more likely to have Internet access than their poorer and less well-educated brethren (Madden, 2003, p. 10). In some respects, the Internet remains a realm where the middle class can ignore the plight of the lower class and rich countries can ignore the problems of poor countries; panhandlers are rarely encountered on the way to an online meeting, and disagreeable voices can frequently be filtered out with the touch of a button (Baym, 1998, p. 40).

7.7 Acknowledgments

I am grateful for the advice and aid of Lorne Dawson, Kieran Bonner, John Goyder, Barry Wellman, and Brent Berry in preparing this chapter. Thank you as well to Patrick Purcell for suggesting the title.

7.8 References

Anderson, B., and Tracey, K. (2001) Digital living: the impact (or otherwise) of the Internet on everyday life. *American Behavioral Scientist,* **45**(3), 456–475.

Anderson, B.R.O'G. (1983) *Imagined Communities: Reflections on the Origin and Spread of Nationalism.* Verso, London.

Bakardjieva, M. (2003) Virtual togetherness: an everyday perspective. *Media, Culture and Society,* **25**, 291–313.

Baym, N. (1998) The emergence of community in computer-mediated communication. In Jones, S.G (ed.), *Cybersociety 2.0: Revisiting Computer-Mediated Communication and Community.* Sage, Thousand Oaks, CA.

Berger, B.M. (1988) Disenchanting the concept of community. *Society,* **25**(6), 50–52.

Bernard, J. (1973) *The Sociology of Community.* Scott, Foresman and Company, Glenview, IL.

Birchmeier, Z., Adam, N.J. and Beth, D.-U. (2005) Storming and forming a normative response to a deception revealed online. *Social Science Computer Review,* **23**(1), 108–121.

Blanchard, A. (2003) Whose community is it anyway? Testing a model of member participation and sense of community in virtual communities. Presented at the Association of Internet Researchers Conference on 16 October 2003 in Toronto, Ontario, Canada.

Blumer, H. (1986) *Symbolic Interactionism: Perspective and Method.* University of California, Berkley, LA.

Brint, S. (2001) Gemeinschaft revisited: a critique and reconstruction of the community concept. *Sociological Theory,* **19**(1), 1–23.

Cahnman, W.J. (1995) *Weber & Tönnies: Comparative Sociology in Historical Perspective.* Transaction Publishers, New Brunswick, NJ.

Calhoun, C. (1998) Community without propinquity revisited: communications technology and the transformation of the urban public sphere. *Sociological Inquiry,* **68** (3), 373–397.

Chan, D., Grand, K.S. and Cheng, H.L. (2004) A comparison of offline and online friendship qualities at different stages of relationship development. *Journal of Social and Personal Relationships,* **21** (3), 305–320.

Chee, F. and Smith, R. (2003) Is electronic community an addictive substance? Everquest and its implications for addiction policy. Presented at the Association of Internet Researchers Conference on 18 October 2003 in Toronto, Ontario, Canada.

Coates, G. (2001) Disembodied cyber co-presence: the art of being there while somewhere else. In Watson, N. and Cunningham-Burley, S. (eds), *Reframing the Body*. Palgrave, Hampshire, UK.

Cohen, A.P. (1985) *The Symbolic Construction of Community*. Tavistock Publications, London.

Correl, S. (1995) The ethnography of an electronic bar: the lesbian café. *Journal of Contemporary Ethnography*, **24**(3), 270–298.

Crang, M. (2000) Public space, urban space and electronic space: would the real city please stand up? *Urban Studies*, **27**(2), 301–317.

Cummings, J.N., Sproull, L. and Sara, B.K. (2002) Beyond hearing: where the real-world and online support meet. *Group Dynamics (Special Issue: Groups and Internet)*, **6**(1), 78–88.

Donath, J.S. (1999) Identity and deception in the virtual community. In Smith, M.A. and Kollock, P. (eds), *Communities in Cyberspace*. Routledge, New York.

Driskell, B.R. and Lyon, L. (2002) Are virtual communities true communities? Examining the environments and elements of community.*City and Community*, **1**(4), 373–390.

Dunham, P.J., Hurshman, A., Litwin, E., Guseall, J., Ellsworth, C. and Peter, W.D.D. (1998) Computer-mediated social support: single young mothers as a model system. *American Journal of Community Psychology*, **26**(2), 281–306.

Durkheim, E. [1893] (1969) *Division of Labor in Society*, translated by George Simpson. Macmillan, New York.

Effrat, M.P. (1974) Approaches to community: conflicts and complementarities. In Effrat, M.P. (ed.), *The Community: Approaches and Applications*. Collier Macmillan Publishers, New York, pp. 1–32.

Etzioni, A. (1993) *The Spirit of Community: Rights, Responsibilities, and the Communitarian Agenda*. Crown Publishers, New York.

Etzioni, A. and Etzioni, O. (1999) Face-to-face and computer-mediated communities: a comparative analysis. *The Information Society*, **15**, 241–248.

Frankfort-Nachimas, C. and John, J.P. (1993) Neighbourhood revitalization and the community question. *Journal of the Community Development Society*, **24**(1), 1–14.

Freie, J.F. (1998) *Counterfeit Community: The Exploitation of our Longings for Connectedness*. Rowman & Littlefield, Lanham, MD.

Gershuny, J. (2003) Web use and net nerds: a neofunctionalist analysis of the impact of information technology in the home. *Social Forces*, **82** (1), 141–168.

Giddens, A. (1990) *The Consequences of Modernity*. Polity, Cambridge.

Giddens, A. (1996) *In Defence of Sociology: Essays, Interpretations and Rejoinders*. Polity, Cambridge.

Goldenberg, S. and Haines, V.A. (1992) Social networks and institutional completeness: from territory to ties. *Canadian Journal of Sociology*, **17**(3), 301–313.

Gotved, S. (2002) Spatial dimensions in online communities. *Space and Culture*, **5**(4), 405–414.

Granovetter, M. (1973) The strength of weak ties. *American Journal of Sociology*, **78**, 1360–1380.

Hale, S. (1995) *Controversies in Sociology: A Canadian Introduction*, 2nd edn. C.C. Pitman, Toronto, Onatario, Canada.

Hern, M. and Chaulk, S. (2000) Roadgrading community culture: why the Internet is so dangerous to real democracy. *Democracy and Nature*, **6**(1), 111–120.

Hillery, G.A., Jr. (1955) Definitions of community: areas of agreement. *Rural Sociology*, **20**, 111–123.

Hornsby, A.M. (2001) Surfing the Net for community: a Durkheimian analysis of electronic gatherings. In Kivisto, P. (ed.), *Illuminating Social Life: Classical and Contemporary Theory Revisited*. Pine Forge Press, Thousand Oaks, CA, pp. 73–116.

Kavanaugh, A.L. and Patterson, S.J. (2001) The impact of community computer networks on social capital and community involvement. *The American Behavioral Scientist*, **45**(3), 496–509.

Kendall, L. (2002) *Hanging Out in the Virtual Pub: Masculinities and Relationships Online*. University of California Press, Berkley, LA.

Kling, R. (1996) Synergies and competition between life in cyberspace and face-to-face communities. *Social Science Computer Review*, **14**, 50–54.

Kollock, P. (1999) The economies of online cooperation: gifts and public goods in cyberspace. In Smith, M.A. and Kollock, P. (eds), *Communities in Cyberspace*. Routledge, New York, pp. 220–239.

Kraut, R., Lundmark, V., Patterson, M., Kiesler, S., Mukopadhyay, T. and Scherlis, W. (1998) Internet paradox: a social technology that reduces social involvement and psychological well-being? *American Psychologist*, **53**(9), 1017–1031.

Loomis, C.P. and McKinney, J.C. (1957) Introduction. In Loomis, C.P. (ed.), *Community and Society (Gemeinschaft und Gesellschaft)*. Michigan State Press, East Lansing, MI, pp. 1–29.

Madden, M. (2003) America's online pursuits: the changing picture of who's online and what they do. Pew Internet and American Life Project. Retrieved 28 February 2004 from <http://www.pewinternet.org/reports/toc.asp?Report=106>

Mahowald, M.B. (1973) Marx's "gemeinschaft": another interpretation. *Philosophy and Phenomenological Research*, **33**(4), 472–488.

Marx, K. [1847] (1978) The German ideology. In Tucker, R.C. (ed.), *The Marx–Engels Reader*, 2nd edn. Norton, New York, pp. 146–200.

McKinney, J.C. and Loomis, C.P. (1958) The Typological Tradition. In Roucek, J.S. (ed.), *Contemporary Sociology*. Philosophical Library, New York, pp. 557–582.

Mesch, G.S. (2003) The family and the Internet: the Israeli case. *Social Science Quarterly*, **84**(4), 1038–1050.

Mouritsen, P. (2003) What's the civil in civil society? Robert Putnam, Italy and the republican tradition. *Political Studies*, **51**, 650–668.

Nie, N.H. (2001) Sociability, interpersonal relations, and the Internet. Reconciling conflicting findings. *American Behavioral Scientist*, **45**(3), 420–335.

O'Brien, J. (1999) Writing in the body: gender (re)production in online interaction. In Smith, M.A. and Kollock, P. (eds), *Communities in Cyberspace*. Routledge, New York.

Parsons, T. (1973) A note on *Gemeinschaft* and *Gesellschaft*. In W.J. Cahnman (ed), *Ferdinand Tönnies. A New Evaluation*. E.J. Brill, Netherlands, pp. 140–150.

Paxton, P. (2002) Social capital and democracy: an interdependent relationship. *American Sociological Review*, **67**, 254–277.

Portes, A. (1998) Social capital: its origins and applications in modern sociology. *Annual Review of Sociology*, **24**, 1–24.

Putnam, R. (1995a) Bowling alone: America's declining social capital. *Journal of Democracy*, **6**, 65–78.

Putnam, R. (1995b) Tuning in, tuning out: the strange disappearance of social capital in America. *Political Science and Politics*, **28**, 664–683.

Puttnam, R. (2000) *Bowling Alone: The Collapse and Revival of American Community*. Simon & Schuster, New York.

Reid, E. (1999) Hierarchy and power: social control in cyberspace. In Smith, M.A. and Kollock, P. (eds), *Communities in Cyberspace*. Routledge, New York, pp. 107–133.

Rheingold, H. (1993) *The Virtual Community: Homesteading on the Electronic Frontier*. Addison-Wesley, Reading, MA.

Rheingold, H. (2000) *The Virtual Community: Homesteading on the Electronic Frontier*, 2nd edn. Addison-Wesley, Reading, MA.

Ryan, A. (1997) Exaggerated hopes and baseless fears. *Social Research*, **64**(3), 1167–1190.

Simmel, G. (1949) The sociology of sociability. *American Journal of Sociology*, **55**, 254–261.

Simpson, R.L. (1974) Sociology of the community: current status and prospects. In Bell, C. (ed.) *The Sociology of Community*. F. Cass, London, pp. 313–334.

Slevin, J. (2000) *The Internet and Society*. Blackwell, Malden, MA.

Stacey, M. (1969) The myth of community studies. *The British Journal of Sociology*, **20**(2), 137–147.

Tönnies, F. [1887] (1957) *Community and Society*, translated and edited by Loomis, C.P. Michigan State Press, East Lansing, MI.

Turkle, S. (1995) *Life on the Screen*. Simon & Schuster, New York.

Turner, J.H. and Dolch, N.A. (1996) Using classical theorists to reconceptualize community dynamics. *Research in Community Sociology*, **6**, 19–36.

Walther, J.B. (1992) Interpersonal effects in computer-mediated interaction: a relational perspective. *Communication Research*, **19**, 52–90.

Walther, J.B. (1996) Computer-mediated communication: impersonal, interpersonal, and hyperpersonal interaction. *Communications Research*, **23**, 3–43.

Wellman, B. (1979) The community question. *American Journal of Sociology*, **84**(5), 1201–1231.

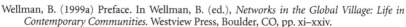

Wellman, B. (1999a) Preface. In Wellman, B. (ed.), *Networks in the Global Village: Life in Contemporary Communities*. Westview Press, Boulder, CO, pp. xi–xxiv.

Wellman, B. (1999b) The network community: an introduction. In Wellman, B. (ed.), *Networks in the Global Village: Life in Contemporary Communities*. Westview Press, Boulder, CO, pp. 1–47.

Wellman, B. (2001) Physical space and cyberplace: the rise of personalized networking. *International Journal of Urban and Regional Research*, **25**(2), 227–252.

Wellman, B. and Gulia, M. (1999) Virtual communities as communities: Net surfers don't ride alone. In Smith, M.A. and Kollock, P. (eds), *Communities in Cyberspace*. Routledge, New York, pp. 167–194.

Wellman, B. and Leighton, B. (1979) Networks, neighbourhoods and communities. *Urban Affairs Quarterly*, **14**, 363–390.

Wellman, B. and Wetherell, C. (1996) Social network analysis of historical communities: some questions from the present for the past. *The History of the Family*, **1**(1), 97–121.

Wirth, L. (1938) Urbanism as a way of life. *American Journal of Sociology*, **44**(1), 1–24.

Part C
The Research Impetus

Connected Lives: The Project[1]

Barry Wellman and Bernie Hogan
with Kristen Berg, Jeffrey Boase, Juan-Antonio Carrasco,
Rochelle Côté, Jennifer Kayahara, Tracy L.M. Kennedy and Phuoc Tran

8.1 Being Networked

8.1.1 Connected Lives Before the Internet

> Only connect, and the beast and the monk, robbed of the isolation that is life to either, will die. (E.M. Forster, *Howard's End*, 1910, Chapter 22)

Barring the odd beast and monk, just about everyone is connected these days – at most by 6 degrees of interpersonal connection and often by less (Milgram, 1967; Kochen, 1989; Watts, 2003). Yet only a tiny fraction of those who are connected ever interact in any meaningful way as friends, relatives, neighbours, workmates, and acquaintances. These ties comprise our individual personal communities, each a solar system of 10–2000 persons orbiting around us (Wellman, 1979).

Such personal networks abounded before the coming of the Internet, and they flourish now. This chapter uses survey and interview information from our new *Connected Lives* project to investigate what information and communication technologies (ICTs) are doing to us and reciprocally, what we are doing to ICTs.[2] We begin with a long-term view of personal networks and work our way toward present day shifts characterized by "networked individualism" (Wellman, 2001). Thereafter we elaborate the substantive areas of inquiry that the Connected Lives project is addressing and present early findings to bolster our claims.

---

[1] Barry Wellman is the principal investigator of the Connected Lives project. He and Bernie Hogan have major responsibility for the overall drafting of this chapter. Phuoc Tran is the project's computer specialist while the other coauthors are doctoral students at NetLab.

[2] This formulation is a slight paraphrase of Rheingold (2005, p. 6).

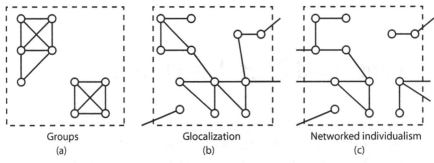

Groups
(a)

Glocalization
(b)

Networked individualism
(c)

Figure 8.1   Three models of personal community.

8.1.1.1 Neighbourhood and Village Groups

Our elders and ancestors tell us that once upon a time personal communities were small and stable. They were rooted in villages and neighbourhoods, with community members changing slowly through the life course via marriage, death, quarrels, and war. People had stable marriages and were members of a single, local, small densely knit group that normally communicated by walking, shouting, or glancing door-to-door (Figure 8.1a). A preindustrial village or an urban village would be exemplars. Such communities often contained many kin and neighbours, with frequent communication among them. Interactions involved much dropping in on people, awareness of the rhythms of daily communal life, and strong community norms and solidarity (Wellman and Leighton, 1979; Wellman, 2001).

Computer scientists – and their acolytes – like to think that it was the Internet that changed bounded village communities to the "global village" (McLuhan, 1964), "the death of distance" (Cairncross, 2001), and "the world is flat" (Friedman, 2005). Yet they are only echoing a long tradition of pastoralist nostalgia that has overstated the stable localness of past times (Marx, 1964). The shift from local, group-based social structures to far-flung, sparsely knit, network-based social structures started well before the advent of the Internet and other ICTs. Even before the advent of the Industrial Revolution, marriages often were short; remarriages and informal liaisons were common. Nor was community always local. For example, Jane Austen's heroines galloped past their neighbours to visit friends and relatives hundreds of miles away; shepherds and nomads wandered long distances; students, soldiers, and camp followers journeyed far to universities and wars (see, for example, the tales recounted in LeRoy Ladurie, 1975, 1997; Davis, 1983).[3]

---

[3] See also the discussions in Thébert (1985/1987), Barthélmy and Contamine (1985/1988), Ward (1999), Wellman and Wetherell (1996), and Wellman and Leighton (1979).

8.1.1.2 Glocalized Place-to-Place Networks

In the last half of the 20th century, the spread of cars, planes, buses, rail, and phones broadened the base and frequency of long-distance connectivity. These technologies enabled ordinary people to keep in touch with friends and relatives, and workers to travel long distances. The result was that by the 1970s, if not before, neighbours were only a small percentage of personal communities. Rather than being born into life-long local community groups, people have been better able to choose their personal community members. Their neighbourhood communities have transmuted into personal community networks: fragmented multiple social networks connected only by the person (or the household) at the center.

Concomitantly, the proliferation of paid opportunities for women to work – in conjunction with postponed marriage and parenthood, accessible birth control, dual-job families, and the prevalence of divorce – affected the extent to which North American households are stable, heavily interacting units where husbands, wives, and children see much of each other (Fagan, 2001; Jacobs and Gerson, 2001; Statistics Canada, 2003, 2005). Even the act of a family eating meals together as a solidary group has been on the decline (Putnam, 2000).[4] Moreover, the number of Americans in "core discussion networks" – people to discuss important matters with – has declined by 29% (2.94–2.08) between 1985 and 2004 (McPherson et al., 2006).

American involvement in some group-oriented activities – such as bowling leagues, civic organizations, and church groups – has also declined. Rather, individuals are extensively involved in less-bounded, less-structured informal networks where they maneuver through multiple sets of ties shifting in saliency and frequency of contact. Each person enacts multiple roles at home, in the community, and at work. Their friends – and even their relatives – are often loosely linked with each other (Wellman and Hampton, 1999). These loose linkages do not imply a complete untethering of social relations: there are few isolates "bowling alone," as Putnam's metaphorical book title asserts (Putnam, 2000). They are bowling in sparsely knit networks rather than in solidary groups (see research by Fischer, 1982; our NetLab, e.g. Wellman, 1979; Wellman and Wortley, 1990; and others reviewed in Wellman, 1999).

Networked relations are no longer confined to neighbourhoods and villages. Yet, until the turn of the 21st century they have been based in a few specific and fixed places although many ties stretch well beyond neighbourhoods. They are "*glocalized*": both far-flung (global) and local.[5] Households

---

[4] We confine our discussion to North American trends, but we suspect that our argument is largely applicable to developed societies and perhaps to societies elsewhere. There are exceptions of course. For example, Catalans continue to live with their parents and adult children, eating most meals together (Wellman, 2002).

[5] We write in the present tense because glocalized interaction patterns remain prevalent in the developed world, even as ICTs proliferate. Indeed, so do densely knit local solidarities. For example, Robert Putnam

remain the preeminent units for organizing marital and community relations. Many friends, relatives, and coworkers travel substantial distances to
get together. Phone calls and even Internet communication are made to
households wired by telephone and cable lines (Wellman, 1982; Wellman
and Wortley, 1989; Wellman and Wellman, 1992). People connect "place-
to-place": aware of local contexts but not dealing with places in between as
they travel, phone, or e-mail sizeable distances to connect with dispersed
friends, kin, and workmates.

Glocalized networks contain overlapping groups of people (Figure 8.1b).
There is much group interaction within local places – homes and offices –
but no overall integration. It is not that there are simply less or more ties,
it is that there are clusters of ties that are really dense, many of which are
affinity groups associated with a particular milieu, such as neighbourhoods,
church, work, old school friends, and kin (Simmel, 1903; Wellman and
Leighton, 1979; Feld, 1982; Kadushin, 1966). Hence, glocalized networks
connect across small clusters, rather than connecting within a large cluster.
They provide diversity, choice, and maneuverability at the probable cost of
cohesion and long-term trust (Wuthnow, 1998; Putnam, 2000; Wellman,
2001; Fischer, 2005).

Recent research into how information flows on the Web has shown
that such intercluster connectivity is an efficient networking structure
(Wellman, 1988; Adamic et al., 2003; Watts, 2003). Most clusters contain
superconnectors – people linked to large numbers of others in multiple social milieus – and these connectors rapidly diffuse information. Although
superconnectors were first identified in studies of links between Web sites,
we believe that they are even more network-efficient for humans, because
people are more likely to connect to multiple other social milieus than are
oft-isolated Web sites (Watts, 2003).

8.1.2 Person-to-Person Networked Individualism

The most recent shift has been to glocalized networks in which the
individual – and not the household, kinship group, or work group – is
the primary unit of connectivity. Such a social structure preserves the aforementioned advantages of glocalization: access to a variety of information-
providing social milieus and rapid linkage by superconnectors. Because the
networks are not confined to one or two solidary groups, they acquire resources from a variety of sources (Merton, 1957; Granovetter, 1973, 1995).
The strength and content of ties vary from situation to situation and from

developed some of his ideas about the persistence of village community in the Italian village of Bellagio (personal communication to Wellman, 2005): a place with densely knit, long-standing internal communication
that serves a glocalized international tourist population including movie star George Clooney (Wellman's
observations, 1999–2005).

day to day in how active or latent they are – as people maneuver through their days and lives. The very presence of a large, active, and resource-filled set of ties has become an important resource in itself.

This individualization of connectivity means that acquiring resources depends substantially on personal skill, individual motivation, and maintaining the right connections. The loss of group control and reassurance is traded for personal autonomy and agility. With networked individualism, people must actively network to thrive or even to survive comfortably. More passive or unskilled people may lose out, as the group (village, neighboood, household) is no longer taking care of things for them (Kadushin et al., 2005). Recall that most of the chains in the small worlds studies were *not* completed (Milgram, 1967; Dodds et al., 2003), presumably because of ignorance about whom to connect with next or a lack of motivation to make the connection.[6] By contrast, hypernetworkers use social networking software to find, connect, and capitalize on thousands of current, former, and potential network members, with one person achieving nearly 8000 connections through *LinkedIn* (Mayaud, 2005a, b).

The shift to networked individualism has happened recently. Up until the 1990s, places were still the main context for interaction for most people. Along came the Internet and its progeny: Usenet and e-mail were followed by a myriad of ICTs: instant messages (IMs), webcams connecting individuals; chat rooms and listservs connecting groups; and blogs, photoblogs, and podcasts broadcasting thoughts, pictures, and sounds. Parallel to the proliferation of ICTs has been a huge global expansion of mobile phone use, carrying both voice and text (Katz and Aakhus, 2002; Ling, 2004). With the Internet and mobile phone, messages come to people, not the other way around. Individuals are connected by their phones, but their phone is not tied to a place and its environment (such as a family or office). Mail is delivered less to a physical box at a family home than to an inbox accessible wherever an individual has an Internet connection.

In short, there has been a shift from place-to-place networking toward person-to-person networking. This is not a shift toward social isolation, but toward flexible autonomy using social networks. It simultaneously implies the responsibility for people to keep up their own networks with more freedom to tailor their interactions. The shift is toward a form of social structure that we call "*networked individualism*" (Wellman, 2001; see also Castells, 2001). Although networked individualism encompasses broader trends in the organization of work and nations, our concern here is with social interactions mediated by modernity and technology. (Figure 8.1c; see also Wellman, 2001, 2007; Wellman and Hogan, 2004; Hennig, 2007).

Our Connected Lives research is investigating the extent and nature of networked individualism. We believe that individualized networks are often larger than glocalized networks and are less densely knit. Networked

---

[6] We are grateful to Charles Kadushin for pointing this out.

individuals know people through individual networking, such as ad hoc meetings over lunch or sending individually tailored e-mail. Their ties are specialized, providing them with different types of support and sociability in a variety of social milieus. Each milieu has limited control over an individual's behavior; each individual has limited commitment to a specific milieu and a low sense of group membership.

Networked individualism can be contrasted with the glocalized situation of networking in which people are involved in a number of specific groups. Networked individuals often have time binds, since they are constantly negotiating plans with disconnected sets of individuals. As they maneuver through their multiple networks, their ties often vary from hour to hour, day to day, and week to week (Menzies, 2005).

The shift to networked individualism has been accompanied by a shift in theorizing about interpersonal behavior. Rather than seeing society as driven by individual norms or by the collective activities of solidary groups, social network analysts focus on how people's connections – to each other, groups, organizations, and institutions – affect possibilities and constraints for their behavior. The social network approach allows for people maneuvering among their relationships, with ICTs providing further maneuverability (Wellman, 1988, 2001).

Hence we – and other analysts – are replacing one-way technological determinism – the assumption that ICTs cause behavior – with a two-way "social affordances" viewpoint that inquires about the opportunities and constraints that ICTs and social systems provide for each other (Bradner et al., 1998; see also Ling, 2004, Wellman, 2004 on the "domestication" of technology).

8.1.3 Plan of Chapter

In this chapter, we use the shift toward networked individualism to help frame and explain the difference ICTs can make to social interaction. We look at social networks and the Internet in the home and beyond the home. We focus on how network structures can influence communication patterns and travel, and how personal dispositions can influence network structures. Our key concerns are:

- How does the shift to individual means of communication – the Internet and mobile phones – affect domestic and community solidarity?

- If people are immersed in the Internet, how does this affect their relations with household members?

- Has the shift from groups to networks affected the ways in which ICTs are being used?

- Do ICTs increase or decrease involvement with community members and more organized forms of civic life?

- How do ICTs affect traveling to see friends and relatives?

- What is the nature of social support – emotional and material aid – in a networked individualized society in which many interactions take place via ICTs?

- How are the ways in which people obtain information related to their ICT use?

This chapter presents the rationale, measurement, and preliminary findings from the Connected Lives project. As our NetLab is itself a team of networked individuals, we present our study in multiple complementary sections. In this, our first reconnoitering of the field, we each use different analytic approaches with variations in sample size, variable definitions, and analytic techniques. We work outward from household, through personal community networks, to finding information through networks or further afield. Our presentation is divided into four principal sections:

- The networking of households.

- The size, composition, and management of personal network communities.

- ICTs and travel to social activities.

- Finding support and information online and offline.

8.2 Doing the Connected Lives Project

8.2.1 Data Gathering

In this first report, we discuss the overall sample of survey respondents and interview participants in order to provide an overview of our concerns and our data. Future research will examine the connected lives of subgroups and individualism in more detail.

The Connected Lives project gathered quantitative and qualitative data through a large survey followed by detailed interviews with a subsample of survey respondents. The study fits between large-scale surveys that provide overall (often national) statistics and ethnographic studies of a small number of cases. The large sample size of the survey provides statistical generalizability while its 1-hour length provides useful detail. The in-depth interviews with a subsample of the same participants provide more detail plus the ability to acquire information about social networks and search processes.

8.2.1.1 Survey

The team collectively developed a lengthy 32-page survey from November 2003 to June 2004. We randomly sampled English-speaking nonfrail adults

(18+) in East York and completed 350 surveys between July 2004 and March 2005. The sampling frame yielded 621 valid names, and we obtained a response rate of 56%. Each survey took between 1 and 2 hours to complete. It was dropped off at each respondent's house and picked up 1–3 weeks later.

The survey makes it possible to establish a fairly good picture of how people in East York are currently using the Internet. It asks about respondents' computers, jobs, household members, personal community networks, community involvement, social attitudes, and the customary demographics. Except where noted, all statistics used in this chapter are from the survey.[7]

8.2.1.2 Interviews

In-home interviews were conducted between February and April 2005 with one-quarter of the survey participants (87 in total). The interview schedule was developed by the Connected Lives team between September 2004 and January 2005, in tandem with the survey deployment. The interviews were conducted by Connected Lives doctoral students and took 2–4 hours – usually in a single evening session. The response rate was 85% of those survey respondents who wrote "yes" or "unsure" when asked at the end of the survey if they would be willing to be interviewed.

The interview starts with a semistructured section on daily routines and moves on to computer and Internet use. The interview obtained detailed information about household relations, Internet use, travel behavior, social networks, and information seeking. It includes a name generator to help describe the personal networks of the respondents (Carrasco et al., 2005).

Participants were questioned during the interviews about their general culture and leisure activities, how they select specific activities to engage in, and the role the Internet plays in their leisure lives. Information about cultural activities was gathered by having people rank a series of cue cards listing leisure activity groups, and then asking them to elaborate on the specific activities people engaged in. This was then followed by a series of questions about how people gather information and make decisions about the culture and leisure activities they identified as being of interest. Participants were then asked about the role that the Internet plays in information gathering, decision making, and engaging in activities. As we are still coding

---

[7] The 32-page survey was designed and typeset in *Adobe InDesign CS* by Bernie Hogan. The cover logo was designed by Phuoc Tran and mirrors a public web page, *http://www.connectedlives.ca*. Given the emphasis on contemporary ICT use, we felt it important for the project to have a public face online. The survey package included the logo, an introductory letter, and a picture of the Connected Lives team. Standard survey procedures were used: an initial letter followed by an in-person follow-up and subsequent pick-up, Tim Hortons™ coffee shop gift certificates to respondents, and extensive attempts to convert refusals and incompletes into completed surveys.

the interviews, no statistics from it are presented, but the interviews do provide interpretive enrichment.

8.2.1.3 Observations

Free and semistructured observations and discussions were used to relate participants' actual behavior to their interview and survey responses. If the interview participant had an Internet connection, we concluded our visit with an in-home observation of how the participant actually uses a computer and searches for health and cultural information. Interview participants were asked to demonstrate how they use the Internet. This included both unstructured demonstrations of everyday uses plus structured demonstrations of specific skills (see also Hargittai, 2005). The observations focused on how the participants obtain health and cultural information. We also photographed the participants' computer setups.

8.2.2 East York and East Yorkers

8.2.2.1 East York

The case study is set in East York, a residential area of Toronto that has played host to NetLab's two previous community studies in pre-Internet days. A distinct self-governing "borough" of Toronto until metropolitan amalgamation in 1998, East York has always prided itself on its local community and small town atmosphere (Davidson, 1976; Cooper, 2004). East York was originally chosen for the first study in 1968 because of its convenient locale (30 minutes drive from the downtown core), atmosphere, cooperative government, and cultural homogeneity. Its selection for the second study (1978–1979) was for longitudinal continuity as 33 original respondents were reinterviewed (Wellman et al., 1997b). While it would not be feasible to do a third wave of a longitudinal study 25 years later, East York retains its value for comparisons with our pre-Internet data, and it provides a fair cross section of the Canadian urban public.

East York sits squarely within the arterial highway system of Toronto. It is bounded on the west by an expressway and on the south by a subway line, and buses frequently travel main routes. Mobile phone and broadband Internet service is widely available throughout Toronto, the largest metropolitan area of Canada. Computer access is good, with telephone and cable companies competing to provide broadband connectivity.

East York is near the heart of metropolitan Toronto, 30–45 minutes travel from Toronto's central business districts (Figure 8.2a). Its population of 114,240 (2001 census) is ethnically and socioeconomically mixed, residing in houses and apartment buildings (Figure 8.2b).

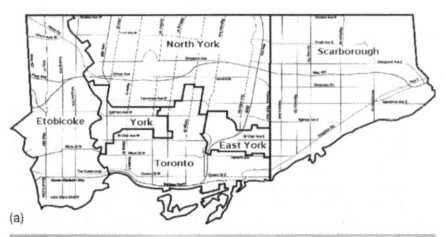

Figure 8.2   (a) East York in Metropolitan Toronto;
(b) Houses and apartment buildings in East York.

8.2.2.2 The East Yorkers

In many respects, East Yorkers reflect Anglophone urban Canada. Fifty-eight percent of the survey respondents are women, with a median age of 45. Fifty-nine percent of the somewhat less representative interview sample

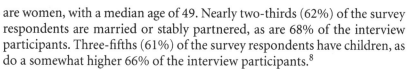
are women, with a median age of 49. Nearly two-thirds (62%) of the survey respondents are married or stably partnered, as are 68% of the interview participants. Three-fifths (61%) of the survey respondents have children, as do a somewhat higher 66% of the interview participants.[8]

The East Yorkers are educated. Forty-three percent of the survey respondents have a university degree, while 27% have a high school education or less.

The bulk of the population is working class and middle class. Median personal income is between $30,000 and $40,000. With most adult household members doing paid work, median household income is substantially higher, between $50,000 and $75,000.[9] Sixty-two percent of the survey respondents are doing paid work. Of the rest, a high percentage (37%) are retired, 16% are students, and 13% are full-time homemakers. The rest are between jobs, on leave, or have other reasons for not working.

Sixty-two percent of the survey respondents are coupled: married, common-law, or in a long-term relationship. Twenty-three percent are single. Compared to the survey respondents, a higher percentage of the interview participants are likely to be coupled (68%) and a lower percentage to be single (15%). A higher percentage (51%) of interview participants than survey respondents have a university degree, while only 20% of the interview participants have a high school education or less.

Recent immigrant migration and high-rise apartment development has made the East York cityscape more complex than its village-like past. When we previously gathered data in East York in 1968 and 1978–1979, almost all residents were Canadian born and of British-Canadian ethnicity. The situation has changed substantially in the past decades. East York is similar to much of the metropolitan Toronto area (and different from many other places in Canada) in its high percentage of foreign-born residents. Fifty-three percent of East York residents were Canadian-born in 2001 (2001 census), similar to the 51% Canadian-born survey respondents and 58% interview participants.

The largest ethnic group remains British-Canadian, comprising nearly half (44%) of the survey sample. However, visible minorities (i.e., nonwhite-Canadians) comprise 27% of the survey sample: principally East Asians and South Asians, with Chinese-Canadians and Indian-Canadians being the largest groups. This is substantially lower than the 2001 Canadian census report according to which visible minorities comprise 36% of the East York population. These ethnic groups are underrepresented in our survey (and subsequent interviews) because of language and cultural barriers. In most

---

[8] Kayahara and Wellman (2005) provide more demographic detail.

[9] Following common survey practice, we asked respondents to report their income within ranges, such as $30,000–$40,000. All dollar amounts are in Canadian dollars, which at the time of our research was equivalent to about 78 US cents, 67 Euro cents, 45 British pence, 87 Japanese yen, and 6.6 Chinese yuan.

other respects, our data reflect census demographics, including gender, age, income, education, and family composition.

8.3 Networked Households[10]

Contemporary household structures are becoming "post-familial families" (Beck-Gernsheim, 2002; Wehner and Abrahamson, 2004). Within this transition, households have become networked in two mutually reinforcing ways.

First, they have become the hubs of communication networks rather than self-contained homes that are penetrated only by doorbells, wired phones, and paper mail. At any one moment, a household member may be talking on a wired phone, another using a mobile phone, while still others – adults or children – are e-mailing, playing online games, or chatting in online groups. With the widespread diffusion of the Internet and mobile phones, patterns of online use have shifted significantly from work and school to the more personal context of the home. ICTs have become key ways in which household members communicate and coordinate with others.[11]

Second, many households – like personal communities – resemble social networks more than solidary groups (Putnam, 2000). Household members keep different schedules, no matter if they are dual career, single parent, married couple, or several friends. Although household members usually take each other's agendas into account, they do not move in solidary lock-step. Women – the historic kinkeepers and networkers within and between households (Wellman, 1982; Rosenthal, 1985; Wellman and Wellman, 1992; Logan and Spitze, 1996) – are spending less time at home doing household chores and more time out of the home doing paid work (Robinson and Godbey, 1997). Moreover, in networked households, individual household members are less able to rely on each other to arrange their social life with friends and kin. This is a major change since we last interviewed East Yorkers in 1979, when one man (typically) reported that his wife "can remember everything except where my socks are" (Wellman, 1985).

8.3.1 ICT Use in Households

With the shift of analytic attention from demographics – who uses the Internet – to dynamics – who do they use it with, where, why, and when – comes a need to understand the Internet's role in households. The majority

---

[10] Tracy L.M. Kennedy has major responsibility for this part of the Connected Lives project and drafted much of this section.

[11] Dickson and Ellison (2000), Cumming and Kraut (2001), Bakardjieva and Smith (2001), Wellman and Haythornthwaite (2002), and Fortunati et al. (2003).

(79%) of the survey respondents have at least one computer at home. Almost all (94%) of these computerized households are connected to the Internet. This means that 75% of all the surveyed households are connected to the Internet, a rate similar to national Canadian and American Internet use (Ekos, 2004; Rideout and Reddick, 2005). Respondents report being online a *median* of 10 hours per week, and sending e-mails a median of 21 times per week. This is similar to the July 2005 Canadian *mean* usage of 12.7 hours per week, once outliers are accounted for (Ipsos-Reid, 2005).[12]

Thus, the Internet is not a part of every home – even in Canada – nor does every Canadian feel that it is the Internet that connects them to the wider world. Yet, the Internet and ICTs permeate Canadian society. Even those who do not have computers at home often have access to them at work, school, cafes, libraries, etc. (Boase et al., 2003). While some use the Internet for a wide variety of things – communication, information, recreation, and commerce, are more focused. Some people still feel hesitant to shop online, while others see it as a tool for work rather than for recreation (Katz and Rice, 2002; Kraut et al., 2002; Wellman and Haythornthwaite, 2002).

Several phenomena play a role in Internet use, including higher levels of education and income, and the presence of children in the household (Statistics Canada, 2002; Chen and Wellman, 2005). That the presence of children makes a difference suggests that parents have a particular understanding of what the Internet is good for, perhaps even before they start to use it. Indeed, in some households age dynamics are reversed, with teenage children helping their parents to use the computer (Kiesler et al., 2000; Ribak, 2001).

The complex lives of household members – coupled with their personal mobility and mobile connectivity – means that household members often want to use ICTs to communicate with each other as well as with community members. Although enthusiasts have seen computer and Internet use as an unalloyed good, in practice, use can create stress. A generation ago, some households argued about who would get the family car. Now, some households argue about who gets to use the family computer (Lenhart et al., 2005).

Despite the Internet's potential for creating conflict and stress in households, little stress and conflict reportedly happens. Only a minority of households have such arguments. Sixty percent of respondents in households with more than one resident and at least one computer say they never argue about who gets to use the Internet, while only 5% say they argue half the time or more. Most disagreements only happen "some of the time" (Table 8.1). Moreover most interview participants report little conflict about computer use. When disagreements do occur, they are about who has access to the computer, what people are doing online (e.g. porn or "goofing-off" rather

---

[12] The 12.7 mean hours per week of Internet usage in 2005 is up 46% from 8.7 hours in 2002. It is slightly less than the 14.3 hours per week that Internet-using Canadians spend watching television and the 11.0 hours they spend listening to the radio (Ipsos-Reid, 2005).

173

Table 8.1 Household disagreement about Internet use ($N = 242$)[a]

	Disagreements about someone using the Internet too much (%)	Disagreements about who gets to use the Internet (%)
Never	64	64
Some of the time	28	32
Half of the time	3	2
Most of the time	5	2
All of the time	<1	<1

[a] Percentages calculated from respondents who reported Internet access at home.

than "serious work" or looking for leisure information), and who is online too much. As one participant reported:

> *Interviewer:* So you think it takes away from things that you like, other shared activities, or other things that you might be doing together. What kinds of things would you be doing together if he wasn't online?
>
> *Participant 608:* It could be a combination of anything from entertaining ourselves together, which could be a physical activity as well as discussion on a personal level, which we could be discussing an article that we might have read...I've accepted the situation. I've tried different methods to get what I want like anyone else and when you finally give up, you go onto something else. Therefore, OK, once you go onto something else, that is no longer shared time.

Families with more than one child at home were 2.2 times more likely to argue than families with one or no children. This could be parents arguing with children about computer use and also children arguing with each other (see also Mesch, 2006b).

Having multiple computers at home does not affect the likelihood of disputes about computer use, perhaps because households with a lot of tension have already purchased several computers. For example, one household we studied has three Internet computers in their living room, bought partially to eliminate disputes (Figure 8.3). Thirty-one percent of the survey respondents have more than one computer at home; this is 39% of those households that have at least one home computer (Table 8.2). Not only is there less competition for use with multiple computers, they are often dispersed in different rooms, giving more privacy and minimizing household members' disapproval of each other's computer use. Our findings are consistent with a large US national survey of teens that found that "increasing numbers of teenagers live in a world of nearly ubiquitous computing and communication technologies that they can access at will" (Lenhart, 2005).

Another contention-reducing option is to use the Internet at different times of the day (see also Mesch, 2006a,b). This works best if some people

Copyright © 2005 Wellman Associates.

Figure 8.3 Networked at home: the three-computer living room.

stay home during daytime hours as the survey data show that the Internet is used most frequently at home between 5 and 11 p.m. (Table 8.3), when many people have returned home from paid work or school. Although interview participants report that they are usually on the Internet without the participation of other household members, they do not feel that their Internet use interferes with household life even though they are connected to people online as well as to household members. Their online networking appears to fit into their networked household lives. While Sherry Turkle's studies of cyberaddicts (Turkle, 1984, 1995) led her to argue that people get so immersed online that they develop "second selves", this is not the case among the ordinary people of East York.

Table 8.2 Distribution of the number of computers in the household ($N = 328$)

Number of home computers	N	%
0	69	21
1	164	50
2	64	20
3	22	7
4	7	2
5	1	<1
7	1	<1

Table 8.3 Period of the day when respondents use the Internet $(N = 263)^a$

Period	N	%
5–8 a.m.	47	18
8 a.m.–12 p.m.	89	34
12 p.m.–5 p.m.	89	34
5 p.m.–11 p.m.	217	83
11 p.m.–5 a.m.	78	30

^{a} Percentages calculated from respondents who reported Internet access at home.

8.3.2 The Place of Computers in Households

The location of computer use in a household can affect how it is used as well as relationships among family members. Placing a computer in an isolated den frames it as a "work" machine whereas placing it in a family room frames it as a "home" machine. Nearly half (46%) of the survey respondents who have a home computer have at least one in an office or study (Table 8.4). At the extreme end, one major Canadian telecommunications company insists that its home-based teleworkers work in a separate lockable room (Salaff, 2002; Dimitrova, 2003). Thus, spatial boundaries become social boundaries, especially for young children. If a computer is in a parent's or child's bedroom, it is difficult for other household members to have access to it when occupants are sleeping or otherwise engaged (Haddon and Skinner, 1991; Frohlich and Kraut, 2002; Aro and Peteri, 2003).

Our interviews reveal that it is not always the case that the home office is work space and the family room is recreational space. While organizations may insist that teleworkers have a separate closed-door room, in practice they are unable to enforce this. Home and work boundaries of these spaces are blurring, with people thinking creatively about reorganizing household spaces to accommodate their Internet use. People decorate their offices, or have an "open door/closed door" policy to indicate availability

Table 8.4 Percentage of respondents who have a computer in specific locations

Location	% of all computer owners $(N = 265)$	% of total sample $(N = 328)$
Office/study	46	38
Living room	24	20
Recreational room/family room	23	19
Child(ren)'s bedroom	18	14
Master bedroom	12	10
Other	7	6
Kitchen	2	2

Figure 8.4 A computer integrated into a family room.

for interruption. For example, Figure 8.4 shows a computer that is well integrated into the living room of one of the participants we interviewed. It is accessible to all household members in this open, recreationally oriented area. Another interview participant told us how the location of his computer affected his relations with other household members:

> We have an open area on the second floor that we designed on the second story. So, it could have been a 4-bedroom, but I wanted it to be open, so it's like a big landing where the computer is. So, when I'm working at home, or doing something at home, I'm available to everybody still. I don't want to be off in a room somewhere. (Participant 232)

8.3.3 Gendered Power over Household Computers

The development of ICTs has resonated with the networking of technology. The networking of households has created a wide demand for personalized, often-mobile ICTs far beyond the early dreams of mobile phone developers that they would be rich persons' toys. The development of ICTs has encouraged household members to go their separate ways while remaining connected and coordinated.

In such networked households, we wonder how gendered power dynamics mediate online behavior. We are investigating patterns of household relations, including divisions of labor, gender ideology, and the valuation of unpaid domestic work. We are tracking the performance of gender within households through the negotiation of technologies and online tasks. Have household patterns of gender ideology and interactions between husbands and wives (and parents and children) that affect domestic divisions of labor expanded to include computer use, with some online activities interpreted according to preexisting gender roles? Consider the oft-demonstrated differences between women and men on time spent on household responsibilities (Robinson and Godbey, 1997). Will the Internet be interpreted as a labor saving device to be used by the woman of the house, or as a toy and tool to be preferred by the man (Cowan, 1983; Wajcman, 1991)? Will this vary by content area, with women responsible for socializing online and finding cultural information and men responsible for playing games and dealing with finances?

The Connected Lives survey shows that women continue to spend more time than men in traditionally gendered tasks such as chores and cleaning, childcare, and cooking and baking, while men continue to spend more time on yard work and home maintenance (Figure 8.5). This gendered division of labor also includes men spending 23% more time on the Internet: 11.9 hours as compared to 9.7 hours. Overall Internet use accounts for more hours per week than chores/cleaning, cooking, yard work, and home repair – for women as well as for men.[13]

Often, when women do go online it is for involvement in what has historically been deemed to be women's work. For example, Japanese homemakers search for advice and emotional support for dealing with children and husbands (Miyata, 2002). Among the East Yorkers, male and female survey respondents specialize in different things (Figure 8.6). Thus, the East York women who are responsible for cooking sometimes use the Internet to search for recipes:

> I got round steaks, so I'll look up recipes for round steak in the slow cooker or you know chicken or whatever. I do that almost on a daily basis, you know get ideas about what am I going to make for supper tonight. (Participant 174)

Women continue their offline role as social networkers by "communicating with others online, while men do more searching for general information". The only anomaly is the tendency for men to do more online shopping. We believe that this is linked to the greater involvement of men in searching for information, and that this may be a diminishing difference as women accumulate greater experience online.

Although Internet use may be gendered in part, interview participants report little conflict about it. The gender gap has disappeared, with women online as much as men, and teenage girls as likely as teenage boys to be computer gurus in their families (see also Kiesler et al., 2000).

---

[13] We caution that these are preliminary data and do not take into account variations in such things as family and work situations.

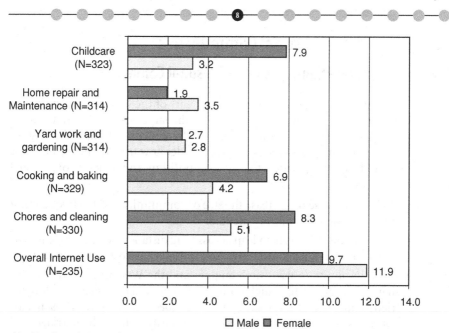

N = Number of respondents

Figure 8.5 Mean number of hours per week spent on household jobs by gender.

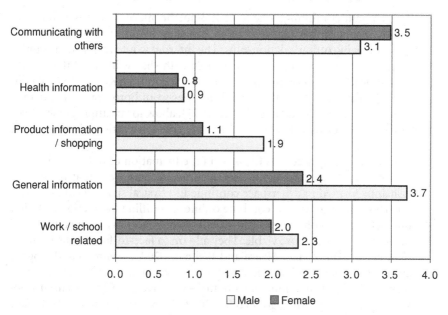

Figure 8.6 Mean number of hours per week spent on Internet activities at home by gender, for those with the Internet ($N = 235$).

8.4 Networking Personal Communities

8.4.1 The Size and Composition of Personal Community[14]

Despite ongoing scholarly and political fears of the loss of community, we now know that community has survived the large-scale social transformations of urbanization, industrialization, bureaucratization, technological change, capitalism, and socialism. We wonder how the Internet – and other forms of ICTs – might affect the size, composition, and structure of personal communities.

- Do ICTs increase or decrease the size of community and the frequency of contact among community members?
- Does the ability of ICTs to leap across long distances with a single mouse click foster far-fling community, and is this at the cost of neighbouring?
- Does the ability to use search engines and the Web find comrades with shared interests foster a high number of "achieved ties" such as voluntary friendships? Is this at the expense of "ascribed ties" with relatives, neighbours, and coworkers that come less voluntarily from birth, marriage, and local juxtaposition?
- Does the person-to-person nature of ICTs lead to less group solidarity and more sparsely knit networks as people maneuver among multiple, often loosely coupled components of their personal community?

Early – and continuing – excitement about the Internet saw it as stimulating positive change in people's lives by creating new forms of online interaction and enhancing offline relationships. The Internet would restore community by providing a meeting space for people with shared interests that would overcome the limitations of space and time (Sproull and Kiesler, 1991; Baym, 1997; Wellman, 2001). Online communities would promote open, democratic discourse (Sproull and Kiesler, 1991), allow for multiple perspectives (Kapor and Berman, 1993), and mobilize collective action (Tarrow, 1999; Kelly, 2005).

Although early accounts focused on the formation of online ("virtual") communities (e.g. Rheingold, 1993), it has become clear that most relationships formed in cyberspace continue in physical space, leading to new forms of community that combine online and offline interactions. Online interactions fill communication gaps between face-to-face meetings and make nonlocal ties more viable. They add on to face-to-face contact, rather than replacing it (Quan-Haase and Wellman, 2002; Wellman and Hogan, 2004). The result probably is that the amount of contact among friends, relatives – and even among neighbours – was greater in 2005 than in 1995 (see Müller, 1999; Rheingold, 2000, 2002, 2005; Hampton and Wellman,

---

[14] Jeffrey Boase and Bernie Hogan have major responsibility for this part of the Connected Lives project and drafted much of this section.

2002). Certainly, more people are writing more, even as keyboards have replaced pen, paper, and postage.

Yet, one continuing fear is that the entrancing possibilities of online communication will pull people away from face-to-face (and even telephone) contact, leading to alienation and depression. This has been a concern not just for community but also for households, where data indicate that Americans eat together at family dinners only 3 days in a week and rarely have family outings (Putnam, 2000). As an irate letter writer asserted in the *New York Times*:

> How about if all those who spend much of their time chattering on their cell-phones stow them somewhere and actually talk to the living, breathing human beings right in front of them. Then maybe they wouldn't have to spend so much time blogging us all senseless. We'd all be truly communicating and we'd have more time to truly accomplish something. (Hunter, 2005, p. A18)

Although early research (Kraut et al., 1998) suggested that Internet use may alienate heavy users from other household members, a follow-up study showed that this is a problem only for newbie computer users that disappears with experience (Kraut et. al, 2002). Even if household psychodynamics are not involved, ICTs may compete with other activities for time in an inelastic 24-hour day (Anderson and Tracy, 2002; Nie et al., 2002; Gershuny, 2003). Moreover, with the shift to networked individualism, people must maintain many ties one by one, as compared to going regularly to kinship gatherings or favorite cafés where the milieu does much of the maintenance work. The work of sustaining individual ties may be easier online where only fingers do the walking on keyboards and multiple friends may be connected at once. This ease, coupled with the narrower interpersonal bandwidth of ICTs (as compared to face-to-face contact), may foster contact with weak ties of acquaintanceship at the expense of socially close ties.

This is not just speculation. Our data show that people believe the Internet generally makes life easier and arguably more social. We asked survey respondents about the perceived ease of doing nine different online activities commonly conducted online (Figure 8.7).[15] Their responses show that virtually all of the Internet-using population report learning new things online and that they are learning more easily. Respondents report that the Internet has made contacting members of their personal community much easier. The highest mean score is for information ("learning new things") – principally the Web – as the easiest activities to do, followed closely by communication ("connecting with friends", "connecting with relatives") – principally via e-mail and IM.

Most respondents have found all the tasks to be easier since they first began using the Internet. There are few negatives. Although this scale includes

---

[15] The perceived ease of doing these online activities was rated on a scale: −2 = "made it much more difficult," −1 = "made it somewhat more difficult," 0 = "has not affected it," 1 = "made it somewhat easier," 2 = "made it much easier." The nine items were summed to obtain an overall score.

181

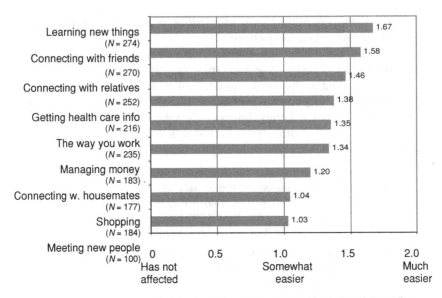

N = Number of people engaged in that activity online

Figure 8.7   Mean level of perceived ease of using the Internet for everyday activities.

"somewhat more difficult" and "much more difficult" response categories, no respondent reports that the Internet makes any task "much more difficult". Only four people say that the Internet makes work somewhat more difficult, and only four say that it makes getting health care information more difficult. As for the rest of the activities, only two or fewer people say that the Internet makes any of these tasks more difficult.

It is possible that these self-reported data are biased with some of those who go online saying that the Internet has made their life easier only out of cognitive dissonance. Hence, we are also investigating behavioral measures such as network size and time spent online.

8.4.2 Measuring Networked Personal Communities[16]

How many people are in a personal community? 5? 15? 150? 1500? We have seen estimates of all these numbers. The size of a personal community network is a difficult question. Academics have evolved some techniques; aficionados of social software (such as *Friendster* and *Orkut*) can count their lists, as can conscientious Rolodexers and databasers. Yet, most people have no idea of the size of their networks. Hence we need some way to ask people about this systematically.

---

[16] Jeffrey Boase had major responsibility for this part of the Connected Lives project and drafted much of this section.

The Connected Lives project uses two methods of ascertaining the size and shape of an individual's personal community: the *summation method* in the survey and the *name generator method* in the interviews (Marsden and Campbell, 1984). We have used the summation method in the survey because it can be self-administered and takes less time. Although it provides less detailed results than the name generator, it still provides reasonable approximations of network size and composition.[17]

The summation method was invented as a way to break up a larger network into separate and more easily estimated pieces (McCarty et al., 2001). Researchers ask respondents to report the number of people they know in a number of roles, such as "relatives outside the home" or "neighbours". In the Connected Lives survey, we further differentiate this measure by having respondents report first on the number of people in roles who are *very close* and then on the number of people in roles who are *somewhat close*. This allows us to test measures that may vary by the strength of ties while simultaneously making the categories more manageable for the respondents.

Very close ties include those with whom people discuss important matters, regularly keep in touch, or provide much help. Very close ties often provide resources that require substantial time, energy, and trust. They are more likely to provide intensive care for those in poor health and they are more likely to provide financial aid (Wellman and Wortley, 1990; Wellman, 1992).

By contrast, somewhat close ties may have some or all of these traits, but to a lesser extent. However, such weaker ties may be more likely than very close ties to provide new ideas and information because they tend to connect to a wider variety of social circles (Granovetter, 1973, 1995).

We asked people to enumerate the number of somewhat and very close ties in the following eight categories:

1. Members of your immediate family living outside of your household (parents, siblings, children)
2. Other relatives
3. Neighbours
4. People you currently work with, or go to school with
5. People you know only online
6. People from organizations (such as church, sports leagues, business associations)
7. Friends not included above
8. Other people not included above.[18]

---

[17] Data from the name generator are still being prepared for analysis.

[18] A similar approach was used in February 2004 for a telephone survey of 2200 American adults in the Social Ties study by the Pew Internet and American Life project (Boase et al., 2006).

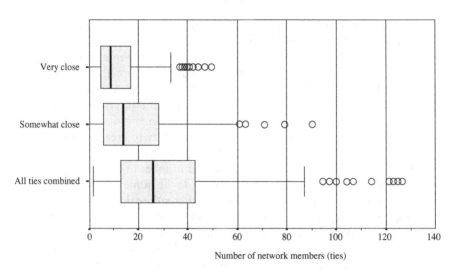

The solid bar inside each box is the median. The boxes themselves show
the interquartile range: the network size of the middle 50% of the sample

Figure 8.8 Variations in network size for very close, somewhat close, and all close ties
($N = 317$).

The personal communities of the respondents in our study vary substantially in size, both for very and somewhat close ties (Figure 8.8). Overall, respondents report a median of 23 ties: 9 that are very close and 14 that are somewhat close.[19] These are roughly comparable to those found in other studies of personal communities (Fischer, 1982; Wellman et al., 1988). We caution that close ties are only the heart of a personal community network. Estimates of the overall size (including weak acquaintances) of such networks range between 200 and 1500 ties (Boissevain, 1974; Pool and Kochen, 1978; Bernard et al., 1990; Kadushin et al., 2005).

These personal communities also vary in composition. Some people's networks contain many kin while others contain many friends. On average, 25% of both very close and somewhat close ties are purely friends (Table 8.5). In addition, many other nonkin are known through work, school, the neighbourhood, and voluntary organizations. Immediate kin (parents, siblings, adult children, in-laws) comprise a much higher percentage of very close ties (38%) than they do of somewhat close ties (10%). By contrast, extended kin (aunts, uncles, cousins, grandparents), workmates/schoolmates, neighbours, and people known only online or through voluntary organizations comprise higher percentages of somewhat close ties.

---

[19] Medians are reported because a small number of respondents report huge networks (the mean +2 standard deviations [equal to 41 very close ties, 65 somewhat close ties, 106 overall ties]) that positively skew the mean.

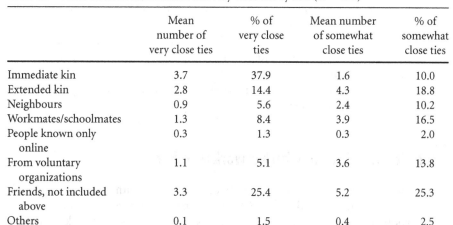

Table 8.5 Mean number of somewhat and very close ties by role ($N = 297$)

	Mean number of very close ties	% of very close ties	Mean number of somewhat close ties	% of somewhat close ties
Immediate kin	3.7	37.9	1.6	10.0
Extended kin	2.8	14.4	4.3	18.8
Neighbours	0.9	5.6	2.4	10.2
Workmates/schoolmates	1.3	8.4	3.9	16.5
People known only online	0.3	1.3	0.3	2.0
From voluntary organizations	1.1	5.1	3.6	13.8
Friends, not included above	3.3	25.4	5.2	25.3
Others	0.1	1.5	0.4	2.5

Plots of the number of ties for each role show a wide level of diversity within the networks (Figure 8.9). First, different roles make up a large share of the network for different people. Twenty percent are close to 8 or more people from voluntary organizations, 20% are close to 9 or more from work, and 20% are close to 11 or more extended kin. Moreover, correlation analysis reveals that different people usually have different mixes of kin, friends, neighbours, and workmates.

Second, the steep decline of the lines in Figure 8.9 suggests that there is a limit to the number of persons in a given role. For example, 96% have at

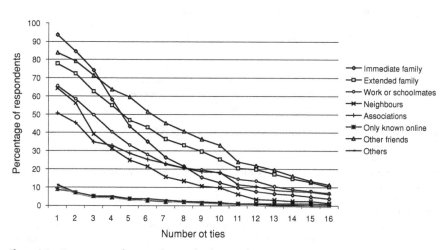

Figure 8.9 Percentage of respondents who have at least one or more network member in each role.

least one immediate kin in their network, but only 40% have five or more. Similarly, 70% of people are close to at least one neighbour, but only 25% are close to five or more neighbours.

Third, people who are only contacted online rarely are socially close. Only 10% of the respondents report being close to people they only know online, and only 2% (four respondents) report being close to 8 or more ties exclusively online. Thus, the Internet is not a separate social system but is embedded in everyday life.[20]

8.4.3 Communicating with Network Members[21]

How do people connect with network members, with and without ICTs? After asking about the raw numbers of people in the network, we asked about the number of very close and somewhat close network members whom respondents contact: (1) at least weekly and (2) between weekly and monthly. Figure 8.10 shows that more ties are interacted with in person than by ICTs. However, the telephone and the Internet are each widely used.

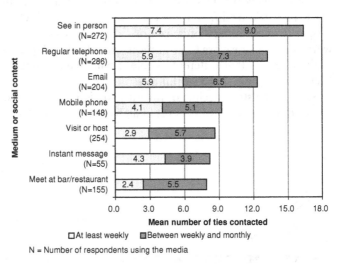

Figure 8.10 Mean frequency of contact with network members by media.[22]

---

[20] See also Chen et al. (2002), Quan-Haase and Wellman (2002), Hampton and Wellman (2003), and Wellman and Hogan (2004).

[21] Jeffrey Boase and Bernie Hogan have major responsibility for this part of the Connected Lives project and drafted much of this section.

[22] Figure 8.10 shows only the means for people who use any of the aforementioned media. For example, the number of people using instant messaging is much lower than the number of people making in-person contacts, because instant messaging is not used by a majority of the sample.

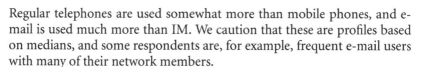

Regular telephones are used somewhat more than mobile phones, and e-mail is used much more than IM. We caution that these are profiles based on medians, and some respondents are, for example, frequent e-mail users with many of their network members.

Some forms of interaction are more suited to a weekly rhythm while others are more suited to a less frequent rhythm. Respondents phone or e-mail about half of their network members at least weekly and the other half less frequently. However, those people who visit or host network members are most apt to do so between once per week and once per month, as do those who meet network members at a bar or restaurant (Figure 8.10).

These findings reinforce our claims about the Internet and sociability. E-mail is a tool for frequent communication, and it is a means to communicate frequently with more people than might be seen at social events (see also Copher et al., 2002). The extent to which this assertion is really the case will be examined in future analyses of the interviews that have gathered more detailed information about communication and socializing among specific network members. For example, the interviews should show us if it is the same network members who are e-mailed at least weekly but only socialized with less often, or if e-mailing and socializing take place with different network members.

Keeping in contact with people requires time. Some media allow people to save time by maintaining contact with small gestures: forwarding jokes and pictures via e-mail as simple bonding gestures. Moreover, e-mail messages can be sent to a large number of people as quickly as they can be sent to a single person. By contrast, some other ICTs, such as mobile phones, require users to make contact one person at a time.

Our concern is not just the amount of time used. E-mail, text messaging, and some IMs are asynchronous, meaning that there can be a time lag between the time that a message is sent and the time that it is received. Therefore, people can communicate around their schedule rather than letting the media dictate their schedule. This suggests that people with more ties than time may benefit from technologies that allow them to contact their community ties efficiently and conveniently.

The social affordances of e-mail – such as asynchronicity, multiple message recipients, and store-and-forward – can be especially useful as network size increases (Bradner et al., 1998; Wellman and Hogan, 2004; Boase and Wellman, 2005). Our survey data highlight these affordances by showing how e-mail scales up much more effectively for large networks than mobile phones, instant messaging, or regular telephones (Figure 8.11). In-person contact also scales up for people with large networks. People with large networks are more likely to drop-in on others, have many neighbours, and participate in voluntary organizations – all social contexts that require little planning ahead of time and are efficient means of getting in contact with other people.

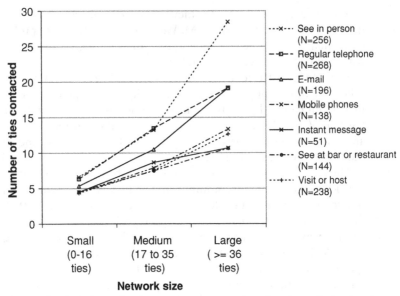

N = Number of pepole who use that context or medium to contact network members in a given months

Figure 8.11 Number of ties contacted monthly by selected media by network size.

8.4.4 Managing Networks[23]

Networked individualism means that people must actively manage their networks. Rather than sitting back and letting densely knit groups provide sociability, support, and control, people must contact their ties and shop for support at relational boutiques rather than at general stores. Ling (2004) has argued that there are two principal forms of coordination: (1) making and revising schedules and arrangements; (2) managing social networks. The two forms intertwine. For example, earlier East York research has shown that people with large networks get more emotional and material aid – not only overall but from *each* network member (Wellman and Gulia, 1999a; Wellman and Frank, 2001). Such people may know how to manage their networks.

Network absorption is one way of managing: the capacity of an individual to bring a new tie into a network or to store information about that individual which will lead to more networking. This can involve participation

---

[23] Bernie Hogan has major responsibility for this part of the Connected Lives project and drafted much of this section.

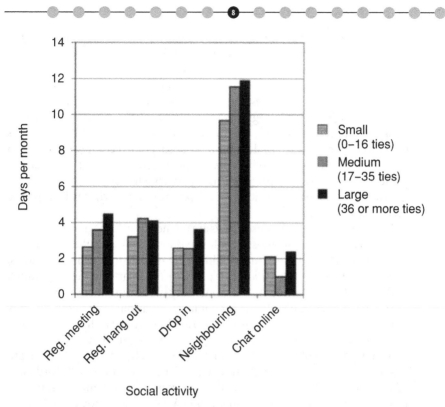

Figure 8.12   Mean days per month of social activity by overall network size ($N = 308$).

in social contexts amenable to adding new ties, using media that lower the cost of interacting with more people, or using tools to remember who is in the network and how to get access to them. The Connected Lives project is measuring involvement in social milieus by asking about people's participation in activities such as neighbouring, visiting/hosting friends, and involvement in voluntary organizations or online chat groups.

People with large networks are on average more active than people with medium and small networks in virtually all the spheres discussed above (Figure 8.12). The only exception is that people with medium-sized networks congregate at regular hangouts slightly more frequently.

We also surveyed the use of personal information management tools such as address books, calendars, and personal digital assistants (PDAs). We hypothesize that one can more easily absorb a new tie if one can remember how to get in contact with that person. People use an average of three tools for personal information management in addition to their own memory: three for recalling telephone numbers and three for remembering occasions.

For *remembering telephone numbers*, the most commonly used tools are the phone book, one's memory, and a written address book of frequently called numbers. For *remembering occasions*, one's memory is the most popular, followed by a wall calendar in the home, and reminders from others. For *remembering e-mail addresses*, the most popular means are features that are embedded in the e-mail program: an existing message, a computer address book, and the auto-complete feature common on many e-mail clients. This bodes well for e-mail as a tool for network absorption. People do not need to recall e-mail addresses, as computers will help them remember. It is easier to get access to a network member's e-mail address while at an e-mail program than to get access to a network member's phone number when using a phone (Table 8.6).

Many of the most widely used tools are not necessarily the most frequently used tools. For example, most people use phone books but they do not use them often. The opposite can be said of PDAs that store addresses, notes, calendars, etc. (e.g. *Palm, BlackBerry, Pocket PC* smart phones). Although only 14% of the respondents use PDAs, they usually use them heavily.

Whatever devices people use for contacts, occasions, and planning, they put that information to use through *network engagement*. Interaction with network members can take place in person or via many media. When in person it can take place in many social milieus, such as casually dropping in on friends or attending a regularly scheduled meeting. The social affordances of communication media can facilitate or constrain how interactions take place (Bradner et al., 1998; Wellman, 2000; Wellman and Hogan, 2004). For example, e-mail allows one person to broadcast messages to many others, whereas telephone calls typically take place between two persons. Answering machines and e-mail allow people to communicate asynchronously, and some technologies are more mobile than others.

The more social milieus that people participate in, the larger their social networks. Nearly two-thirds (63%) of respondents report dropping in on their friends unannounced or only calling just ahead of time. Those respondents with drop-in privileges report having networks that are on average one-third larger: fully 12 more ties. Similarly, both the frequency of neighbouring and of attending regularly scheduled meetings are directly related to increases in network size.

It is probable that various aspects of network management can be related to the ideal types of community discussed earlier. Densely knit, village-like groups would have significantly more neighbours and kin. Hence, people in such solidary milieus would experience more dropping in, less planning, and perhaps less mediated contact. By contrast, in a network individualistic situation, activity and passivity are important for how people engage with their networks: do they invite or get invited; do they call or get called? People in such situations most likely rely more on ICTs because of their far-flung relationships, asynchronous schedules, and greater need to maintain their relationships one-by-one.

Table 8.6 Methods/tools to remember personal information: days per month

Phone numbers	Days	Occasions	Days	E-mail addresses	Days
Phone book ($N = 303$)	5.4	Memory ($N = 310$)	24.4	Existing message ($N = 262$)	18.5
Memory ($N = 299$)	22.3	Wall calendar ($N = 290$)	16.2	Computer program ($N = 211$)	17.3
Rolodex ($N = 253$)	12.9	Reminders ($N = 219$)	8.2	Auto-complete ($N = 206$)	21.1
Ask someone ($N = 246$)	3.9	Post-its ($N = 190$)	11.9	Memory ($N = 201$)	15.6
Stored on phone ($N = 219$)	20.7	Agenda ($N = 175$)	18.7	Ask someone ($N = 183$)	4.2
Post-its ($N = 202$)	7.4	Computer program ($N = 113$)	15.1	The Internet ($N = 161$)	5.5
The Internet ($N = 194$)	6.4	Pocket calendar ($N = 106$)	11.9	Post-its ($N = 141$)	4.4
Computer program ($N = 151$)	10.2	PDA ($N = 49$)	20.0	Rolodex ($N = 127$)	8.5
PDA ($N = 49$)	18.4	Assistant ($N = 27$)	15.4	PDA ($N = 45$)	15.3

N = Number of respondents using the method or tool.

8.5 Networks, ICTs, and Travel[24]

8.5.1 Linking Travel to Social Networks

Patterns of activity and travel are getting more complex throughout the world. Resonating with the shift toward networked individualism, there is a tendency toward increasing suburbanization of homes, shopping, and employment; increasing car ownership; and an increasing number of trips with only one person in the car (Miller and Shalaby, 2003). As in many cities, personal travel in Toronto is becoming more mobile and car oriented, and mass transit systems are moving proportionately less of the population.

Much travel is for socializing with network members. This is especially true for long-distance trips. In addition to the temporal and spatial constraints normally considered relevant for understanding social travel, analysts must take into account the nature of personal communities and the impact of ICTs. Individual trips are becoming even more prevalent as household members live on separate schedules and as communities become spatially dispersed networks of individuals rather than local groups.

Yet, analysts have not studied the relationship of ICT use to traveling in order to engage in social activities (Mokhtarian et al., 2003). Earlier studies have focused on *substitution* effects: trade-offs between ICT and transportation in areas such as telework and shopping. This assumes that higher ICT use will mean less travel (Niles, 1994; Johnson, 1999). However, there has been scant empirical evidence for the presence of these substitution effects (Salomon, 1998). Analysts are coming to recognize that ICTs can have a potentially broader effect and play multiple roles in social travel (Senbil and Kitamura, 2003):

- *Complementary*: ICT use increases the number of trips to social activities through communication and coordination.

- *Modification*: ICT use leads to changes in the characteristics of travel or social activities, such as the duration of trips and activities, the location of activities, and the planning horizon.

- *Neutral*: ICT has no effect on either travel or social activities.

The Connected Lives project has pursued such concerns by linking for the first time the study of social networks, ICTs, and activity and travel behavior (see also Carrasco et al., 2005). We ask if the quantity and type of ICT use – and the characteristics of personal community networks – enhance or diminish the frequency of social activities and the nature of travel to such activities. As people increasingly rely on ICTs for entertainment and communication, do they travel less? Or, does their virtual connectivity actually

---

[24] Juan-Antonio Carrasco has major responsibility for this part of the Connected Lives project and drafted much of this section.

create more need for travel, as people arrange trips online or acquire information about new cultural or entertainment venues (Wellman and Gulia, 1999b; Wellman and Haythornthwaite, 2002)? What are the trade-offs between travel and communication in an increasingly ICT-pervaded world? What are the implications of changing travel behavior for households, networks, and the societies in which they are embedded?

Analysts currently try to explain the generation and spatial distribution of the social activities of individuals and households by

- time and space opportunities and constraints;
- individual and household characteristics, including stage in the lifecycle, ethnicity, psychological characteristics, and socioeconomic status;
- the intrinsic attributes of the activity.

We believe that the generation and spatial distribution of social activities can be better understood by also knowing aspects of social networks, such as

- who are the members of a personal community?
- where are the members of this personal community physically located?
- what is the level of ICT use?
- What is the association between ICT use and ties with community members?

Although there has been some recent discussion of relationships between social networks and travel, such discussion is still hypothetical (Axhausen, 2005). There is practically no evidence about the interplay of social networks, ICT use, and travel. Hence, one thrust of our analysis is to see if ICT use is associated with the physical distance of social activities. At the dawn of the Internet age, hopes flourished that Internet communication would foster a global village with far-flung friends and neighbours, and fears arose that these dispersed communications would diminish local community activities (Wellman and Gulia, 1999b; Kayahara, 2006). It now appears that while there has been an increase in spatially dispersed communication, ICT use has not substituted for neighbouring and may facilitate it.[25] It is time to develop more nuanced analyses: Does the nature of a person's social network intersect with ICT use to influence the physical distance and spatial distribution of social activities?

Has ICT use traded off with travel? We are finding that ICT use increases the number of social network ties, the amount of contact with those ties, and the spatial dispersion of such ties. But, how does it affect travel for social activities? One possibility is substitution: ICT use and travel could be fungible,

---

[25] See Chen et al. (2002), Quan-Haase and Wellman (2002), Hampton and Wellman (2003), and Boase and Wellman (2005).

so that as ICT use increases, travel decreases. We suspect the opposite possibility is more common: increased Internet communication synergistically leads to increased travel, as the exchange of information on the Internet provides more reasons for physical encounters: socializing, emotional support, and the exchange of goods and services. Online contact is best when intermittently reinforced and enhanced by physical contact (Wellman and Haythornthwaite, 2002).

8.5.2 E-mail, Spatial Location, and Social Activities

The relationship between three phenomena is an example of how ICTs, social networks, and travel interact. We have focused on three variables:

- *E-mail use* (number of people with whom the individual interacts by e-mail within a month): none, light, and heavy e-mail use.
- *Spatial location of network members* (number of neighbours and number of people living at more than 1 hour of distance): none, low, and high number.
- *Social activities* (number of people with whom the individual performs social activities and travel, hosting, and visiting): none, low, and high number.[26]

E-mail proves to be complementary to social activities (Table 8.7). Heavy e-mail use is related to a high level of social activity, and low e-mail contact is related to little or no social activity. Thus, e-mail acts as a facilitator to perform social activities, and not as a substitute, suggesting that there is no trade-off between communicating by e-mail and face-to-face social activities, but that there is more of a complementary relationship (see also Copher et al., 2002; Quan-Haase and Wellman, 2002).

Having more network members living relatively far away is also positively associated with being involved in more social activities (Table 8.8). Most of the people with a high number of network members living far away (i.e., more than 1 hour's travel) also have a high number of social activities. The opposite happens for those with a few or no network members living far away.

A similar phenomenon happens locally, although less strongly. The more neighbours in a network, the more social activities in which respondents engage. These similarities in two geographical scales – neighbour (local) and people at more than 1-hour travel (global) – resonate with the glocalization concept.

---

[26] "Light and heavy" as well as "low and high" levels are defined by dividing the sample in two approximately equal groups from those individuals who have some e-mail use and those who have any network member in each category, respectively.

Table 8.7 E-mail use and social activity

# of ties sent or received e-mail	Social activities (# of ties hosted/visited within a month)			Row percentage
	None (A)	Low: 1–6 (B)	High: 7+ (C)	
None	33.3%	32.9%	28.9%	31.4%
Low: 1–7	51.1% (C*)	42.9% (C*)	14.9%	33.5%
High: 8+	15.6%	24.2%	56.2% (A*B*)	35.1%
Total	100.0%	100.0%	100.0%	100.0%

$\chi^2 = 43.9^*$, Tau b $= 0.21$, $^* = p < 0.05$
Note: Labels (A), (B), (C), and their combinations in a given column, indicate a proportion that is statistically significantly higher in that column with respect to the corresponding labeled column(s).

Table 8.8 Network spatial location and social activities by residential distance

Model 1: Far away [a]

# of ties living >1 hour's travel	Social activities (# of ties hosted/visited within a month)			Row percentage
	None (A)	Low: 1–6 (B)	High: 7+ (C)	
None	6.8% (C*)	7.0% (C*)	0.9%	5.1%
Low: 1–7	63.7% (C*)	53.0% (C*)	28.1%	44.6%
High: 8+	29.5%	40.0%	71.1% (A*B*)	50.3%
Total	100.0%	100.0%	100.0%	100.0%

Model 2: Neighbours [b]

# of ties who are neighbours	Social activities (# of ties hosted/visited within a month)			Row percentage
	None (A)	Low: 1–6 (B)	High: 7+ (C)	
None	62.8% (B*C*)	38.6% (C*)	21.5%	35.8%
Low: 1–4	27.9%	34.1%	31.8%	32.3%
High: 4+	9.3%	27.3% (A*)	46.7% (A*B*)	31.9%
Total	100.0%	100.0%	100.0%	100.0%

[a]$\chi^2 = 35.2$, Tau b $= 0.31$, $^* = p < 0.05$
[b]$\chi^2 = 30.7$, Tau b $= 0.29$, $^* = p < 0.05$
Note: Labels (A), (B), (C), and their combinations in a given column, indicate a proportion that is statistically significantly higher in that column with respect to the corresponding labeled column(s).

The combined effect of e-mail use, spatial dispersion, and social activities indicates that the complementary relationship between e-mail and social activities is not mediated by the spatial dispersion of the social network. People with low e-mail use tend to have little or no involvement in social activities. This is true both for networks that are mainly local or have much

Table 8.9 Combined effects of e-mail use, network spatial location, and social activities

# ties living >1 hour's travel	# ties sent or received e-mail	Social activities (# of ties hosted/visited within a month)			Row percentage
		None (A)	Low: 1–6 (B)	High: 7+ (C)	
Low[a]: 1–7	None	32.1%	38.4%	40.6%	37.6%
	Low: 1–7	53.6% (C*)	46.5% (C~)	21.9%	42.1%
	High: 8+	14.3%	15.1%	37.5%	20.3%
	Total	100%	100%	100%	100%
High[b]: 8+	None	38.5%	17.9%	23.5%	22.7%
	Low: 1–7	38.5% (C~)	41.1% (C*)	12.3%	25.3%
	High: 8+	23.0%	41.1%	64.2% (A*B*)	52.0%
	Total	100%	100%	100%	100%

[a] $\chi^2 = 10.8^*$, Tau $b = 0.05$, $^* = p < 0.05$, $^\sim = p < 0.10$.
[b] $\chi^2 = 19.8^*$, Tau $b = 0.20^*$, $^* = p < 0.05$, $\sim = p < 0.10$.
Note: Labels (A), (B), (C), and their combinations in a given column, indicate a proportion that is statistically significantly higher in that column with respect to the corresponding labeled column(s).

spatial dispersion. At the same time, people with high e-mail use tend to be involved in much social activity, regardless of the spatial dispersion of their networks (Table 8.9). These findings suggest that ICTs are catalysts for social activities regardless of the spatial dispersion of social networks.

8.6 Finding Support and Information in Networks

8.6.1 Finding Social Support[27]

With the move from groups to networks, social support – emotional and material aid from others – has become more contingent on the nature of separate relationships. Where the village/neighbourhood once controlled and provided social support – as Hillary Clinton (1996) says, "It takes a village to raise a child" – such support is now provided by spatially and socially dispersed network members. Previous studies of East York have looked at the types of social support exchanged, identifying what types of people are more likely to give and get support. These studies have shown that support is largely supplied in discrete relationships rather than in groups, with different relationships specializing in the kind of support that they

---

[27] Rochelle Côté made a major contribution to this part of the Connected Lives project and drafted much of this section.

provide. For example, parents provide financial support, sisters provide emotional support, while spouses provide a wide range of support.[28]

With the proliferation of ICTs, timely questions include determining the effect of emerging technologies such as the Internet and cell phones. Of equal importance are the evolving relationships between men and women, the potential impact of an aging baby-boomer generation and their relationship to other cohorts, and the effects of the high level of cultural diversity of East York, Toronto, and indeed, cities across Canada and abroad.

The social support questions in our survey ask respondents to identify the support they gave and received from a list of seven types. These combine into three overarching categories of support – emotional aid, minor services, and major services, a result similar to the second East York study (Wellman and Wortley, 1990). Respondents also reported which of the nine groups of people they gave and received support from: household members, immediate kin, extended kin, neighbours, workmates and schoolmates, people known only online, from voluntary organizations, other friends, and "others". The measures of social support used here are the number of groups (household members, etc.) that respondents gave or received for emotional support, minor services, and major services.[29]

The data show that East Yorkers continue being supportive and supported people, as they have been for at least the 36 years since our first study (Wellman, 1979; Wellman and Wortley, 1990). They exchange different types of support with about one-third of available network resources. The relationship between giving and receiving is quite equal: People give to about as many sources in the network as they receive from (Figure 8.13). For example, respondents give and receive social support from about three different relationships in their network or 33%. For the exchange of major and minor services, people give and receive support from roughly two sources in their network or 22%. The low score of two out of a possible nine groups giving support shows role specialization in the provision of support that is similar to what NetLab found in its 1979 study (Wellman and Wortley, 1990). We caution that this is the number of roles providing support, and not the number of persons.

---

[28] See Wellman (1979), Wellman (1985), Wellman and Wortley (1990), Wellman and Wellman (1992), Wellman and Frank (2001), and Plickert et al. (2007).

[29] The seven types of support used were advice on important matters, job advice, care for a serious health condition, help with home renovations, help with looking for health information, help with computer, and talk/listen about the day with someone else. These forms of support were worded on the questionnaire to determine support *given* and support *received*. The summed total of support is a reflection of support given within the network. To illustrate the construction of the summed variables, "emotional aid given" used two variables. Therefore, the overall emotional aid given variable was created by summing the nine role types that could have given each of the two types of emotional support. It was then divided by two to get a total out of nine possible sources of support. This ensures comparability between the three categories of support.

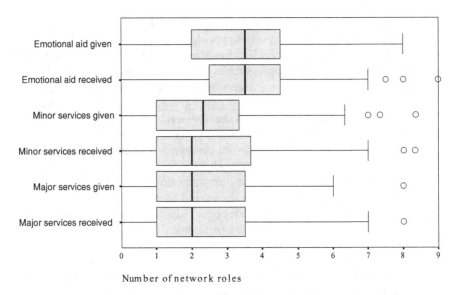

The solid bar inside each box is the median. The boxes themselves show
the interquartile range: the number of ties contact monthly by the middle 50% of the sample

Figure 8.13   Type of social support by number of roles.

8.6.1.1 Emotional Aid Given

Multiple regressions identified several variables contributing to giving and
receiving each type of support: gender, age, education, marital status, level
of income, e-mail frequency, and network size (Table 8.10).

The data reveal several discrepancies from previous research that showed
women receiving more emotional support than men, and men more likely
to exchange services than emotional support (Wellman, 1985; Perlman and
Fehr, 1987; Wellman and Wortley, 1990; Wellman and Frank, 2001; Liebler
and Sandefur, 2002). Our study does not find this gender gap. Men and
women in East York do not differ significantly in the mean amount of
emotional or material support they give or receive.

Consistent with the theory of networked individualism, it is the social
network characteristic of the number of network members – and not the
personal characteristics of people – that contributes to more support – and
more diverse support. Larger numbers of very close and somewhat close
ties increase support by increasing awareness and communication about
needs, and coordination and social control to foster the delivery of aid.[30]
Respondents with larger overall networks (very close + somewhat close)

---

[30] House et al. (1988), Wellman and Wortley (1990), Erickson (1996), Wellman and Gulia (1999a), Hurlbert
et al. (2000), Molm et al. (2000), and Kadushin (2002).

Table 8.10 Regressions of social support by demographic and network characteristics

	Emotional support received		Emotional support given		Minor services received		Minor services given		Major services received		Major services given	
	B^+	β	B	β	B	β	B	β	B	β	B	β
Female	0.15	0.05	0.01	0.00	−0.07	−0.02	−0.34	−0.11	0.04	0.01	−0.17	−0.06
Age	−0.01^a	−0.12	−0.01	−0.07	−0.03**	−0.21	−0.03***	−0.24	−0.01	−0.11	−0.02*	−0.16
Education	0.01	0.01	−0.02	−0.02	−0.04	−0.04	−0.09	−0.08	0.04	0.04	−0.11	−0.10
Married	0.33	0.10	0.06	0.02	−0.05	−0.02	0.17	0.05	−0.19	−0.06	0.15	0.04
Personal income	0.01*	0.14	0.01*	0.16	0.00	0.05	0.00	0.07	0.00	0.05	0.01	0.11
Frequency of e-mail use	−0.05	−0.02	−0.01	−0.02	−0.02	−0.01	0.28	0.09	−0.19	−0.02	0.03	0.01
# of very close ties	0.35	0.12	0.04***	0.26	0.04*	0.20	0.04**	0.22	0.06***	0.32	0.06***	0.34
# of somewhat close ties	0.81***	0.28	0.02*	0.17	0.01	0.11	0.01	0.10	0.01	0.07	0.00	0.05
Constant	1.84***		3.11***		3.24***		3.09***		2.23***		2.69***	
Adjusted R^2	0.154		0.137		0.085		0.137		0.118		0.143	

$N = 203$; ^{a} $p < 0.10$; * $p < 0.05$; ** $p < 0.01$; *** $p < 0.001$. B^+ B = unstandardized; β = standardized

tend to provide and receive support from a wider range of "role types" (immediate kin, etc.) than those with smaller networks. When giving or receiving emotional support, respondents with large networks give to 14% more role types and receive from 13% more role types. Minor services are much the same: respondents with large networks give support to 9% more role types and receive support from 7% more role types. They also provide major services to 10% more role types and receive major services from 9% more role types. Higher income levels correlate with network diversity (Erickson, 2003) and may explain why people with large networks exchange more support.

Large *very close* networks give and receive the most major and minor services, followed by large *somewhat close* networks. By contrast, it is large *somewhat close* networks that give and receive the most emotional support, followed by large *very close* networks. Thus, people are more apt to exchange emotional aid than services with network members who are only somewhat close to them.

In addition to the effects of social networks on support, some personal characteristics also affect the exchange of support. For example, age is especially associated with the exchange of services. Younger and early middle aged respondents (18–40) are more likely to give and receive major and minor services from many more individuals in their network: on average, from 30% to 50% of nine possible role types as compared to only 10% to 29% for those aged 60+ (see also Campbell and Lee, 1992; Haines et al., 1996). Younger adults also obtain support from more diverse role types in their network, whereas older adults rely more heavily on a smaller number of network members.

In the 1990s, theorists speculated that the limited "social presence" of text-based e-mail would limit its use for emotional support (see Rice's review, 1993). Then it became obvious through experience and research that ICTs are frequently used to provide emotional support, both interpersonally (e-mail, IM, mobile phone) and via online support groups (Barrera et al., 2002; Fogel et al., 2002).

To investigate this, we use e-mail frequency as an indicator of Internet use and compare the supportiveness of nonusers, low and high frequency users. To our surprise, e-mail use is not significantly associated with the provision of social support, when compared to other factors, except for a marginal association of ICT use with the provision of minor services ($p < 0.10$). While nonusers access a mean of only 20% of possible role types (1.8 role types actually accessed out of a possible 9), light users access 26% (2.3), and heavy users access 31% (2.8).[31] On the other hand, contrary to those who feared that time online would suck life out of relationships, e-mail use does not diminish supportiveness.

---

[31] We used the median frequency of e-mailing, 20 e-mails per week to differentiate "light" and "heavy" users. Nonusers are those who do not e-mail at all.

Thus, e-mail use diversifies access to different sources of support within social networks. It is an easy way of communicating bits of information and making arrangements such as job information or computer help. The more people are online, the more arranging and information providing they do. The argument that e-mail provides less social presence than in-person contact is borne out to some extent by the comparatively greater association of ICTs with services rather than emotional support. It is not that ICTs provide less emotional support; it is that they facilitate more services rather than more emotional support.

8.6.2 Finding Health Information[32]

ICTs convey information as well as communication. Communicating, like finding information, is affected by social phenomena. People differ markedly in their ability to find information online. Unlike in the early years of the Internet, women row search as often as men. Older people have less skill in doing Web searches as do people with little technological experience (Hargittai, 2002a, b). Such differences among individuals are more important in networked individualistic situations where people may be less apt to have someone physically present to help them with their searches than they would in active neighbourhood groups.

How do people search for information online and offline? Is the turn away from groups and institutions to social networks associated with much reliance on interpersonal ties? Is the combination of widespread Internet use and an abundance of information on the Web leading to a reliance on online sources of information? Our concern is with how people's social networks intersect with their ICT networks to provide them with information. We focus on two areas: health and culture. This section describes our health focus.

When dealing with a health issue for oneself or others, people look for relevant information and support from their families and social networks; health care providers (doctors, homeopaths); specialized government, pharmaceutical, and nongovernmental organizations (such as Cancer Care Canada); and published sources (such as *Prostate Cancer for Dummies*, Lange and Adamec, 2003).[33] ICTs can amplify many of these information sources, by providing e-mail information from friends and relatives and Web information from organizations. It is also easy to find new sources of health information and social support from strangers via chat rooms and listservs (Fox, 2003; Gold, 2003).

The rise of ICTs makes it tempting to take for granted that people will use such tools extensively to inform themselves about health and discuss their

---

[32] Kristen Berg has major responsibility for this part of the Connected Lives project and drafted much of this section.

[33] See also Cohen and Syme (1985), House (1985), Pescosolido (1992), Pearlin et al. (1995), Thoits (1995), Wellman et al. (1997a), and Statistics Canada (2003).

concerns with interested others. With ICTs probably expanding the number and availability of social network members, ICTs may play key roles in providing information or support about health concerns, especially for those who are technologically comfortable (Miyata, 2002; Legris et al., 2003). Two large national US surveys have found that health information is one of the most frequently searched areas online (Fox 2003, 2005). Women are especially apt to go online to look for health information and to participate in discussion groups concerned with health (Pandey et al., 2003; Fox, 2005). The most recent US survey found that as more people gain prolonged Internet experience and use broadband connections, the amount and diversity of their health searching increases. In addition to searching for information about specific diseases, they are doing more expansive searches using the Internet to find out about well-being, nutrition, and alternative forms of medical care (Fox, 2005).

With going online for health information a popular activity, are there patterns of ICT use for health concerns? We are investigating the importance of social network composition and structure, personal characteristics, and perceived ease of use (see Figure 8.7). To learn how respondents communicate about health, the Connected Lives survey asked:

Do you communicate about health concerns with either: a) a doctor or other health care professionals? b) friends or family members, or c) with individuals who share a similar health concern?

Nearly one-third (31%) of the respondents communicate about health issues with friends or family members.[34] By contrast, only 9% communicate with health care professionals and 11% communicate with people who have similar health concerns. As Figure 8.7 showed (page 178), people consider it easy to find some sort of health information on the Internet.

Large networks, high levels of e-mailing, and high interest in health issues are all associated with high levels of online communication about health. More specifically, the more very close and somewhat close friends that people have, the more they communicate with doctors and health care professionals. Correlation analysis suggests that the more e-mail people send per week from their home and work, the more they communicate online with friends or family members about health issues ($r = 0.243$, $p < 0.001$). Behavior and attitudes are similar: a positive attitude toward the Internet is related to more online communication with family and friends about health ($r = 0.320$, $p < 0.001$). Furthermore, the more respondents find that the Internet makes tasks easier, the more likely they are to communicate about health issues with health care professionals.

In sum, there are general associations between having larger networks, communicating online with friends and family, and communicating online

---

[34] See Wellman (1979), Wellman et al. (1988), and Wellman and Wortley (1990) for results from previous surveys in East York.

about health. Our findings are consistent with what social scientists have called "the buffering model" (House, 1985): There is a relationship between the number of network ties, the total number of e-mails, and communicating health concerns online as well as offline. Moreover, the positive attitudes toward the Internet operate in conjunction with network size and Internet use to foster high levels of seeking information online and discussing health. That our data do not show male–female differences in communicating and seeking information about health online suggests that ICTs may be lessening the longstanding specialization of women in this area.

8.6.3 Finding Culture[35]

Cultural knowledge and activities are strongly related to success in both school and jobs (e.g. DiMaggio, 1982, 1997; Bourdieu, 1984). The rise of ICTs, particularly the Internet, has been accompanied by a massive increase in potential access to cultural information. Yet, such access is only meaningful if people actually use the Internet for such purposes. Culture is a broad term that can encompass a vast array of concepts. For the purposes of our analysis, culture will be limited to leisure-type activities from both "high culture" and "popular culture" categories, including reading and writing, television and film, music, fine art, performing arts, and games and sports (Gans, 1974).

Studies of the relationship between culture and life outcomes suggest that the types of cultural knowledge people possess are also important in determining outcomes, although the relationship is more contingent and less straightforward than was once believed (Erickson, 1996). Given the importance of cultural knowledge and the ability of ICTs to expand access to information of all types, it is important to investigate how people are taking advantage of this new access for cultural purposes.

The popularity of the *Internet Movie Database, iTunes,* and the *ESPN* sports Web site; online book vendors such as *Amazon* and *Chapters/Indigo*; and the Web sites of public libraries indicate that people are going online to engage with culture (defined broadly). However, relatively little research has been done on where they are going and what they are looking for. Much research on the connection between leisure, culture, and the Internet has tended to focus on exclusively online activities such as multiplayer games, virtual communities, and online gambling (Reid, 1999; Rheingold, 2000; Bryce, 2001; Griffiths and Parke, 2002; Kendall, 2002; Chee and Smith, 2003). Others have looked at behaviors perceived as deviant, such as cyberporn (Mitchell et al., 2003; Stack et al., 2004) and its more mainstream cousin, cyberdating (Whitty, 2004; Baker, 2005; Whitty and Carr, 2006).

---

[35] Jennifer Kayahara has major responsibility for this part of the Connected Lives project and drafted much of this section. For more information, see Kayahara and Wellman (2005).

This research, while valuable, is limited because it treats going online as a leisure activity unto itself and ignores the interplay between online and offline activities in everyday life. Interacting directly online is only one way that people can use the Internet to access culture. In addition to providing a location for engaging in cultural and leisure activities, the Internet can also facilitate access to information about new cultural activities through features such as book reviews; offer information that enables people to access culture offline, such as movie times; enable people to manufacture and share with others their own cultural activities such as photoblogs; and improve ease of communication about culture through e-mail and instant messaging.

Gaps in the literature suggest some important questions. In general, we are concerned with where people get cultural information from and how they decide which cultural activities to consume. Our more specific, Internet-focused questions are:

Who goes online in search of culture? What types of people are most likely to go online, and what types of people are least likely to do so?

For what kinds of cultural information are people searching? Included in this is the question of what types of online cultural activities people are engaging in: Are people interested in online cultural experiences, such as games and podcasts? Are they interested in supplementary information, such as biographies of musicians and reviews of books and movies? Or are they using the Internet for access to offline experiences, as a gateway to learn about – and buy tickets for – concerts and galleries?

To address these questions, interview participants were asked about how they use the Internet to engage with their two favorite cultural activities. In addition, our interviewers also observed users as they navigated through their favorite cultural sites (for more details, see Kayahara and Wellman, 2005).

The answer to the first question – who goes online for culture – is a majority of all interview participants and the great majority of Internet users. Overall, 69% of interview participants use the Internet for gathering information about cultural and leisure activities. This rises to 81% if the sample is limited to people who go online. This high participation rate means that Internet users who use the Internet for culture are almost indistinguishable from Internet users who do not. They tend to be younger and better educated than the interview participants as a whole, but that is a product of being Internet users, and the effect disappears once Internet use is controlled for. No demographic factor we checked is statistically significant: gender, employment status, relationship status, or the presence of children.

On the issue of what people are looking for online, we have learned a few things. First, people go online for a variety of cultural and leisure information, reflecting their diverse interests. The topics participants search for include gardening tips, bird watching locations and sightings; online

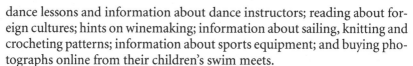

dance lessons and information about dance instructors; reading about foreign cultures; hints on winemaking; information about sailing, knitting and crocheting patterns; information about sports equipment; and buying photographs online from their children's swim meets.

Second, information related to books and movies is relatively popular. This is consistent with an earlier study that found a positive correlation between Internet use and pleasure reading, based partly on the fact that people sometimes go online to seek pleasurable reading (Griswold and Wright, 2004). Our data show that 8% of participants go online to look for book reviews or purchase books from *Amazon* or *Chapters/Indigo*, and 13% go in search of movie times, locations, or tickets. It is likely that even more people engage in these activities, since each participant was only questioned about their top two cultural and leisure activities. The popularity of looking up books and movies online reflects the interests of the participants: 74% mention reading and writing as an interest and 68% percent mention television and film. This behavior may also be influenced by the structure of Web sites that serve as portals to cultural and leisure information, as many sites feature popular culture items such as movie listings more prominently than high culture items such as fine art shows (Hargittai, forthcoming).

Third, when deciding what cultural activities to engage in, people often turn first to sources other than the Internet for inspiration. The most frequently cited source of recommendations for new cultural activities is personal networks, mentioned by 71% of interview participants (Figure 8.14). Many value suggestions from friends and family. They can be personalized to individual tastes. This suggests that recent moves toward "social bookmarking" – automatically sharing information with others about popular Web sites – might be popular (Hargittai, 2005). As one participant explains:

(Participant 810): What people tell me [is more important than ads]. Like The Incredibles. We rent this movie.... But they [the children] didn't like it. The movie was okay for us, but not for children. It's about government . . . they were waiting for something to happen and finally they get tired.

Not all participants value the recommendations from friends and family quite so much. One explains:

(Participant 274): I don't quite like everything [my sister] reads. Even though it's nice, I am not into that genre like Nora Roberts. I have read her books, they're nice but I'm not really into it. She likes Wicca and witch stuff; I've already been through that period.

Participant 498 put it more succinctly:

To each his own. Your tastes might be different from mine.

Those who prefer not to take advice from their social networks have a variety of other sources to which they turn, including existing knowledge

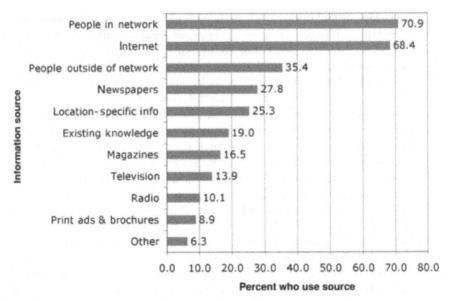

Figure 8.14 Sources of cultural information ($N = 79$).

of genres, authors, actors, or musicians (19%); location-based information gathered by scanning the bookshelves at the store or driving to the theater to look at the listings (25%); listening to the radio (10%); or consulting print sources such as newspapers (28%).

It is after people have a recommendation or suggestion – from their network or from elsewhere – that they often turn to the Internet for information. They usually seek specific information, such as upcoming performances by a favorite band, book reviews, or hotel prices for a summer vacation. This suggests that ICT-involved individuals are going beyond the longstanding theory of the "two-step flow of communication" (Lazarsfeld et al., 1948; Katz and Lazarsfeld, 1955). The initial conception of this theory stated that most people are not directly influenced by messages from the mass media. Instead, opinion leaders filter the messages and influence their followers through social networks (see also Weimann's critique, 1982). With the proliferation of ICTs, our findings suggest that while social networks remain influential in spreading the word about cultural matters, a large number of people are adding a step by taking the recommendations they receive from their social networks and going online to research these recommendations further. Such behavior can result in a feedback spiral, where people learn something online and share it with friends who then go research it further online before sharing the information with others. It could also suggest an interruption in the traditional two-step pattern if people are going online and finding opinions that contradict the recommendations they receive from network members.

8.7 Connected Lives – On and Offline

This, our initial reconnoitering of Connected Lives, has found that ICTs have become part of everyday life in East York, Toronto, Canada – from mobile phones to the Internet. Rather than the separate, often kinky, online-only virtual communities so beloved of the media,[36] we have discovered that most people use ICTs easily and routinely to find information and to contact family, friends, and neighbours. Rather than special household shrines to personal computers, we have found computers sharing domestic space in living rooms, family rooms, and bedrooms. Even home offices – home to computers in nearly half of the households – are usually accessible to all household members.

The most popular time to use home computers (and the Internet) is during traditional evening family hours. Even though all household members are not as likely to be at home as yesteryear for family dinners or gatherings around the television, people use mobile phones, IM, and e-mail extensively in order to contact them – be they across the continent or in the next room. Indeed, using the Internet to communicate with family, friends, and acquaintances is second only to using it for work and school. Contrary to the pre-Internet era, men do as much online communication at home as do women. Indeed, working online from home now takes a bit more of the average woman's time than does doing household chores. Men do about the same level of online work from their homes as do women but, as usual, they do less household chores.

The high level of ICT-based communication reflects the networked lives of household members and the continued strength of personal communities. East Yorkers have an average of 9 very close members of their personal communities and 14 somewhat close members. These are substantially higher numbers than when NetLab last measured network size in 1979 and 1968, although we caution that different network generators were used then to estimate network size (Wellman, 1979; Wellman and Wortley, 1990). Moreover, given the ability of the Internet to support even weaker ties (Boase et al., 2006), we suspect that the size of personal community networks is larger than it has been since the post-World War II move away from street corner neighbourhoods to castle-like detached suburban homes. Our data suggest a situation similar to Japan where mobile phones are used extensively to keep in touch with extremely close ties – household members, friends, immediate kin, work partners – and to make local arrangements while the Internet is used to keep in contact with ties ranging from the extremely close to acquaintances and strangers (Miyata et al., 2005). E-mail scales up more effectively than the mobile phone to support more contact with more network members. In addition, rather than substituting for in-person contact,

---

[36] "Have webcam, will copulate" reads a recent newspaper headline (Friesen, 2005).

e-mail lubricates and increases in-person contact, both locally and via long-distance travel. And the data show that the larger the network, the more social activities.

Consistent with the theory of networked individualism, people get a variety of social support – major and minor goods and services as well as emotional support – but that support may be as specialized in 2004 as the second East York study found in 1979 (Wellman and Wortley, 1990). On average, only two or three role relations (friend, neighbour, etc.) give any one type of support (although multiple friends, etc. may be supportive). Emotional support flows as copiously to heavy e-mail users as it does to non- or light users (see also Copher et al., 2002). Contrary to early fears (detailed in Wellman and Gulia, 1999b; Kayahara, 2006), the Internet does not turn people away from supportive ties. Moreover, the facilitative affordances of e-mail appear to be associated with the greater extent of supportive services that heavy e-mail users exchange. Where both the Internet doomsayers (e.g. Stoll, 1995) and the community doomsayers (e.g. Putnam, 2000) have argued that things are falling apart, we believe that things are becoming more complicated and lively with the help of ICTs.

ICTs are information technologies as well as communication technologies. "We're entering an era in which people are participating rather than just receiving information," said Jonathan Swartz, president of Sun Microsystems (Knowledge@Wharton, 2005, p. 2). Our Connected Lives data agree, showing that the Internet is used extensively for finding a good deal of diverse information about health and culture. (We only asked about these two areas.) For example, the Internet is second only to network members for providing cultural information, well more than any other means of providing information. The very nature of ICTs as both information and communication technologies means that these two domains are interpenetrating more than before. People discuss with network members what they have found on the Internet. Similarly, people go to the Internet (and mobile phones) to check out what they have heard from network members. Our interview participants describe multistep feedback spirals between network information and interpersonal information – communicated online and offline – that goes far beyond the traditional model of the two-step flow of information.

In short, as computer, communication, and social networks have intertwined, ICTs have become part of the household and community. ICTs are increasingly being taken for granted. They are becoming part of the furniture, like the living room couch, and when they get old, they may hang around as coffee tables (Richtel and Markoff, 2005).

8.8 Acknowledgments

We are grateful for the financial support provided by the Social Science and Humanities Research Council, the Joint Program in Transportation of the University of Toronto, Heritage Canada, the Intel Research Council, and Microsoft Research. We have benefited from the advice of Shyon Baumann, Wenhong Chen, Dimitrina Dimitrova, Bonnie Erickson, Keith Hampton, John Horrigan, Charles Kadushin, William Michelson, Eric Miller, John Miron, Anabel Quan-Haase, Carsten Quell, Lee Rainie, Covadonga Robles, Inna Romanovska, Irina Shklovski, Beverly Wellman, Sandy Welsh, and East York's Neighbourhood Information Centre. We gratefully appreciate the assistance of Monica Prijatelj, Grace Ramirez, Inna Romanovska, Esther Rootham, Julia Weisser, Lee Weisser, Sandra Wong, Natalie Zinko, and our many surveyors, data enterers, and transcribers (listed at www.chass.utoronto.ca/∼ wellman). We thank the Centre for Urban and Community Studies at the University of Toronto for providing our research base. Photographs and diagrams copyright © 2005 Wellman Associates and are used by permission.

8.9 References

Adamic, L., Buyukkokten, O., and Eytan, A. (2003) A social network caught in the web, *First Monday*, **8**(3), <http://www.firstmonday.org/issue8-6/adamic/index.html>

Anderson, B. and Tracy, K.T. (2002) Digital living: the impact (or otherwise) of the Internet in everyday British life. In Wellman, B. and Haythornthwaite, C. (eds), *The Internet in Everyday Life*. Blackwell, Oxford, pp. 139–163.

Aro, J. and Peteri, V. (2003) Constructing computers at home. Presented at the Association for Internet Researchers Conference, Toronto, October.

Axhausen, K.W. (2005) Activity spaces, biographies, social networks and their welfare gains and externalities: some hypotheses and empirical results. Presented at the PROCESSUS Colloquium, Toronto, June.

Bakardjieva, M. and Smith, R. (2001) The Internet in everyday life: computer networking from the standpoint of the domestic user. *New Media and Society*, **3**(1), 67–83.

Baker, A. (2005) *Double Click: Romance and Commitment Among Online Couples*. Peter Lang, New York.

Barrera, M., Glasgow, R.E., McKay, H.G., Boles, S.M. and Feil, E.G. (2002) Do Internet-based support interventions change perceptions of social support? An experimental trial of approaches for supporting diabetes self-management. *American Journal of Community Psychology*, **30**, 637–654.

Barthélmy, D. and Contamine, P. (1985 [1988]) The use of private space [Goldhammer, A., trans.]. In Duby, G. (ed.), *A History of Private Life*. Vol. II, Revelations of the Medieval World. Belknap Press, Cambridge, MA, pp. 395–505.

Baym, N. (1997) Identity, body, and community on-line. *Journal of Communication*, **47**(4), 142–148.

Beck-Gernsheim, E. (2002) *Reinventing the Family: In Search of New Lifestyles*. Polity Press, Malden, MA.

Bernard, H.R., Killworth, P., Johnsen, E., Shelley, G.A., McCarty, C. and Robinson, S. (1990) Comparing four different methods for measuring personal networks. *Social Networks*, **12**, 179–216.

Boase, J., Chen, W., Wellman, B. and Prijatelj, M. (2003) Is there a place in cyberspace? The uses and users of the Internet in public and private places. *Culture et Geographie*, **46**(Été), 5–20.

Boase, J., Horrigan, J., Wellman, B. and Rainie, L. (2006) The strength of Internet ties. Pew Internet and American Life Project, Washington, January. Retrieved from <http://www.pewinternet.org>

Boase, J. and Wellman, B. (2005) Personal relationships: on and off the Internet. In Perlman, D. and Vangelisti, A.L. (eds), *Handbook of Personal Relations*. Blackwell, Oxford.

Boissevain, J. (1974) *Friends of Friends: Networks, Manipulators, and Coalitions*. Blackwell, Oxford.

Bourdieu, P. (1984) *Distinction* [Nice, R., trans.]. Routledge and Kegan Paul, London.

Bradner, E., Kellogg, W. and Erickson, T. (1998) Social affordances of BABBLE. Presented at the European Computer Supported Cooperative Work Conference, Copenhagen, October.

Bryce, J. (2001) The technological transformation of leisure. *Social Science Computer Review*, **19**(1), 7–16.

Cairncross, F. (2001) *The Death of Distance: How Communications Revolution is Changing Our Lives*. Harvard Business School Press, Boston, MA.

Campbell, K.E. and Lee, B.A. (1992) Sources of personal neighbour networks: social integration, need, or time. *Social Forces*, **70**(4), 1077–1100.

Carrasco, J.A., Hogan, B., Wellman, B. and Miller, E.J. (2005) Collecting social network data to study social activity-travel behavior: an egocentric approach. Working Paper. Connected Lives Study, NetLab, Centre for Urban and Community Studies, University of Toronto. Retrieved from <http://www.chass.utoronto.ca/~wellman/publications/index.html>

Castells, M. (2001) *The Internet Galaxy*. Oxford University Press, Oxford.

Chee, F. and Smith, R. (2003) Is electronic community an addictive substance? Everquest and its implications for addiction policy. Presented at the Association of Internet Researchers Conference, Toronto, Ontario, October.

Chen, W., Boase, J. and Wellman, B. (2002) The global villagers: comparing Internet users and uses around the world. In Wellman, B. and Haythornthwaite, C. (eds), *The Internet in Everyday Life*. Blackwell, Oxford, pp. 74–113.

Chen, W. and Wellman, B. (2005) Charting digital divides: within and between countries. In Dutton, W., Kahin, B., O'Callaghan, R. and Wyckoff, A. (eds), *Transforming Enterprise*. MIT Press, Cambridge, MA, pp. 467–497.

Clinton, H.R. (1996) *It Takes a Village*. Simon & Schuster, New York.

Cohen, S. and Syme, S.L. (1985) *Social Support and Health*. Academic Press, New York.

Cooper, J. (2004) Stiffing East York. Retrieved 30 December 2004 from <http://www.nowtoronto.om/issues/2004-12-30/news_story4.php>

Copher, J.I., Kanfer, A.G. and Walker, M.B. (2002) Everyday communication patterns of heavy and light e-mail users. In Wellman, B. and Haythornthwaite, C. (eds), *The Internet in Everyday Life*. Blackwell, Oxford, pp. 263–298.

Cowan, R.S. (1983) *More Work for Mother: The Ironies of Household Technology from the Open Hearth to the Microwave*. Basic Books, New York.

Cumming, J. and Kraut, R. (2001) Domesticating computer and the Internet. *The Information Society*, **18**(3), 221–231.

Davidson, T. (1976) *The Golden Years of East York*. Centennial College Press, Toronto.

Davis, N.Z. (1983) *The Return of Martin Guerre*. Harvard University Press, Cambridge, MA.

Dickson, P. and Ellison, J. (2000) Plugging in: the increase of household Internet use continues into 1999. Statistics Canada Report, Connectedness Series. Retrieved from <http://www.statcan.ca/english/research/56F0004MIE/56F0004MIE2000001.pdf>

DiMaggio, P. (1982) Cultural capital and school success: the impact of status culture participation on grades of U.S. high school students. *American Sociological Review*, **47**(2), 189–201.

DiMaggio, P. (1997) Culture and cognition. *Annual Review of Sociology*, **23**, 263–287.

Dimitrova, D. (2003) Controlling teleworkers: supervision and flexibility revisited. *New Technologies, Work and Employment*, **18**(3), 181–195.

Dodds, P.S., Muhamad, R. and Watts, D. (2003) An experimental study of search in global social networks. *Science*, **301**(8), 827–829.

Ekos (2004) *The Dual Digital Divide IV*. Ekos Research Associates, Ottawa.

Erickson, B.H. (1996) Culture, class and connections. *American Journal of Sociology*, **102**, 217–251.

Erickson, B.H. (2003) Social networks: the value of variety. *Contexts*, **2**(1), 25–31.

Fagan, C. (2001) The temporal reorganization of employment and the household rhythm of work schedules: the implications for gender and class relations. *American Behavioral Scientist*, **44**(7), 1199–1212.

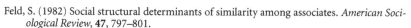
Feld, S. (1982) Social structural determinants of similarity among associates. *American Sociological Review*, **47**, 797–801.

Fischer, C. (1982) *To Dwell Among Friends*. University of California Press, Berkeley, CA.

Fischer, C. (2005) Bowling alone: what's the score? *Social Networks*, **27**(2), 155–167.

Fogel, J., Albert, S.M., Schabel, F., Ditkoff, B.A. and Neuget, A.I. (2002) Internet use and social support in women with breast cancer. *Health Psychology*, **21**, 398–404.

Forster, E.M. (1910) *Howard's End*. Edward Arnold, London.

Fortunati, L., Katz, J. and Riccini, R. (eds) (2003) *Mediating the Human Body: Technology, Communication, and Fashion*. Lawrence Erlbaum, Mahwah, NJ.

Fox, S. (2003) Internet health resources. Pew Internet and American Life Project, Washington, July. Retrieved from <http://www.pewinternet.org>

Fox, S. (2005) Health information online. Pew Internet and American Life Project, Washington, May. Retrieved from <http://www.pewinternet.org>

Friedman, T. (2005) *The World is Flat*. New York: Farrar, Straus and Giroux.

Friesen, J. (2005) Have web cam, will copulate. *Toronto Globe and Mail*, 15 July, p. F2.

Frohlich, D. and Kraut, R. (2002) The Social Context of Home Computing. Retrieved from <http://www.hpl.hp.com/techreports/2003/HPL-2003-70.pdf>

Gans, H. (1974) *Popular Culture and High Culture*. Basic Books, New York.

Gershuny, J. (2003) *Web-Use and Net-Nerds: A Neo-Functionalist Analysis of the Impact of Information Technology in the Home*. Institute for Social and Economic Research, University of Essex, Colchester, UK.

Gold, G. (2003) Rediscovering place: experiences of a quadriplegic anthropologist. *Canadian Geographer*, **47**, 467–479.

Granovetter, M.S. (1973) The strength of weak ties. *American Journal of Sociology*, **78**, 1360–1380.

Granovetter, M.S. (1995) *Getting a Job: A Study of Contacts and Careers*, 2nd edn. University of Chicago Press, Chicago.

Griffiths, M.D. and Parke, J. (2002) The social impact of Internet gambling. *Social Science Computer Review*, **20**(3), 312–320.

Griswold, W. and Wright, N. (2004) Wired and well read. In Howard, P.N. and Jones, S. (eds), *Society Online: The Internet in Context*. Sage, Thousand Oaks, CA, pp. 203–222.

Haddon, L. and Skinner, D. (1991) The enigma of the micro: lessons from the British home computer boom. *Social Science Computer Review*, **9**(3) 435–449.

Haines, V.A., Hurlbert, J.S. and Beggs, J.J. (1996) Exploring the determinants of support provision. *Journal of Health and Social Behavior*, **37**(3), 252–264.

Hampton, K. and Wellman, B. (2002) The not so global village of Netville. In Wellman, B. and Haythornthwaite, C. (eds), *The Internet in Everyday Life*. Blackwell, Oxford, pp. 345–371.

Hampton, K. and Wellman, B. (2003) Neighbouring in Netville: how the Internet supports community and social capital in a wired suburb. *City and Community*, **2**(3), 277–311.

Hargittai, E. (2002a) Beyond logs and surveys: in-depth measures of people's online skills. *Journal of the American Society for Information Science and Technology*, **53**(14), 1239–1244.

Hargittai, E. (2002b) Second-level digital divide: differences in people's online skills. *First Monday*, **7**(4). Retrieved from <http://www.firstmonday.org/issues/issue7_/hargittai>

Hargittai, E. (2005) Social bookmarking goes mainstream. *Crooked Timber*. Retrieved 29 June 2005 from <http://crookedtimber.org/2005/06/29/social-bookmarking-goes-mainstream-or-attempts-to-anyway/>

Hargittai, E. (forthcoming) Content diversity online: myth or reality? In Napoli, P. (ed.), *Media Diversity and Localism: Meanings and Metrics*.

Hennig, M. (2007) Re-evaluating the community question from a German perspective. *Social Networks*, forthcoming.

House, J.S. (1985) *Structures and Sentiments of Social Support*. American Sociological Association, Washington, DC.

House, J.S., Umberson, D. and Landis, K.R. (1988) Structures and processes of social support. *Annual Review of Sociology*, **14**, 293–318.

Hunter, L.R. (2005) Alternative to blogging [Letter to the editor]. *New York Times*, 9 August, p. A18.

Hurlbert, J., Haines, V. and Beggs, J. (2000) Core networks and tie activation. *American Sociological Review* 65, 598–618.

Ipsos-Reid, (2005) Online news and information seeking: what the future holds. Report. Retrieved 9 August 2005 from <www.ipsos-na.com>

Jacobs, J.A. and Gerson, K. (2001) Overworked individuals or overworked families? Explaining trends in work, leisure, and family time. *Work and Occupations*, **28**(1), 40–63.

Johnson, L. (1999) Bringing work home: developing a model residentially-based telework facility. *Canadian Journal of Urban Research*, **8**(2), 119–142.

Kadushin, C. (1966) The friends and supporters of psychotherapy: on social circles in urban life. *American Sociological Review*, **31**, 786–807.

Kadushin, C. (2002) The motivational foundation of social networks. *Social Networks*, **24**, 77–91.

Kadushin, C., Killworth, P., Bernard, H.R. and Beveridge, A. (2005) Scale-up methods as applied to estimate of heroin use. Working paper, Brandeis University, East Waltham, MA.

Kapor, M. and Berman, J. (1993) A superhighway through the wasteland? *New York Times*, 24 November, p. A15.

Katz, E. and Lazarsfeld, P. (1955) *Personal Influence: The Part Played by People in the Flow of Mass Communications*. Free Press, Glencoe, IL.

Katz, J. and Aakhus, M. (2002) *Perpetual Contact: Mobile Communications, Private Talk, Public Performance*. Cambridge University Press, Cambridge, UK.

Katz, J. and Rice, R. (2002) *Social Consequences of Internet Use: Access, Involvement and Interaction*. MIT Press, Cambridge, MA.

Kayahara, J. (2006) Community and communication: a rounded perspective. In Purcell, P. (ed.), *Networked Neighbourhoods*. Springer-Verlag, Berlin, pp. 123–154.

Kayahara, J. and Wellman, B. (2005) *Finding Culture Online and Offline*. Heritage Canada, Ottawa.

Kelly, K. (2005) We are the web. *Wired*. Retrieved 13 August 2005 from <http://www.wired.com/wired/archive/13.08/tech.html>

Kendall, L. (2002) *Hanging Out in the Virtual Pub: Masculinities and Relationships Online*. University of California Press, Berkeley, CA.

Kiesler, S., Zdaniuk, B., Lundmark, V. and Kraut, R. (2000) Troubles with the Internet: the dynamics of help at home. *Human–Computer Interaction*, **15**, 323–351.

Knowledge@Wharton (2005) Supernova 2005: it's a whole new, connected world. Retrieved 1 August 2005 from <http://knowledge.wharton.upenn.edu/index.cfm?fa=viewArticle&id=1244&specialId=38>

Kochen, M. (ed.) (1989) *The Small World*. Ablex, Norwood, NJ.

Kraut, R., Kiesler, S., Boneva, B., Cummings, J., Helgeson, V. and Crawford, A. (2002) Internet paradox revisited. *Journal of Social Issues*, **58**(1), 49–74.

Kraut, R., Patterson, M., Lundmark, V., Kiesler, S., Mukhopadhyay, T. and Scherlis, W. (1998) Internet paradox: a social technology that reduces social involvement and psychological well-being? *American Psychologist*, **53**(9), 1017–1031.

Lange, P.H. and Adamec, C. (2003) *Prostate Cancer for Dummies*. Wiley, Seattle, WA.

Lazarsfeld, P.F., Berelson, B. and Gaudet, H. (1948) *The People's Choice*. Columbia University Press, New York.

Legris, P., Ingham, J. and Collerette, P. (2003) Why do people use information technology? A critical review of the technology acceptance model. *Information and Management*, **40**, 191–204.

Lenhart, A. (2005) Pew Internet report on teens and technology: press release. Pew Internet and American Life Project, Washington, July, 25. <http://www.pewinternet.org/PPF/r/109/press_release.asp>

Lenhart, A., Madden, M. and Hitlin, P. (2005) Teens and technology. Pew Internet and American Life Project, Washington, July. Retrieved from <http://www.pewinternet.org>

LeRoy Ladurie, E. (1975 [1978]) *Montaillou: The Promised Land of Error* [Bray, B., trans.]. Braziller, New York.

LeRoy Ladurie, E. (1997) *The Beggar and the Professor: A Sixteenth-Century Saga* [Goldhammer, A., trans.]. University of Chicago Press, Chicago.

Liebler, C.A. and Sandefur, G.D. (2002) Gender differences in the exchange of social support with friends, neighbours, and co-workers at midlife. *Social Science Research*, **31**(3), 364–391.

Ling, R. (2004) *The Mobile Connection: The Cell Phone's Impact on Society*. Morgan Kaufmann, San Francisco.

Logan, J. and Spitze, G. (1996) *Enduring Ties*. Philadelphia: Temple University Press.

Marsden, P. and Campbell, K.E. (1984) Measuring tie strength. *Social Forces*, **63**, 482–501.

Marx, L. (1964) *The Machine in the Garden*. Oxford University Press, New York.

Mayaud, C. (2005a) Cheater's guide to *LinkedIn*, v. 01. *Sacred Cow Dung*, 1 July (revised version). Retrieved from <http://www.sacredcowdung.com/archives/2005/05/cheaters_guide.html>

Mayaud, C. (2005b) "Right sizing" your PANs, CANs, and FANs. *Sacred Cow Dung*, 11 March. Retrieved from <http://www.sacredcowdung.com/archives/2005/03/right_sizing_yo.html>

McCarty, C., Killworth, P.D., Bernard, H.R., Johnsen, E.C. and Shelley, G.A. (2001) Comparing two methods for estimating network size. *Human Organization*, **60**(1), 28–39.

McLuhan, M. (1964) *Understanding Media: The Extension of Man*. McGraw-Hill, New York.

McPherson, M., Smith-Lovin, L. and Brashears, M.E. (2006) Social isolation in America: changes in core discussion networks over two decades. *American Sociological Review*, **71**(3), forthcoming.

Menzies, H. (2005) *No Time: Stress and the Crisis of Modern Life*. Douglas & McIntyre, Vancouver.

Merton, R.K. (1957) Patterns of influence: Cosmopolitans and Locals. In Merton, R.K., *Social Theory and Social Structure*. Free Press, Glencoe, IL, pp. 387–420.

Mesch, G. (2006a) Family characteristics and intergenerational conflicts over the Internet. *Information, Communication and Society*, **9**(4), forthcoming.

Mesch, G. (2006b) Family relations and the Internet: exploring a family boundaries approach. *Journal of Family Communication*, **6**(2), xxx–xxx.

Milgram, S. (1967) The small-world problem. *Psychology Today*, **1**, 62–67.

Miller, E.J. and Shalaby, A. (2003) Evolution of personal travel in Toronto area and policy implications. *Journal of Urban Planning and Development*, **129**, 1–26.

Mitchell, K.J., Finkelhor, D. and Wolak, J. (2003) The exposure of youth to unwanted sexual material on the Internet. *Youth and Society*, **34**(3), 330–358.

Miyata, K. (2002) Social supports for Japanese mother online and offline. In Wellman, B. and Haythornthwaite, C. (eds), *The Internet in Everyday Life*. Blackwell, Oxford, pp. 520–548.

Miyata, K., Boase, J., Wellman, B. and Ikeda, K. (2005) The mobile-izing Japanese: connecting to the Internet by PC and webphone in Yamanashi. In Ito, M., Matsuda, M. and Okabe, D. (eds), *Portable, Personal, Pedestrian: Mobile Phones in Japanese Life*. MIT Press, Cambridge, MA, pp. 143–164.

Mokhtarian, P., Salomon, I. and Handy, S. (2003) A taxonomy of leisure activities, with implications for the choice of ICT alternatives. Presented at the International Conference on Travel Behaviour Research, Lucerne, August.

Molm, L.D., Takahashi, N. and Peterson, J.T. (2000) Risk and trust in social exchange: an experimental test of a classical proposition. *American Journal of Sociology*, **105**, 1396–1427.

Müller, C. (1999) Networks of "personal communities" and "group communities" in different online communication services. Presented at the ECS Conference, Newcastle, UK, October.

Nie, N., Hillygus, D.S. and Erbring, L. (2002) Internet use, interpersonal relations and sociability: a time diary study. In Wellman, B. and Haythornthwaite, C. (eds), *The Internet in Everyday Life*. Blackwell, Oxford, pp. 215–243.

Niles, J. (1994) *Beyond Telecommuting: A New Paradigm for the Effect of Telecommunications on Travel*. US Department of Energy, Washington, DC.

Pandey, S.K., Hart, J.J. and Tiwary, S. (2003) Women's health and the Internet. *Social Science and Medicine*, **56**(1), 179–191.

Pearlin, L., Aneshensel, C., Mullan, J. and Whitlatch, C. (1995) Caregiving and its social support. In Binstock, R. and George, L. (eds), *Handbook of Aging and the Social Sciences*. Academic Press, San Diego, CA, pp. 283–302.

Perlman, D. and Fehr, B. (1987) The development of intimate relationships. In Perlman, D. and Duck, S. (eds), *Intimate Relationships*. Sage, Newbury Park, CA, pp. 13–42.

Pescosolido, B. (1992) Beyond rational choice: the social dynamics of how people seek help. *American Journal of Sociology*, **97**, 1096–1138.

Plickert, G., Wellman, B. and Côté, R. (2007) It's not who you know, it's how you know them: who exchanges what with whom? *Social Networks*, forthcoming.

Pool, I.D.S. and Kochen, M. (1978) Contacts and influence. *Social Networks*, **1**, 5–51.

Putnam, R. (2000) *Bowling Alone*. Simon & Schuster, New York.

Quan-Haase, A., Wellman, B. with Witte, J. and Hampton, K. (2002) Capitalizing on the Internet: network capital, participatory capital, and sense of community. In Wellman, B. and Haythornthwaite, C. (eds), *The Internet in Everyday Life*. Blackwell, Oxford, pp. 291–324.

Reid, E. (1999) Hierarchy and power: social control in cyberspace. In Smith, M.A. and Kollock, P. (eds), *Communities in Cyberspace*. Routledge, New York, pp. 107–133.

Rheingold, H. (1993) *The Virtual Community: Homesteading on the Electronic Frontier*. Addison-Wesley, Reading, MA.

Rheingold, H. (2000) *The Virtual Community*, revised edn. MIT Press, Cambridge, MA.

Rheingold, H. (2002) *Smart Mobs*. Perseus, New York.

Rheingold, H. (2005) On and offline communities. [Letter to the Editor], *Contexts*, **4**(2), 6.

Ribak, R. (2001) "Like immigrants": negotiating power in the face of the home computer. *New Media and Society*, **3**(2), 220–238.

Rice, R. (1993). Media appropriateness. *Human Communication Research*, **19**(4), 451–484.

Richtel, M. and Markoff, J. (2005) Home for corrupted PC's is often the dumpster. *New York Times*, 17 July, p. A16.

Rideout, V. and Reddick, A. (2005) Sustaining community access to technology: who should pay and why. *Journal of Community Informatics*, **1**(2). Retrieved from <http://www.ci-journal.net/viewarticle.php?id=39>

Robinson, J.P. and Godbey, G. (1997) *Time for Life: The Surprising Ways Americans Use Their Time*. Pennsylvania State University Press, State College.

Rosenthal, C. (1985) Kinkeeping in the familial division of labor. *Journal of Marriage and the Family*, **47**, 965–974.

Salaff, J. (2002) Where home is the office: the new form of flexible work. In Wellman, B. and Haythornthwaite, C. (eds), *The Internet in Everyday Life*. Blackwell, Oxford, pp. 464–495.

Salomon, I. (1998) Technological change and social forecasting: the case of telecommuting as a travel substitute. *Transportation Research C*, **6**(1/2), 17–45.

Senbil, M. and Kitamura, R. (2003) Simultaneous relationships between telecommunications and activities. In *International Conference on Travel Behaviour Research*, Lucerne, August.

Simmel, G. (1903) The metropolis and mental life. In Wolff, E. (ed. and trans.), *The Sociology of Georg Simmel*. Free Press, Glencoe, IL, 409–424.

Sproull, L. and Kiesler, S. (1991) *Connections*. MIT Press, Cambridge, MA.

Stack, S., Wasserman, I. and Kern, R. (2004) Adult social bonds and use of Internet pornography. *Social Science Quarterly*, **85**(1), 76–88.

Statistics Canada (2002) Characteristics of household Internet users. Retrieved from <http://www40.statcan.ca/101/cst01/comm10a.htm?sdi=internet>

Statistics Canada (2003) The Daily – Births. Retrieved from <http://www.statcan.ca/Daily/English/050712/d050712a.htm>

Statistics Canada (2005) The Daily – Divorces. Retrieved from <http://www.statcan.ca/Daily/English/050309/d050309b.htm>

Stoll, C. (1995) *Silicon Snake Oil: Second Thoughts on the Information Highway*. Doubleday, New York.

Turkle, S. (1984) *The Second Self: Computers and the Human Spirit*. Simon & Schuster, New York.

Turkle, S. (1995) *Life on the Screen: Identity in the Age of the Internet*. Simon & Schuster, New York.

Tarrow, S. (1999) Fishnets, Internets and catnets: globalization and transnational collective action. In Hanagan, M.P., Moch, L. and TeBrake, W. (eds), *The Past and Future of Collective Action*. Russell Sage, New York, pp. 228–424.

Thébert, Y. (1985 [1987]) Private life and domestic architecture in Roman Africa (Goldhammer, A., trans.). In Veyne, P. (ed.), *A History of Private Life*. Belknap Press, Cambridge, MA, pp. 313–410.

Thoits, P. (1995) Stress, coping, and social support process. *Journal of Health and Social Behavior*, **35**(Extra Issue), 53–79.

Wajcman, J. (1991) *Feminism Confronts Technology*. University Press, Pennsylvania.

Ward, P. (1999) *A History of Domestic Space: Privacy and the Canadian Home*. UBC Press, Vancouver.

Watts, D.J. (2003) *Six Degrees: The Science of a Connected Age*. Norton, New York.

Wehner, C. and Abrahamson, P. (2004) Individualisation of family life and family discourses. Presented at the ESPAnet Conference, Oxford, September 2004. Retrieved from <http://www.apsoc.ox.ac.uk/Espanet/espanetconference/papers/ppr.10C.PA.pdf.pdf>

Weimann, G. (1982) On the importance of marginality: one more step into the two-step flow of communication. *American Sociological Review*, **47**, 764–773.

Wellman, B. (1979) The community question. *American Journal of Sociology*, **84**(5), 1201–1231.

Wellman, B. (1982) Studying personal communities. In Marsden, P. and Lin, N. (eds), *Social Structure and Network Analysis*. Sage, Beverly Hills, CA, pp. 61–80.

Wellman, B. (1985) Domestic work, paid work and net work. In Duck, S. and Perlman, D. (eds), *Understanding Personal Relationships*. Sage, London, pp. 159–191.

Wellman, B. (1988) Structural analysis: from method and metaphor to theory and substance. In Wellman, B. and Berkowitz, S.D. (eds), *Social Structures: A Network Approach*. Cambridge University Press, Cambridge, UK, pp. 19–61.

Wellman, B. (1992) Which types of ties and networks give what kinds of social support? *Advances in Group Processes*, **9**, 207–235.

Wellman, B. (ed.) (1999) *Networks in the Global Village*. Westview Press, Boulder, CO.

Wellman, B. (2000) Changing connectivity: a future history of Y2.03K. *Sociological Research Online*, **4**(4). <http://www.socresonline.org.uk/4/wellman.html>

Wellman, B. (2001) Physical place and cyber-place: changing portals and the rise of networked individualism. *International Journal for Urban and Regional Research*, **25**(2), 227–252.

Wellman, B. (2002) Redes de sociabilidad y redes de internet [Social networks and computer networks]. In Castells, M. (ed.), *La societat xarxa a catalunya: informe de recerca.* [*The Network Society in Catalonia: Research Report.*] Barcelona, Spain. <http://www.uoc.edu/in3/pic/cat/pdf/PIC_escoles_1_3>

Wellman, B. (2004) The three ages of Internet studies: ten, five and zero years ago. *New Media and Society*, **6**(1), 108–114.

Wellman, B. (ed.) (2007) Personal Networks. Special issue of *Social Networks*.

Wellman, B., Carrington, P. and Hall, A. (1988) Networks as personal communities. In Wellman, B. and Berkowitz, S.D. (eds), *Social Structures: A Network Approach*. Cambridge University Press, Cambridge, UK, pp. 130–184.

Wellman, B. and Frank, K.A. (2001) Network capital in a multilevel world: getting support from personal communities. In Lin, N., Burt, R.S. and Cook, K. (eds), *Social Capital*. Aldine De Gruyter, Hawthorne, NY, pp. 233–273.

Wellman, B. and Gulia, M. (1999a) A network is more than the sum of its ties: the network basis of social support. In Wellman, B. (ed.), *Networks in the Global Village*. Westview Press, Boulder CO, pp. 83–118.

Wellman, B. and Gulia, M. (1999b) Net surfers don't ride alone: virtual communities as communities. In Wellman, B. (ed.), *Networks in the Global Village*. Westview Press, Boulder, CO, pp. 331–366.

Wellman, B. and Hampton, K. (1999) Living networked on and offline. *Contemporary Sociology*, **28**(6), 648–654.

Wellman, B. and Haythornthwaite, C. (eds) (2002) *The Internet in Everyday Life*. Blackwell, Oxford.

Wellman, B. and Hogan, B. (2004) The immanent Internet. In McKay, J. (ed.), *Netting Citizens: Exploring Citizenship in the Internet Age*. Saint Andrew Press, Edinburgh, pp. 54–80.

Wellman, B. and Leighton, B. (1979) Networks, neighbourhoods and communities. *Urban Affairs Quarterly*, **14**, 363–390.

Wellman, B. and Wellman, B. (1992) Domestic affairs and network relations. *Journal of Social and Personal Relationships*, **9**, 385–409.

Wellman, B., Wellman, B. and Lloyd, D. (1997a) *Describing Social Support Networks: Development, Findings and Techniques*. Statistics Canada, Ottawa.

Wellman, B. and Wetherell, C. (1996) Social network analysis of historical communities: some questions from the present and the past. *History of the Family*, **1**(1), 97–121.

Wellman, B., Wong, R., Tindall, D. and Nazer, N. (1997b) A decade of network change: turnover, mobility and stability. *Social Networks*, **19**(1), 27–51.

Wellman, B. and Wortley, S. (1989) Brothers' keepers: situating kinship relations in broader networks of social support. *Sociological Perspectives*, **32**, 273–306.

Wellman, B. and Wortley, S. (1990) Different strokes from different folks: community ties and social support. *American Journal of Sociology*, **96**, 558–588.

Whitty, M. (2004) Cyber-flirting: an examination of men's and women's flirting behaviour both offline and on the Internet. *Behaviour Change*, **21**(2), 115–126.

Whitty, M. and Carr, A.N. (2006) *Cyberspace Romance: The Psychology of Online Relationships.* Palgrave Macmillan, Basingstoke, UK.

Wuthnow, R. (1998) *Loose Connections: Joining Together in America's Fragmented Communities.* Harvard University Press, Cambridge, MA.

9

The Impact of the Internet on Local and Distant Social Ties

A. Kavanaugh, T.T. Zin, M.B. Rosson and J.M. Carroll

9.1 Preamble

People use various modes of communication to maintain their social networks, both local and distant. In the United States, Internet use has been growing steadily, and electronic mail has consistently been the most popular online activity. We investigate the effect of online communication with social ties in the highly networked community of Blacksburg, VA, and surrounding rural Montgomery County. We conducted a random stratified household survey to residents in two rounds (2001 and 2002) as part of a larger study on the Internet and community. Our findings provide further evidence that computer networking helps to strengthen and cultivate different types of ties and support within a person's social network at both the local and distant levels. We found significant differences in online communication based on type of social tie (close, somewhat close, and acquaintances), gender, and type of Internet user (heavy versus light, experienced versus novice). Our findings clearly support claims that overall the Internet is used to support and strengthen sociability and social interaction. This finding also holds for social circles at the local level, suggesting that local community is not undermined by Internet use. Finally, people who use the Internet more heavily (more hours per day) also show more social interaction than people who use the Internet less or not at all.

9.2 Introduction

Research on the use and impact of the Internet for sociability and developing and maintaining social relationships shows mixed results on whether communication technology tends to isolate people or to bring them closer together (Patterson and Kavanaugh, 1994; Kraut et al., 1996, 2002; Cohill and Kavanaugh, 1997, 2000; Nie, 2001; Wellman et al., 2001; Kavanaugh,

2003; Turow and Kavanaugh, 2003; Kavanaugh et al., 2005, among many others). Does Internet use with social network members tend to strengthen distant ties at the expense of local ones (Fischer, 1992; Wellman 1992, 1997; Wellman et al., 1996; Fischer et al., 1977)? What are the opportunity costs of using the Internet – is time spent online the time that would otherwise have been spent in face-to-face social interaction (Fischer, 1992; Nie, 2001)? What Internet usage and social and demographic attributes affect social interaction?

In this chapter we report on findings from survey research that was part of a larger 3-year study (2001–2003) we called "Experiences of People, Internet, and Community" (EPIC). The EPIC project comprised a triangulation of three data sources: (1) a three-phase qualitative data collection, most notably household interviews, (2) logging of participant Internet activities, and (3) two rounds of survey data collection, spaced 1 year apart. As part of the qualitative data collection we conducted group interviews (with all members of a given household forming each group) with a subset of households. For the log data, we configured the network connections of a subset of households (all that were possible) so that we could monitor household Web use (hits, time of use, etc.) and e-mail exchange (headers only). Focusing at the household level allowed us to capture interaction and usage patterns related to Internet use in the home. This chapter reports findings from the survey primarily, although we consulted interview data where clarification to survey data was useful. From an original random sample of over 800 households, we created a stratified set of respondents from 100 households, stratified on the categories of location (town of Blacksburg or surrounding rural Montgomery County), user type (Internet access at home, work, or other locations), and education to ensure a representative sample for the area.

The town of Blacksburg (home of the land grant university, Virginia Tech) and surrounding Montgomery County in southwest Virginia is an area directly served since 1993 by the Blacksburg Electronic Village (BEV; http://www.bev.net). BEV is a mature, well-established community network project initiated and subsidized by Virginia Tech, in partnership with the Town of Blacksburg and the local telephone company, Bell Atlantic of Virginia (now known as Verizon). Blacksburg was named by *The Reader's Digest* "the most wired town in America" in 1996. A critical mass of users had formed; local government, schools, and libraries were online, as well as many local businesses and community groups (Cohill and Kavanaugh, 1997, 2000; Ehrich and Kavanaugh, 1997; Kavanaugh and Cohill, 1997; Kavanaugh et al., 2000). By 2001, 87.7 % of Blacksburg residents and 78.9 % of Montgomery County residents reported having Internet access at either home, school, work, or public location (Kavanaugh et al., 2003). This is among the world's highest Internet densities. Over 150 community groups and more than 450 local businesses (more than 75%) maintained Web sites. All of the 20 Montgomery County schools (encompassing Blacksburg) had

T1 Ethernet or Token Ring local area networking since 1996, compared to about 65% nationally; all of the middle and high schools had T1 Internet connectivity since 1995.

The BEV has hosted standard Internet content since its inception in 1993 and has managed local services such as Web space, listservs, and e-mail accounts, as well as specific community-oriented services like information and Web-based forums for local town and county government, social services, public education, libraries, and health care. It has also provided some support for the commercial sector, by linking from its "Village Mall" listing to merchant Web sites hosted elsewhere. Many community-oriented initiatives are maintained, including community newsgroups, organization lists, a senior citizens' nostalgia archive, and video streaming of Town Council meetings. Free public access and a variety of ongoing training classes have been available through the local libraries and town recreation center since the early 1990s.

9.3 Prior Research

The analysis of social networks (a key feature of social capital) has roots in sociology and structural analysis. Social network analysis investigates the concrete social relations among specific social actors and the ordered arrangements of relations that are contingent upon exchange among members of social systems – whether people, groups, or organizations (Wellman and Berkowitz, 1988). Members of social networks garner or mobilize resources through a process of exchange, competition, dependency, or coalition.

A person's social network is composed of the friends, family, and acquaintances with whom that person stays in contact and exchanges resources of friendship and/or aid, including information (Fischer et al., 1977; Wellman and Berkowitz, 1988). Our research interest in social networks focuses on closeness networks (Milardo, 1988). Reported "closeness" of a tie (e.g. labeling someone an acquaintance versus a close friend) is preliminary evidence of the level of intimacy defining the strength of a tie. Additional standard methods to measure strength of social ties are to ask respondents to generate names according to certain criteria or circumstances. For example, for strong ties, respondents are asked to name someone with whom they discuss personal matters; someone, whom, if *they* were in trouble, the respondent would do anything she or he could to help them out and who would do the same for the respondent; someone with whom the respondent enjoys spending a lot of time, doing (or talking about) a variety of activities (or subjects). Intimacy, frequency of contact, the sources and length of the relationship, and the symmetry of the exchange are all important attributes of a tie that indicate its strength or weakness. Among the traditional factors that shape bonds are similar social, economic, and life cycle stage (Fischer, 1992).

9.3.1 Local Versus Distant Social Ties

While the telephone has traditionally been an important communication technology for maintaining contact with social network members who are far away, it was used predominantly for communication with social ties at the local level (Fischer, 1992). Primarily due to the greater cost of long distance telephone calls, especially before the 1990s, there was greater use of this communication technology with local rather than distant social ties. Deriving from the works of Fischer (1992), some have argued that the telephone both counteracted social distance and reinforced local ties. There is a substantial body of literature on the telephone showing that the telephone is a stimulant, preserver, and enhancer of community (Aronson, 1971; Pool, 1983; Fischer, 1992; Dimmick et al., 1994; Katz, 1999). Wellman et al. (2001) found that most Internet communication occurs between people who live less than an hour's drive from each other. Further, their study participants exchanged more e-mail with local friends than with distant friends.

With the advent of the Internet, distant communication (e.g. by e-mail) became no more expensive than local communication. A social shaping of technology approach in recent empirical studies (Katz and Aspden, 1997; Ball-Rokeach et al., 2000; Rainie and Kohut, 2000; Hampton, 2001) shows that people who connect to the Internet are more likely to use it for cultivating proclivities, both social and cultural. Matei and Ball-Rokeach (2001) found that communication technology has been used for reinforcing preexisting social, political, and cultural patterns (Dutton, 1996; Rabby and Walther, 2002; Stafford et al., 1999; Winner, 1997).

9.3.2 Differences by Internet Use

Using the Internet has been associated in various studies with both decreased social interaction (isolation and depression) and increased social interaction. Kraut et al. (1996, 1999) found in an early study of the Internet that users were significantly more likely to become more isolated socially, as well as more depressed, than nonusers. However, their follow-up study was not able to replicate these findings, and the authors have attributed the early findings to the fact that the users were new to the Internet. The isolating effects diminished and disappeared as users became more experienced and as members of their social networks also came online. As part of a larger study (the Syntopia Project), Katz and Rice (2001) found that Internet usage is associated with increased social interaction. They analyzed random telephone survey data they collected on a national scale from 1995 to 2000 and compared it with data from the Pew Internet and American Life Project findings in 2000 (Howard et al., 2001). Their findings regarding social interaction online and offline were consistent in both studies: there were many statistically significant relationships between being an Internet user and greater offline as well as online social interaction. Internet use was positively associated with sociability and interaction. That is, Internet users have more, not less, social interaction on average than nonusers, even among introverts. They also found that the Internet was helping to create new forms of social support (e.g. health-related groups, monitoring and tutoring groups) as well as sustaining long distance relationships and maintaining personal social networks of friends, family, and acquaintances.

Some researchers (see Nie, 2001, among others) have suggested "because of the inelasticity of time, Internet use may actually reduce interpersonal interaction and communication" (p. 420). Nie suggested that data would not lead to conflicting conclusions if researchers standardized their analyses through "parallel measures and replication of multivariate analyses on each of the data sets" (p. 421). Researchers should partition their data, looking at the characteristics of heavy Internet users (more than 10 hours per week) separately from those of light Internet users (less than 10 hours per week). (Ten hours per week was the average at the time of the studies under review, roughly 2000–2001.) Nie's analysis of multiple studies indicated that heavy Internet users may be predisposed to higher levels of sociality than are light users due to their higher education levels. The Nie analysis further suggested that heavy users should be, intrinsically, more gregarious and availed of greater social support than light users.

9.3.3 Differences by Gender

Gender has emerged as an important variable in e-mail communication for social purposes; using e-mail to communicate with friends and family replicates preexisting gender differences (Duck, 1994; Pew, 2000; Boneva et al., 2001, among others). For example, women are more likely than men to maintain kin relationships by e-mail, and women find e-mail with friends and family more gratifying than men do. Women are more likely than men to use e-mail to keep in touch with people who live far away and the contents of their messages sent to people far away are more likely to be filled with personal information and are more likely to be exchanged in intense bursts. Finally, women seem to be expanding their distant social networks due to the fit between women's expressive styles and the features of e-mail.

9.3.4 Social Support, Significant Life Changes, and Internet Use

Social support is considered to be the exchange of verbal and nonverbal messages conveying emotion, information, or referral, to help reduce someone's uncertainty or stress, and to communicate that the individual being supported is valued and cared for (Barnes and Duck, 1994, among others). Conventional social support typically takes place in small intimate dyadic personal relationships and in organized semistructured support groups.

Since the advent of the Internet, many support groups meet online (e.g. for individuals with special health problems, learning disabilities, or family members with special needs and interests). People with larger social circles tend to report a higher sense of social support, since they tend to have more people they can call on for emotional and logistical aid.

Significant life changes whether positive (such as getting married, having a child) or negative (such as a death in the family) are often associated with stress. Even positive changes, like a job promotion, typically require some adjustments and a period of transition before a person is back into a daily routine. As such, these life changes are associated with some level of anxiety and stress for varying lengths of time. Staying in close communication with friends and family and drawing on other forms of social support can play an important role in easing transitions and reducing stress associated with significant life changes (Wellman, 1992). Various studies have found that the Internet has sometimes been a factor in facilitating communication and support during these circumstances (Kraut et al., 2002; Walther and Boyd, 2002).

9.4 Methodology

We used a subset of our EPIC survey data to investigate the relationship between social ties and online communication. The larger EPIC questionnaire investigated the effect of the Internet upon the local community in six areas: Community Involvement, Personal Interests, Internet Activities, Collective Efficacy, Social Networks, and Psychological scales. Questions were largely drawn from previous BEV surveys (Patterson and Kavanaugh, 1994; Kavanaugh et al., 2000) and from the HomeNet study (Kraut et al., 1996, 2002). Led by underlying theory derived from the literature and previous instruments, we posited composites (sets of variables) and modified the set members when necessary to ensure reliable alpha levels (see full description of EPIC composites at http://java.cs.vt.edu/~epic). Most composites are each individual's mean of the set. In order to calculate a composite to represent the total number of significant life-changing events a participant reported during the 6 months preceding the survey (positive and/or negative), we calculated the sum of the individual events reported by each individual.

Data for this set of analyses were drawn from the first and second rounds of the study, with data collected between Fall 2001 and Fall 2002. While the sample was stratified at the household level (100 households), all members of each household were asked to complete a survey (adults and children over 10 years old). We focus in this chapter on data collected from adults only (age 17 and older). The unit of analysis is the individual, with 158 adults (18 or older) in the first round and 146 in the second. We had some attrition between rounds one and two, and this also affected some of the analyses we were able to conduct on the social circles data in round two. Our research team hand-delivered and collected the surveys and we paid participants for completing the surveys. In keeping with the earlier studies, we defined Internet users as those who access the Internet from any location and deleted case information for non-Internet users (33 individuals) from this analysis.

The Social Circles section was divided into subsections, "local social circle" and "distant social circle". Within both of these subsections, we asked about close friends, somewhat close friends, and acquaintances. In this chapter we focus on local social circle because we are interested in usage and effects in the geographic community, and contrast them with communication with distant social circles.

We adapted Marsden and Campbell's (1984) three-part measure of closeness into a set of name generator questions, asking participants to estimate the number of close friends, somewhat close friends, and acquaintances in their local and distant social networks. We provided the following definitions in the survey for local and distant social circles: (1) Local: the people you know *not in your immediate family* who live close enough that you could easily visit them in a single day; and (2) Distant: the people you know *not in your immediate family* who live too far away to visit easily in a single day. We provided closeness measures as follows:

- *People close to you*: People with whom you confide, spend a lot of time doing a variety of activities, and have a deep and reciprocated relationship.
- *People somewhat close to you*: People with whom your relationship is not as deep or involved, but who are significant to you and with whom you stay in touch.
- *Your acquaintances*: People you know but to whom you do not feel close and with whom your contact may be infrequent or casual.

For each type of social tie (close, somewhat close, acquaintance) and for each level of proximity (local and distant) we asked respondents about the number of such ties, the percentage of each with whom they exchanged e-mail and with whom they exchanged instant messages.

Additional questions were borrowed from Wellman and Berkowitz (1988) to elaborate on the nature of the social relationships within each closeness subnetwork. For practical reasons, these follow-up questions focused on a single local and a single distant social circle member at any level of closeness (close, somewhat close, acquaintance). We created sets of questions about prompts called "critical incidents" in order to elicit patterns of associations and reports of behavior instead of suppositions. Rather than depend upon participants' judgments of the effect of the Internet on these social ties, we asked participants to consider the last time they chatted online, sent an instant message, or e-mailed someone in their social circle (using separate sections for local and distant ties).

Specifically, we asked respondents about their current relation to the selected tie (e.g. neighbour, friend, coworker), and whether they were a close friend, somewhat close friend, or an acquaintance. What was the purpose of the communication (e.g. no set topic, exchanging information, personal problem, school or work, organize an activity)? We asked about the mode of communication with this person (e-mail, online chat, instant message, or other online format) and whether that mode was the respondents' preferred way of communicating with this person. How effective was this mode in accomplishing the purposes of their communication? How long had they been corresponding online (e.g. since they met, or later once either one started using the Internet)? Finally, how important were online modes of communication for developing or maintaining their relationship with this person? Had the respondent met in person less, more, or about the same, since using the Internet to communicate with this person? Had the respondent talked on the phone less, more, or about the same, since using the Internet with this person?

In a separate section, we asked respondents with how many *new* people had they become close, somewhat close, or acquaintances (local and distant) in the past year, and how instrumental the Internet had been in helping the respondent develop these new relationships. In subsections branching off each type of tie (close, somewhat close, acquaintance) and each proximity (local, distant), we asked about the relationship and communication patterns. If the respondent had not met a new person in the past year that fit a particular category (for example, local close tie), we asked them to skip the section and go to the next subsection in the series: for example, new somewhat close local tie. Respondents only answered questions that were relevant to their circumstances in terms of having met someone in the past year for a particular category for type and tie and proximity.

We asked respondents to classify the selected *new* tie in terms of their current relation to the respondent (e.g. neighbour, coworker, relative, romantic partner), the source of the tie (e.g. friend, school or work associate, fellow member of a voluntary association), the frequency with which they engaged in eight different activities together (discuss community issues, leisure, hobbies, information exchange, political discussion, support or advice, practical favors or help, participation in community events). We asked about the frequency with which they used six various modes of communication (in person, letters, telephone, e-mail, instant message, and online chat) and whether that mode was the respondents' preferred way of communicating with the selected person. We asked about the purpose of the communication (e.g. no set topic, exchanging information, personal problem, school or work, organize an activity). How effective was this mode in accomplishing the purposes of their communication? How long had they been corresponding online (e.g. since they met, or later once either one started using the Internet)? Finally, we asked respondents how important online modes of communication had been in developing their relationship with the selected person.

Had the respondent met in person and/or used the telephone less, more, or about the same, since using the Internet with the selected person? Each respondent completed all of these questions for each type of relation (close, somewhat close, and acquaintance) for both types of proximity (local and distant).

We investigated patterns of online and offline communication with respondents' selected close friends, somewhat close friends, and acquaintances by geographic proximity. We also tested various constructs in the study to identify predicting variables for EPIC respondents' types of relationships and communication modes with different social ties. The independent variables we investigated included demographics, psychological attributes, levels and purposes of Internet use, interests, significant life changes, and social support. In addition to descriptive statistics and paired sample *t* tests, we conducted one-way repeated measures ANOVAs to test the relationship between the eight different activities (within subject factor) and the respondents' ratings on the frequency of each activity (dependent variable). We ran correlations and regression analyses to test the predictive power of various factors in the survey (including demographics, psychological attributes, level of Internet experience and usage, social support, and significant life changes) on social ties and online communication. Pearson correlations and nonparametric association measures were calculated. We conducted most of the statistical analyses on the social circles data for round one only, since about half of the respondents did not answer these questions in round two.

We measured level of Internet usage as simply the number of hours of Internet usage on an average day reported by respondents. We also measured level of Internet use as a binary variable for heavy versus light Internet users. We distinguished between heavy and light users by dividing our participants at the median of 10.5 hours per week (about half of the sample was in each category for both rounds). Our Social Support construct is derived from the average level of agreement (range from 1 to 5) on 12 statements describing if the respondent has received social and emotional support from people with whom she or he is dealing. The Significant Life Changes construct is the sum of significant life changes the respondent marked in a list of 27 potential changes in the past 6 months (for example, the birth of a child, change of job, getting married, a death in the family, and so on). Significant life changes can be positive or negative and are often associated with some level of stress and adjustment.

9.5 Results

The number of close ties, somewhat close ties, and acquaintances that respondents report for each level of proximity (local and distant) depends on the type of tie and the proximity. Specifically, like most people, EPIC respondents report having fewer close ties (8 and 6 on average in rounds one and two, respectively) than somewhat close ties (between 11 and 15, and between 6 and 10 on average in rounds one and two, respectively) and acquaintances (between 26 and 50 on average in both rounds one and two, respectively). They also report having more distant close ties (9 and 8 on average in rounds one and two, respectively) than local ones. Distant ties that are somewhat close and acquaintances are similar in range to local ones.

The percentage of social ties with whom respondents exchanged e-mail also varied by type of tie and proximity. Not surprisingly, respondents report exchanging e-mail with a higher proportion of close ties than with acquaintances, and with a higher proportion of distant than local close ties. Among those ties with whom respondents reported *no e-mail exchange*, the greatest proportion was with distant acquaintances (44 and 39% in rounds one and two, respectively). By contrast, e-mail exchange with *all* of the respondents' local close ties increased from 11 to 18% between rounds one and two. The frequency with which respondents "used the Internet to communicate with friends" was higher (about once a week) for distant friends compared with "less than weekly" for local friends, in both rounds.

9.5.1 Online Communication with Local Versus Distant Social Circle

Of those reporting on the last time they chatted online, sent an instant message, or e-mailed someone in their local and distant social circle, respondents were more likely to report they had contacted *close friends* rather than relatives or acquaintances. Compared to the person in the respondent's distant social circle with whom the respondent had the most recent online communication, the person in his or her local social circle is more likely to be a neighbour or fellow member of a local club or church (Table 9.1).

Table 9.1 Online communication with local versus distant social circle

Variable names/labels	Means		t test (df)
	Local circle	Distant circle	
Current relation of selected person is:			
Neighbour[a]	0.06	0.00	2.29* (89)
Fellow member of local group[a]	0.23	0.09	2.81** (89)
How close the person is[b]	1.78	2.11	−3.08** (89)
Preferred mode of communication[a]	0.41	0.70	−5.24** (87)
Importance of Internet for relationship[c]	2.90	3.66	−4.71** (87)

$*p < 0.05$; $**p < 0.01$. Response categories codes: [a]$0 = $no, $1 = $yes; [b]$1 = $close, $2 = $somewhat close, $3 = $acquaintances; [c]$1 = $completely unimportant, $5 = $extremely important.

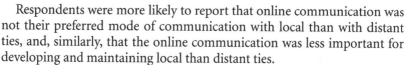
Respondents were more likely to report that online communication was not their preferred mode of communication with local than with distant ties, and, similarly, that the online communication was less important for developing and maintaining local than distant ties.

Regression models show that using the Internet to communicate with friends in one's local social circle is predicted by a linear combination of being female, using the Internet for political purposes, having a larger local social circle, and being involved in the local community ($R^2 = 0.318\%$, or showing these independent variables explain 32% of the variance in online communication). Using the Internet to communicate with *distant friends* is predicted by a linear combination of using the Internet for political purposes, being female, having higher level of education, being extroverted, and using the Internet a greater number of hours per day ($R^2 = 0.348$, or explaining 35% of the variance in online communication). The use of the Internet for political purposes may be acting as a proxy for higher education, although it is not clear.

9.5.2 Activities and Communication with New Social Ties

Most of the new relationships respondents reported making in the previous year were acquaintances – either local acquaintances (18 or 19 people on average) or distant acquaintances (6 or 7 people on average). Fewer relationships had become close. The average number of new close friends was only 2; the average number of somewhat close new friends was about 2 or 3 people. Regression models showed that the number of new people met in the past year who had become part of the respondent's local social circle was explained by a linear combination of having a larger social circle already, age (being younger), and to a lesser extent being involved and interested in politics ($R^2 = 0.389$ and 0.429 for close and somewhat local close friends, respectively). These same variables were also significant in predicting the number of new *distant* relationships ($R^2 = 0.354$ and 0.35 for close and somewhat close new ties). For the number of distant new acquaintances, a linear combination of social circle size, age-related factors such as lower level of education and income and using the Internet for politics explain 45% of the variance ($R^2 = 0.453$). In addition, social support is a lesser, but still significant predictor (change in $R^2 = 0.03$), meaning that people with lower levels of social support report having met a greater number of people in the past year who had become distant acquaintances.

Respondents generally reported that the Internet had not been instrumental in developing new relationships, especially with local close ties. The Internet was just a bit more instrumental (2 on a 5-point scale, with 1 being not at all and 5 being extremely instrumental) for distant ties of all levels of closeness. Regression models showed that the extent to which the Internet was instrumental in developing new relationships was predicted by factors such as using the Internet more heavily, being involved in the local community, and being younger (R^2 ranged from 0.271 to 0.38 for both local and distant social circles. (Please see the EPIC research site for a complete listing of the regression model results: http://java.cs.vt.edu/~epic.)

9.5.2.1 Online Communication with New Ties

When asked to consider one person *whom they met in the past year* who had become *close* with them in their local social circle, most respondents described their current relationship with *new close* ties as friends (57%) followed by coworkers (23%). The predominant source of new relationships that had become *close* was through work or school (45% compared with 39% for acquaintances). "Work or school" appeared as one response category in the survey. Respondents described the current relation of new local *acquaintances* predominantly as being a fellow member of their church, club, or other group (32%) followed by being a neighbour (20%). As such, the source of new *weak* ties (acquaintances) was predominantly through local voluntary associations.

9.5.2.2 Activities with Selected New Social Ties

As would be expected, there were significant differences in the amount of time people spent with *new* close friends versus acquaintances doing various activities. Paired sample *t* tests (Table 9.2) show significant differences

Table 9.2 Frequency (means) of activities with close friends versus acquaintances

	Mean[a]			
	Round 1[b]		Round 2[c]	
Types of activities	With close friends	With acquaintances	With close friends	With acquaintances
1. Leisure activities	3.17	1.69	2.94	1.67
2. Discuss hobbies	3.25	1.91	3.11	2.18
3. Give support or advice	3.92	2.20	3.78	2.25
4. Give/receive favors or help	3.81	2.17	3.47	2.18
5. Exchange information/ideas	4.06	2.57	3.89	2.64
6. Discuss politics	2.47	1.48	2.14	1.49
7. Discuss community issues	2.79	1.83	2.66	1.86
8. Participate in local events	2.08	1.51	2.11	1.37

[a] Mean ratings were calculated from the scale 1 (not at all often) to 5 (very often).
[b] Number of valid cases N range from 142 to 145 for round 1 data.
[c] Number of valid cases N range from 69 to 84 for round 2 data.

Table 9.3 Type and frequency of activity by type of social tie

Tests	Wilks' lambda[a] Λ	f values and p	Partial eta squared η^2
Round 1		$f(7132) = 57.393$	
With close friends	0.247	$p < 0.01$	0.753
		$f(7134) = 25.94$	
With acquaintances	0.425	$p < 0.01$	0.575
Round 2		$f(762) = 28.481$	
With close friends	0.237	$p < 0.01$	0.763
		$f(775) = 17.652$	
With acquaintances	0.378	$p < 0.01$	0.622

[a] Multivariate results are reported.

($p = 0.000$) for all eight activities in the survey, meaning that respondents were likely to do all these eight activities more often with new close friends than with new acquaintances.

One-way repeated measures ANOVA tests show the relationship between the eight different activities (within subject factor) and the respondents' ratings on the frequency of each activity (dependent variable). Although the lower response rate in the second round of these questions (reported by the same respondents) makes it inappropriate to compare across the two rounds for this section of the survey, respondents' ratings on the frequency of eight activities differed consistently for both rounds and for close friends versus acquaintances (Table 9.3).

The ANOVA tests indicate that the differences among the eight means were significant for acquaintances and for close friends in both rounds. Furthermore, the order of the frequency (means) with which respondents report doing the eight activities stays the same in both rounds, with close friends as with acquaintances.

9.5.2.3 Gender Differences in Types of Activities and Online Communication

There were a number of significant gender-based differences in the types of activities and the modes of communication respondents used with different types of ties (Table 9.4). Women engaged in all the following activities with close friends significantly more often than men: participate in leisure activities, discuss hobbies, give or receive support, give practical advice, and exchange information. Since they started using the Internet to communicate with close friends, women were more likely than men to report that their communication with that friend had increased. Reporting on somewhat close friends, women were likely to give or receive

Table 9.4 Activity and communication by type of tie by gender

How often do you:	Mean		t test results and p values
	Male	Female	
1. With local *close* friend: How often do you participate in leisure activities?	2.86	3.42	$t(140) = -2.61$** $p = 0.01$
How often do you discuss hobbies?	2.89	3.54	$t(140) = -3.18$** $p = 0.002$
How often do you give or receive support?	3.55	4.21	$t(142) = -3.89$** $p = 0.000$
How often do you give practical favors or advice?	3.55	4.03	$t(141) = -2.96$** $p = 0.004$
How often do you exchange information?	3.80	4.28	$t(142) = -3.23$** $p = 0.002$
How has communication changed (since using Internet together)	2.27	2.48	[a]$t(93.47) = -2.10$* $p = 0.038$
2. With *somewhat close* friend: How often do you give or receive support?	2.81	3.25	$t(143) = -2.58$** $p = 0.011$
How often do you give practical advice?	2.73	3.09	$t(142) = -2.00$* $p = 0.047$
How often do you discuss politics?	2.11	1.63	$t(142) = 2.65$** $p = 0.009$
3. With *acquaintances*: How often do you discuss politics?	1.68	1.31	[a]$t(135.72) = 2.64$** $p = 0.009$
How often do you organize local events?	1.34	1.64	[a]$t(139.04) = -2.2$* $p = 0.029$
How frequently do you communicate by phone?	2.65	2.01	[a]$t(114.312) = 2.86$** $p = 0.005$

*$p < 0.05$; **$p < 0.01$.
[a]Equal variance is not assumed for t tests.

support and give practical advice more often than men. Men discussed politics with somewhat close friends and acquaintances more often than women. Men communicated by telephone with acquaintances more often than women. Women organized local events with acquaintances more often than men.

Round two showed similar trends, with the addition that women were likely to use more instant messages than men with close friends, and that women were more likely than men to report that the Internet was instrumental for communicating with close friends.

Table 9.5 Communication modes with local versus distant new close ties[a]

Variable	Mean		t (df)
	Local circle	Distant circle	
Frequency of communication ...			
°In person	4.81	1.75	21.12* (125)
°By letter	1.26	1.45	−2.81* (119)
°By telephone	4.28	2.53	10.86* (124)
°Via online chat	1.67	1.18	3.12* (115)

* $p < 0.01$.
[a] Values shown are from round 1.

Table 9.6 Modes of communication with local versus distant new acquaintances[a]

Variable	Mean		t (df)
	Local circle	Distant circle	
Frequency of communication ...			
In person	3.52	1.56	14.22** (119)
	3.44	*1.71*	*7.2** (40)*
By telephone	2.23	1.52	5.29** (116)
	2.00	*1.60*	*2.10* (41)*

* $p < 0.05$; ** $p < 0.01$.
[a] Values from round 2 are in the second row of each cell and in italic.

9.5.2.4 Modes of Communication with Selected New Social Ties

The frequency (means) with which respondents communicate in various *modes* (in person, telephone, e-mail, Internet chat, instant messenger, and letters) also differed by the type of tie and proximity. Communication modes were all significantly higher for local ties (except for letters) and for close friends rather than acquaintances, according to paired *t* tests (Tables 9.5 and 9.6).

Comparisons of frequencies of each communication mode with new *somewhat close* ties in the local versus distant social circle show similar trends as with close ties. Respondents communicate most frequently in person with new somewhat close ties at the local level, followed by telephone, e-mail, and online chat. All modes are significantly higher for local than distant communication, in both rounds one and two (with the exception of online chat in the second round).

For communication with local and distant *acquaintances* (Table 9.4), only in person and telephone communication are significantly higher with local acquaintances. That is, the frequency of respondents' e-mail and other online communication with local and distant acquaintances is about the same *regardless* of proximity.

9.5.3 Predicting Communication: Social Support

We tested specifically for correlations between the construct Social Support and related other variables in our study, such as gender, age, education, income, extroversion, Internet experience and level of usage (hours per day), and activities and interests (Table 9.7). Not surprisingly, higher levels of social support are associated with being younger, extroverted, and having a larger social circle. In addition, greater frequency of interaction through various activities (leisure, discuss hobbies with social ties) is associated with higher levels of social support. While not all variables are significant in both rounds, the results point to a pattern that can guide future research. The correlation between social support and using the Internet to communication with social ties (through instant messaging, e-mail, and even letters) suggests further that online interaction supports and strengthens sociability and interpersonal relationships of varying types, both local and distant.

Other variables that we tested were not significant and are not included in the presentation of results above.

9.6 Discussion

Our survey results indicated that people use various modes of communication to maintain their social networks, both local and distant. The findings show significant differences in online communication based on type of social tie (close, somewhat close, and acquaintances), gender, and type of Internet user (heavy versus light). People are using the Internet to support and strengthen sociability and social interaction, especially women and younger people who are involved in the community and use the Internet more heavily. The findings also suggest that local community is not undermined by Internet use.

In personal interviews with a subset of our survey respondents from 20 households, there were many people who when asked about how they communicate with friends and family locally, many of them said by telephone. They indicated, for example, that they needed to hear more of the expression in a friend's voice or that they wanted to see the friend face to face. It is still easier and more enjoyable for some people just to pick up the phone and chat, or wait until the next time they see their friends in person. Most people communicate with friends and family online, especially with friends at a distance or who have moved away. One interviewee said that when his

Table 9.7   Correlations between social support and selected variables

Variable	Correlation (N)[a]	
	Round 1	Round 2
Age	−0.19*[b]	−0.20*
	(155)	(143)
Extroversion	0.31**	0.30**
	(155)	(136)
With number of close people exchanged	0.20*	NS
instant messages	(138)	
Number of people considered to be close	NS	0.17*
		(136)
Number of people considered to be somewhat close	0.19*	0.20*
	(152)	(137)
Number of new people had become somewhat close	NS	0.20*
		(129)
With number of somewhat close people e-mail	0.20*	NS
exchange	(146)	
With number of somewhat close people instant	0.21*	NS
messages	(138)	
With number of acquaintances exchange instant	0.19*	NS
messages	(137)	
With a selected close person, frequency of:	0.22**	NS
participate in leisure activities	(141)	
discuss hobbies	0.26**	0.37**
	(141)	(70)
give or receive support	0.26**	0.37**
	(143)	(69)
give practical advice	0.22**	NS
	(142)	
exchange information	0.24**	0.40**
	(143)	(70)
With a selected somewhat close person, frequency of:	NS	0.24*
give practical advice		(71)
exchange information	NS	0.24*
		(71)
With an acquaintance, frequency of:	−0.20*	NS
discuss local community issues	(143)	
With a selected close person, frequency of:	0.20*	NS
communication by letter	(133)	
communication via instant message	0.20*	NS
	(131)	

* $p < 0.05$; ** $p < 0.01$.
[a] Spearman Rho values were in italic. Otherwise cell entries are Pearson correlation values. The numbers in parentheses are valid number of cases.
[b] Correlation coefficients are rounded into two decimal places.

friend moved out of town, they continued to meet online to play online blackjack – a game they played in person before he moved.

Some researchers have expressed concern that time spent online might reduce interpersonal networks by stealing time that might otherwise be spent on social interaction. Contrary to those expectations, our study participants indicate that their interpersonal relationships are enhanced by Internet communications. Heavy Internet users report that use of e-mail to communicate increases the amount of face-to-face contact. Additionally, they find the Internet important in maintaining distant relationships. Thus, time spent online is often time spent in social interaction and support.

9.7 Acknowledgments

This research was supported, in part, by the National Science Foundation, IIS 0080864. We would like to thank our collaborators Albert Bandura, Ann Bishop, Daniel Dunlap, Philip Isenhour, Robert Kraut, Debbie Denise Reese, Wendy Schafer, Jennifer Thompson, Steven Winters, and Katie Worley.

9.8 References

Aronson, S. (1971) The sociology of the telephone. *International Journal of Comparative Sociology*, **12**: 153–157.

Ball-Rokeach, S., Gibbs, J., Jung, J.Y., Kim, Y.C. and Qui, J. (2000) The globalization of everyday life: visions and reality (White Paper 2) [Online]. Annenberg School for Communication, Metamorphosis Project, Los Angeles. Retrieved March 24, 2006 from <http://www.metamorph. org/vault/globalization.html>

Barnes, M.K. and Duck, S. (1994) Everyday communicative contexts for social support. In Burleson, B., Albrecht, T. and Sarason, I.G. (eds), *Communication of Social Support: Messages, Interactions, Relationships and Community*. Sage, Thousand Oaks, CA, pp. 175–194.

Boneva, B., Kraut, R. and Frohlich, D. (2001) Using email for personal relationships: the difference gender makes. *American Behavioral Scientist*, **45**(3): 530–549.

Cohill, A. and Kavanaugh, A. (eds) (1997, 2000) *Community Networks: Lessons from Blacksburg, Virginia*, 1st and revised edns. Artech House, Norwood, MA.

Dimmick, J.W., Jaspreet, S. and Patterson, S.J. (1994) The gratifications of the household telephone: sociability, instrumentality, and reassurance. *Communication Research*, **21**(5): 643–661.

Duck, S. (1994) Steady as she goes: relational maintenance as a shared meaning system. In Canary, D. and Stafford, L. (eds), *Communication and Relational Maintenance*. Academic Press, San Diego, CA, pp. 45–60.

Dutton, W.H. (ed.) (1996) *Information and Communication Technologies: Visions and Realities*. Oxford University Press, Oxford.

Ehrich, R. and Kavanaugh, A. (1997) Managing the evolution of a virtual school. In Cohill, A. and Kavanaugh, A. (eds), *Community Networks: Lessons from Blacksburg, Virginia*. Artech House, Norwood, MA, pp. 89–116.

Fischer, C., Jackson, R., Stueve, C.A., Gerson, K., Jones, L. and Baldassare, M. (1977) *Networks and Places: Social Relations in the Urban Setting*. The Free Press, New York.

Fischer, C.S. (1992) *America Calling: A Social History of the Telephone to 1940*. University of California Press, Berkeley, CA.

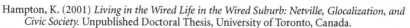

Hampton, K. (2001) *Living in the Wired Life in the Wired Suburb: Netville, Glocalization, and Civic Society.* Unpublished Doctoral Thesis, University of Toronto, Canada.

Howard, P., Rainie, L. and Jones, S. (2001) Days and nights on the Internet: the impact of a diffusing technology. *American Behavioral Scientist,* 45(3): 383–404.

Katz, E. and Aspden, P. (1997) A nation of strangers. *Communications of the ACM,* **40**(12): 81–86.

Katz, E.J. (1999) *Connections: Social and Cultural Studies of the Telephone in America Life.* Transaction, New Brunswick, NJ.

Katz, E.J. and Rice, R.E. (2001) *The Internet and Health Communication: Experiences and Expectations.* Sage, Thousand Oaks, CA.

Kavanaugh, A. (2003) When everyone is wired: the impact of the Internet on families in networked communities. In Turow, J. and Kavanaugh, A. (eds), *The Wired Homestead: An MIT Press Sourcebook on the Internet and the Family.* MIT Press, Cambridge, MA, pp. 423–437.

Kavanaugh, A. and Cohill, A. (1997) Conducting business in a networked community. In Cohill, A. and Kavanaugh, A. (eds), *Community Networks: Lessons from Blacksburg, Virginia,* 1st edn. Artech House, Norwood, MA, pp. 149–158.

Kavanaugh, A., Cohill, A. and Patterson, S. (2000) The use and impact of the Blacksburg Electronic Village. In Cohill, A. and Kavanaugh, A. (eds), *Community Networks: Lessons from Blacksburg, Virginia,* 1st edn. Artech House, Norwood, MA, pp. 77–98.

Kavanaugh, A. and Patterson, S. (2001) American Behavioral Scientist, 45(3): 496–509.

Kavanaugh, A., Reese, D.D., Carroll, J.M. and Rosson, M.B. (2003) In Huysman, M. and Wulf, V. (eds) Communities and Technologies. The Netherlands: Kluwer Academic Publishers, pp. 265–286. (Reprinted 2005 in The Information Society, 21(2): 119–131).

Kavanaugh, A., Carroll, J.M., Rosson, M.B., Zin, T.T. and Reese, D.D. (2005). Community networks: Where offline communities meet online. *Journal of Computer Mediated Communication* 10(4) http://jcmc.indiana.edu/vol10/issue4/kavanaugh.html.

Kraut, R., Kiesler, S., Boneva, B., Cummings, J., Helgeson, V. and Crawford, A. (2002) Internet paradox revisited. *Journal of Social Issues,* **58**(1): 49–74.

Kraut, R., Scherlis, W., Mukhopadhyay, T., Manning, J. and Kiesler, S. (1996) The HomeNet field trial of residential Internet services. *Communications of the ACM,* **39**: 55–63.

Marsden, P.V. and Campbell, K.E. (1984) Measuring tie strength. *Social Forces,* 63(2): 482–501.

Matei, S. and Ball-Rokeach, S. (2001) Real and virtual social ties: connections in the everyday lives of seven ethnic neighbourhoods. *American Behavioral Scientist,* 45(3): 550–564.

Milardo, R.M. (ed.) (1988) *Families and Social Networks.* Sage, Newbury Park, CA.

Nie, N.H. (2001) Sociability, interpersonal relations, and the Internet: reconciling conflicting findings. *American Behavioral Scientist,* 45(3): 420–435.

Patterson, S. and Kavanaugh, A. (1994) Rural Users' Expectations of the Information Superhighway. Media Information Australia, 74 (November): 57–61.

Pew Internet and American Life (2000) Tracking online life: how women use the Internet to cultivate relationships with family and friends. Retrieved from <http://www.pewinternet. org>

Pool, I. de Sola (1983) *Forecasting the Telephone: A Retrospective Technology Assessment.* Ablex Publication, Norwood, NJ.

Rabby, M.L. and Walther, J. (2002) Computer mediated communication effects on relationship formation and maintenance. In Canary, D. and Dainton, M. (eds), *Maintaining Relationships Through Communication: Relational, Contextual, and Cultural Variations.* Lawrence Erlbaum, Mahwah, NJ, pp. 141–162.

Rainie, L. and Kohut, A. (2000) Tracking online life: how women use the Internet to cultivate relationships with family and friends (Internet Life Report 1) Retrieved March 24, 2006 from <http://www.pewinternet.org>

Stafford, L., Kline, S.L. and Dimmick, J. (1999) Home email: relational maintenance and gratification opportunities. *Journal of Broadcasting and Electronic Media,* 43(4): 659–669.

The Reader's Digest (1996) The most wired town in America: a small Virginia town is leading the way to a whole new age. July, pp. 54–58.

Turow, J. and Kavanaugh, A. (eds) (2003) *The Wired Homestead: An MIT Press Sourcebook on Internet and the Family.* MIT Press, Cambridge, MA.

Walther, J. and Boyd, S. (2002) Attraction to computer mediated social support. In Lin, C.A. and Atkins, D. (eds), *Communication Technology and Society: Audience, Adoption and Uses.* Hampton Press, Cresskill, NJ, pp. 153–188.

Wellman, B. (1992) Which types of ties and networks provide what kinds of social support? *Advances in Group Processes,* **9**: 207–235.

Wellman, B. (1997) An electronic group is virtually a social network. In Kiesler, S. (ed.), *Culture of the Internet.* Lawrence Erlbaum, Mahwah, NJ, pp. 179–205.

Wellman, B. and Berkowitz, S.D. (1988) *Social Structures: A Network Approach.* Cambridge University Press, New York.

Wellman, B., Haase, A.Q., Witte, J. and Hampton, K. (2001) Does the Internet increase, decrease or supplement social capital? *American Behavioral Scientist,* **45**(3): 436–455.

Wellman, B., Salaff, J., Dimitrova, D., Garton, L., Gulia, M. and Hythornthwaite, C. (1996) Computer networks as social networks: collaborative work, tele-work, and virtual community. *Annual Review of Sociology,* **22**: 213–239.

Winner, L. (1997) *Autonomous Technology: Technics-out-of-Control as a Theme in Political Thought.* MIT Press, Cambridge, MA.

10

The Magic Lounge: Connecting Island Communities Through Varied Communication Services

Thomas Rist, Niels Ole Bernsen and Jean-Claude Martin

10.1 Preamble

The increasing convergence of networks with a range of today's digital media together with assured public access to a global information and communication infrastructure have transformed the character of both personal and social interactions in today's cyberspace. Given this situation, and in the context of a European research project, we created the so-called Magic Lounge concept in order to investigate the possibilities of having virtual encounters on a future, reliable high-bandwidth, low-cost network infrastructure that would allow combined spoken and text chat exchanges between multiple speakers. Thus, "Magic Lounge" refers to a virtual meeting space where members of geographically dispersed communities can come together to carry out joint meetings and other less formal encounters.

The typical Magic Lounge users are assumed to be groups of lay people with little knowledge of the underlying technology. This is a rather new point of view as most previous work on computer-mediated communication has been dealing with well-defined groups of professional users. Hence, the motivations for entering the Magic Lounge are as diverse as are the interests of the potential users themselves. Some may join just to chat, to make new acquaintances or to carry out joint and goal-directed activities. Others may engage in exchanging and sharing ideas, experience and knowledge on matters that relate to their professions or to their hobby. Yet others may share common cultural or political interests and even use virtual meeting places as arenas of civil discourse.

In our chapter we focus on the question of how emerging technologies generate new, technically complex communication scenarios in which groups of non-professional and sometimes mobile users communicate using multiple modalities and a multitude of devices with very different capabilities. Following a participatory design approach, the research has been conducted in close collaboration with several groups of real users, such as habitants of small and remote Danish islands, and the Youth Service Association of the French town Villejuif. The involvement of these user groups revealed a number of interesting insights on how advanced communication services might be actually used by people with a potential need to collaborate on everyday tasks and group activities.

10.2 Introduction

A basic prerequisite for any living community is the ability of its members to communicate with one another. Traditionally, we tend to think of communities as being local in nature, consisting of the neighbours, the school, the shops, the sports clubs, the pub, etc. In most local communities, communication is a simple matter of walking over, meeting in the street or turning up where and when community members meet. Thinking about the communities to which we belong, however, it turns out that most of us belong, or would like to belong, to communities that are much more geographically distributed than that. The national society for naval history, the political party, the grass-roots movement, the colleagues at different branches of the company are all examples of widely distributed communities, not to mention international communities of like-minded enthusiasts. Moreover, the number of distributed communities to which one would like to belong is not just a function of one's interests but also of where one lives. For instance, big cities cater for more communitarian needs than do remote rural areas or even small islands.

Today, telephone, mail and e-mail are the standard communication channels available to the members of geographically dispersed communities. The emergent fusion of telecommunication and information technologies and the emergence of portable, connected and ubiquitous computing devices are about to remove current limitations, such as unimodal (see Box 10.1) communication channels-only, bilateral connections-only and stationary communication partners. Chat corners on the Internet, whilst still unimodal (using text-only communication) and stationary, are already enabling multiparty text chat communication. Future advanced virtual meeting places on the Internet will remove the unimodal constraint and offer mobile access as well. It may be envisioned, in other words, that future communities will be able to connect through modality-rich connections even when some or all of their members are on the move.

Before such systems are able to spread as widely as they evidently deserve, several issues need to be addressed. For instance, despite optimistic expectations, most users are still waiting to get reliable high-bandwidth, low-cost network infrastructure. Similarly, it remains difficult to exploit

Box 10.1   On modalities

The term "multimodal system" is becoming widespread in the world of computers and telecommunication technologies. A modality is simply a way of presenting or exchanging information at the human–computer interface. Thus, multimodal systems are systems that use several modalities for this purpose. The component modalities are called unimodal modalities for information (re-) presentation and exchange. Examples of unimodal modalities are graphical images, written text, speech, non-speech sounds, data-graphics, haptic or touch (Braille) written text, etc. It is important to distinguish between input (from the user) and output (from the system) modalities. Modality Theory is the theory of the properties of modalities (Bernsen, 2001).

---

multi-speaker connections over the Internet. In addition to issues of base technology and infrastructure, there is the multi-facetted question of how the new opportunities could be exploited by the various user communities out there. The emerging technologies generate new, technically complex communication scenarios in which groups of non-professional and some-times mobile users communicate using multiple modalities and a multitude of devices with very different properties, such as speech-only (traditional telephone), speech and small screen (mobile phone, personal digital assistant or PDA), or speech, large screen, keyboard and mouse (workstation). To hide as far as possible the technical complexity from the communication participants, and to enhance naturalness and intuitiveness of use, new intelligent communication services are required which non-intrusively "broker" between different devices and modalities. Moreover, issues of actual use and user preferences are in need of theoretical or experimental clarification. For instance, how will people use combined text chat and speech if both modalities are available? Under which circumstances will they prefer text chat over speech or vice versa? How will the style of communication change when people use communication devices with different bandwidth and different modalities for exchanging information? Or, to help a user who has missed a previous meeting, might it be possible and useful to provide structured views of the exchanges that took place in that meeting? Answers to these questions are needed to guide development of new communication systems that could make life easier for existing, and possibly remote, communities and foster the formation of new communities even when their members live far from one another.

The Magic Lounge concept was created in order to investigate the possibilities of having virtual encounters on a future, reliable high-bandwidth, low-cost network infrastructure that would allow combined spoken and text chat exchanges between multiple speakers. Thus, "Magic Lounge" refers

© Odense Turist Bur

Figure 10.1   Summer on the scenic island of Funen; Faaborg from the sea.

to a virtual meeting space where members of geographically dispersed communities can come together to carry out joint meetings and other less formal encounters.

Initially, we aimed at the broad user population and wanted to involve people from that population from the beginning and thereafter throughout the system development process. The typical Magic Lounge users are assumed to be groups of lay people with little knowledge of the underlying technology but with a potential need to collaborate on everyday tasks and group activities. This is a rather new point of view as most previous work on computer-mediated communication has been dealing with well-defined groups of professional users. Hence, the motivations for entering the Magic Lounge are as diverse as are the interests of the potential users themselves. For instance, a core group of users of the system are people from small and remote Danish islands including, for instance, the scenic island of Funen (cf. Figure 10.1). Many of these islanders wish to meet in virtual space to exchange and share ideas, experience and knowledge on matters relating to their hobbies or professions, or to problems in their local communities. A French version of the Magic Lounge system has been tested by a small team of young women and men from the Youth Service of Villejuif, an association concerned with social problems in the suburbs of large French towns. Last but not least, the Magic Lounge developer team itself constituted a distributed community par excellence as the team members were located in different cities in several European countries. This made it obvious to use the Magic Lounge for discussions on software development and other project-related matters during our work.

During the first year of work, a catalogue of desirable features of virtual meeting spaces was compiled in close cooperation with the user group of people from the smaller Danish isles. This group was recruited early on with help from the Association of the Smaller Danish Isles. Given our technical goals, it is perhaps not surprising that the group was all-male, consisting of six 30–50-year-old amateur computer enthusiasts from all over the Danish island archipelago of more than 200 islands: rocky Bornholm in the Baltic Sea, closer to Sweden than to Denmark, isolated Anholt in the middle of the Kattegat, Ærø in the mild and beautiful archipelago south of Fuenen and Årø and Tunø off Jutland, so small that few Danes even know that these islands exist. In the Danish, "ø" parsimoniously serves as the standard noun meaning "island". Professionally, the Magic Lounge islanders were otherwise as different as can be, from a bank manager and a tourist office manager to a garbage collector, a male nurse and a sailor working in the local ferry company. Incidentally, the garbage collector from the user group is now in training as IT software instructor, perhaps as a result of the encouraging experience of being a very constructive member of the Magic Lounge user group for 3 years.

We wanted the user group members to envision the kinds of activities for which they would use a system that allows them to meet in virtual space. Their feedback clearly demonstrated the conviction that such a system might serve as a means to overcome the physical and social distance between themselves and the rest of the world. More particularly, the rich information produced by the system's would-be users included use of the system for carrying out (a) professional activities, such as solving problems in local fishery and agriculture, meeting with customers, distance working, teaching and training; (b) hobby-related activities, such as remote participation in club meetings and sharing collections with others; and (c) community-related activities, such as remote participation in church contact forum meetings, labour union meetings and political meetings, tele-shopping and communicating with public administrations (Bernsen and Dybkjær, 1998b).

This wealth of ideas reflects the underlying reality of living in remote locations in today's world. For instance, Erik, a bank manager living on the island of Ærø, is a keen student of naval history. However, most meetings in the Danish Society for Naval History take place on workday evenings in Copenhagen, more than 3 hours away by boat and train from Ærø. With Magic Lounge, Erik could participate to all those meetings without leaving his island! It is straightforward to generalise Erik's predicament: shopping, business contacts, social activities and contacts all become much more complex enterprises when the nearest city is several hours away. On the other hand, life on a beautiful but remote island carries its own rewards, which of course is why there is a predicament in the first place. Emerging technologies promise unprecedented opportunities for combining remote living with being part of many different virtual communities.

Figure 10.2 Magic Lounge system setting. A group of geographically dispersed people attend a meeting in the Magic Lounge using heterogeneous communication devices.

To optimally support all those activities (a through c above) would require an omnipotent communication and collaboration tool which, of course, was beyond what could be developed within our relatively small project. Our team, therefore, had to focus system development on a limited number of communication services taking into account criteria such as added value for the users, potential for commercial exploitation and feasibility with regard to the project resources available. The focusing strategy led to an overall system design that adopts a client–server architecture (cf. Figure 10.2) and includes the following key features:

- Support of interleaved text and audio communication
- Provision of a structured meeting memory
- Access through heterogeneous communication devices, such as workstations, PDAs or mobile phones.

As illustrated in Figure 10.2, users (technically called "clients") can register either by PC or mobile phone with Internet connection to the Magic Lounge system. The heart of the system is formed by a centralised component (technically called "server") that is responsible for the distribution of

text and audio messages to the clients. In addition, this component keeps a record of all messages as the basis for the Magic Lounge "meeting memory". In the following sections, we describe these key features of the Magic Lounge system in more detail and then report on the feedback gained from the system's users.

10.3 Text Chat and Spoken Communication in the Magic Lounge

There is an ongoing discussion among experts as to whether text-based communication, such as text chat or short messaging, will become less important or even disappear when better services for audio conferencing via the Internet and the mobile nets become widely available and cheaper. For instance, some providers of Internet telephony services have already set up what they call "phone-chat corners" in which Internet users can verbally chat with each other. In Magic Lounge we were particularly interested in exploring how ordinary people will use combined multi-user text chat and audio conferencing. This aspect of our common telecommunications future, although predictable from current technology trends, represents an important but neglected topic in research on collaborative technologies. One probable reason is that, although predictable, this aspect has been massively overshadowed by the prospects of video conferencing and videophones.

10.3.1 Text Communication

A central component of the Magic Lounge system is a multi-user tool for exchanging text messages. During the past decade, text chat systems have become popular among online communities. In a conventional text chat system, a message is basically a string of characters that does not receive any further interpretation by the system. Such systems provide little support for structuring conversations and collaborations.

In developing the text chat part of Magic Lounge, we wanted to go beyond mere exchange of text messages. An important idea was that a structured communication framework would be needed to realise many of the envisaged memory and communication support functions. It was decided to experimentally adopt a framework based on conversations, communicative acts and the notion of referable objects (objects to which one can refer). An exchange of messages among the meeting participants on a certain topic might be clustered to form a *conversation*. The notion of *communicative acts* (see Box 10.2) serves to distinguish between the different kinds of activities

---

> **Box 10.2   Communicative acts**
>
> Communicative acts have been used in the fields of speech and natural language processing, in particular for the development of systems that conduct spoken or written natural language dialogue with users (Bernsen et al., 1998). The idea is that if the system is not only able to capture the particular utterance made by the user but also the act(s) that the user intends to perform in producing the utterance, the system will gain additional high-level information on the communicative by the sender of a particular message. The labels might facilitate identification of intention(s) of the user, information which can be used to improve its processing of the utterance and ultimately help the system produce more adequate responses during the dialogue. In Magic Lounge we decided to use performative verbs for labelling messages according to the act performed key messages that refer to
> informing about a subject matter e.g.,
> "I would like to tell you that ... "
> making a proposal, e.g., *"I suggest to ... "*
> posing a question, e.g., *"May I ask ... "*
> agreeing on a proposal, e.g., *"I agree to ... "*
> objecting to a proposal, e.g., *"I disagree with ... "*

(or acts) that a particular text message may represent. Based on this notion, a more detailed internal structure of the messages exchanged by the participants can be defined. For instance, labels such as *accept* or *inform* can be used to explicitly mark the intended communicative act of issuing a particular message. By means of these labels it is possible to reconstruct the flow of activity between the clients (humans and system components) at a higher level of abstraction, allowing the users to gain an overview on the structure of a whole conversation and to recognise various relations that may exist between individual messages.

Ideally, the Magic Lounge system would automatically extract or derive the appropriate label from the messages entered by the users. For the time being, however, automated label extraction is only possible in very restricted domains. To avoid overly restricting communication in the Magic Lounge, it was decided to pass on the labelling task to the users by providing an interface for message labelling. As shown in Figure 10.3, the text chat tool interface does not have a send button as in standard text chat systems. Rather, to send a message, the user has to choose a button labelled with a particular communicative act.

Selecting the recipients of a particular message is another way of adding meta-information to a text chat message. In Magic Lounge, although all logged users receive all messages sent to the server, specifying one or several recipients helps with *floor control* in meetings by emphasising who

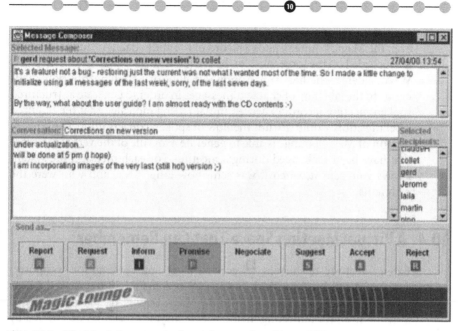

Figure 10.3 The Magic Lounge text chat message composition window on a PC. The user must click on a keyword label to send a message marked with the corresponding communicative act.

among the logged participants is being particularly addressed by a certain message.

A third aspect of the Magic Lounge text messaging structure is the notion of a conversation. A conversation is defined by its name as provided by the user, and denotes a set of messages sharing the same topic or aiming at the same goal, such as a conversation about making a hotel reservation. Each text message exchanged in the Magic Lounge includes a reference to the conversation to which it belongs.

10.3.2 Spoken Communication

The second communication tool of the Magic Lounge system builds on existing technology for multi-party audio conferencing on connected computers. Magic Lounge incorporates a tool for audio communication over the Internet that has been developed by the Network Research Group of the Information and Computing Sciences Division at Lawrence Berkeley National Laboratory. In order to achieve smooth integration of this audio tool into the Magic Lounge system, a new graphical user interface has been developed which has the same look-and-feel as all other system components.

Similar to a telephone with conferencing facilities, the audio tool enables the users to send and receive verbal messages either bilaterally or broadcast to all connected participants. By contrast with telephone conferencing, however, the Magic Lounge builds a record of the audio data. All audio messages are recorded on the server and can be retrieved and replayed after a meeting. We considered whether to incorporate labelling of verbal contributions analogous to the labelling of chat text messages in order to generate a structured spoken meeting record. However, initial testing indicated that the required labelling effort would disrupt the flow of spoken conversation. The current version of Magic Lounge is able to generate a log-file of the verbal messages that have been exchanged during a meeting. For each message, the system knows who sent it, when it was sent, how long it was and who were the recipients.

10.4 A Virtual Meeting Space That Can Remember

Extensive study of current teleconferencing and systems for computer-supported collaborative work (or CSCW systems for short) as well as our own participatory design work with the Magic Lounge user groups revealed a strong need for a meeting memory that may, for instance, be queried by newcomers or latecomers who want to know what has happened in the meeting so far.

In a telecommunications context, the telephone answering machine is an example of an add-on service that, to a modest extent, at least, relates to the concept of remembering. Although useful for many purposes, the mere recording of audio streams is a rather low-level memory function that, moreover, has the drawback of allowing tedious sequential access only.

We were particularly interested in memory functions which go beyond the mere recording and storage of data, and which may support the participants of virtual meetings. Our hypothesis was that a variety of added-value communication services might emerge from a system's capability to memorise information units of the kinds that are actually being exchanged by people in virtual meeting spaces. Moreover, a system like the Magic Lounge could serve as a collective memory for an electronically connected community. This collective memory may be conceived as an information system based on a growing pool of information and knowledge produced or collected by the community members. To be used effectively, the information system should provide functions for the collection, structuring, conservation, retrieval, distribution and customised dynamic presentation of information contents. Information structure presentation, in particular, requires that links among messages be made explicit to users querying the memory.

In the following we sketch a few scenarios to illustrate how the inhabitants of virtual meeting places might benefit from memory services as addressed by the Magic Lounge system. For the purpose of illustration the scenarios draw on three fictive persons, Susan, Lars and Hank, who once met by chance

in the Magic Lounge, eventually teamed up and since then usually enter the Lounge after leaving off work to see what is going on there. Throughout the scenarios we will use the abbreviation VMM to refer to a virtual meeting memory, and the terms "Magic Lounge" and "Lounge" will be used also as synonyms for a virtual meeting place.

Scenario 1: *Briefing oneself before joining an ongoing conversation*

Susan, Lars and Hank know each other from a number of previous meetings in the Magic Lounge. As they have become friends over the time, they plan to meet in the real world too, and therefore try to find a good place to do so. Susan and Lars already started to work on this joint travel-planning task a few days ago, unfortunately without Hank who was on a business trip. Today Hank is also connected to the Magic Lounge. When entering the Lounge, he gets informed about who else is there, and the other inhabitants of the Lounge are made aware of Hank being there. Susan greets Hank and informs him that she is searching the Web for a meeting place. Furthermore, she tells Hank that he may consult the VMM to get informed what she and Lars have discovered and discussed during the last sessions. Though not designed especially to support travelling planning, the VMM provides Hank with information that allows him to get an overview on the current state of the planning task at hand. For example, Hank can query the VMM which messages were exchanged between the inhabitants of the virtual meeting place, and use traditional e-mail functions such as sorting items according to various criteria, such as the temporal order of occurrence, the sender or addressee(s), or the type of media (e.g. text, voice and graphics); or such as plain text searching on the content of exchanged messages. Also, the VMM keeps track of information units, which are brought in from external sources, such as a web page. Also, Hank can browse through the record of messages. However, there are more advanced memory functions, which will help Hank to get an overview much faster. It is important to note that Hank can choose among several different views when accessing memory contents. Some of these views are basic in the sense that they do not require deep processing of the information units; e.g., a certain view may just list contributions to a conversation on a time axis. Other views on memory contents, however, can only be realised if contents and intention of a message are understood to a certain extent by the system.

Scenario 2: *Accessing the meeting memory during a conversation*

While the first scenario intended to illustrate how a VMM can be exploited by a "late-comer", the second scenario focuses on intelligent assistance for Magic Lounge users during a meeting, i.e. while they communicate with each other or perform activities, such as web surfing. Suppose, for instance, Susan has a web browser, which can be controlled using speech input. A spoken command, such as "*Show me again the web page with the hotels in London that Hank has found,"* can only be successfully executed by her web browser if context information is available concerning the current meeting or even a previous meeting. Besides the mere provision of such context information on demand by the user, however, the VMM can take on a more active role, too. There are often situations in which a user would appreciate a helpful comment or a hint without asking for it explicitly. For example, while Susan moves from one web page to the other, the VMM notices that Hank has visited some of these pages just a few minutes before. In order to make Susan aware of this fact, and to provide her with the additional option to follow Hank's navigation path through these web pages, her browser now changes the appearance of exactly those hyperlinks that were selected by Hank and displays a brief message to inform her about the situation and the new option. In this case the VMM provides information for the recognition of a situation in which a suggestion would be appropriate as well as the trace of Hank's previous navigation path. After a while, Susan finds a list of some recommended restaurants, and would like to know which of those Hank and Lars would prefer. Unfortunately, Lars has already left the Lounge. While discussing to which restaurant they will go, again the Magic Lounge system recognises that there is a situation in which the conversation partner should be provided with a comment. This time, it reminds Susan and Hank about Lars preference for Asian food that he stated in an earlier meeting. Further services of this kind include remembering stated arguments, previously made agreements as well as remembering the participants' goals that have been inferred from their communications and actions.

Scenario 3: *Assisting users when using heterogeneous communication devices*
For the purpose of the third scenario, let us assume Lars is walking through a shopping promenade while being connected with the Lounge through his mobile phone. While walking along the street, he passes a travel agency that offers a cheap flight and accommodation package to a place he would like to go with his friends. Lars takes the chance to inform his friends about the information he found. In response to this, Susan and Hank both sitting at their PCs at home start a search on the World Wide Web in order to check how good the offer actually is. Later at home, Lars will not only find all the references to the pages found by his friends but will also be able to trace back the conversation they had on the issue. This scenario illustrates that a VMM should abstract from the actual devices used for communication.

As illustrated by the scenarios above it should be kept in mind that the design of memory functions is strongly dependent upon the particular purposes for which they will be used. Today, for instance, larger enterprises take an increasing interest in so-called organisational memories. They are motivated by the idea that productivity can be increased if employees could draw on the past experiences of their colleagues, rather than

repeating past mistakes or having to rediscover past solutions. Company organisational memories may incorporate, for instance, a company's current understanding of its business processes, work flow and information flow. Moreover, a company often already has certain guidelines for the format and media in which to represent the relevant information. The Magic Lounge system, on the other hand, is a general-purpose communication tool, which makes it less obvious which kinds of memory function to provide. However, the Magic Lounge memory concept is based on the following key ideas:

- Memory content emerges from communication and related activities, such as accessing information sources and making them available to other communication partners.

- There is no best way to structure memory contents. Rather, it should be possible to view memory content from different perspectives corresponding to different cooperative tasks. To this end, it should be possible to reorganise memory contents on the fly.

- Memory content may emerge from the activity of a single user as well as from collaborative activity.

- In terms of content and media format, memory entries can be as diverse as the contents and formats of the messages and information units available to Magic Lounge users.

- Access to memory contents can be restricted at various levels to ensure privacy needs.

The Magic Lounge conversational memory thus includes spoken and text chat utterances as well as other interaction events, such as exchanged references to electronic documents, database access references and results or accessed Web links. In addition, the memory includes asynchronously produced items, such as a meeting agenda that has been uploaded by a single participant before a virtual meeting. As a whole, the memory constitutes the story of the virtual community as made up of synchronous and asynchronous activities. Contributions are recorded in the memory together with structural information, such as sender and recipient(s), the declared communicative act and the conversation to which a contribution belongs.

In designing the meeting memory, a central hypothesis is that the memory should support the user's cognitive activity of remembering. In this view, it is the user who remembers while the system only provides assistance in the process, e.g. by providing various views on information collected by the system. For instance, when running on a PC the Magic Lounge offers the meeting text chat memory viewing tools shown in Figures 10.4 and 10.5. These viewers complement each other, having been tailored for different information needs.

The *Message Viewer* (cf. back screenshot of Figure 10.4) is the basic tool for displaying text chat contributions. Messages can be viewed in three different styles. The screenshot of the Message Viewer in Figure 10.4 shows the display

Figure 10.4 Two of the Magic Lounge tools to access the meeting memory using a PC. Screenshot in the back: *Message Viewer*. Front screenshot: *Tree Viewer*.

of messages as a simple list in chronological order. Each list item has a header that provides the following information about the message:

- the type of the communicative act;
- sender and recipient(s), or *all* if no specific recipient is specified;
- name of the conversation to which the message is a contribution;
- time stamp indicating when the message was sent.

As a further alternative to the chronological display of messages, the user may choose the *Tree View* (cf. front screenshot shown in Figure 10.4) for inspecting relationships between messages and responses to messages. The Tree View is well suited for inspecting hierarchical structures of turns in conversations. It allows the user to trace who contributed what to whom in a conversation. For instance, the hard copy of the Tree View window shows an excerpt of a conversation between two persons, Gerd and Jerôme, one trying to assist the other in a software installation task. The Tree View allows the user to browse quickly through the messages by using the scroll bar

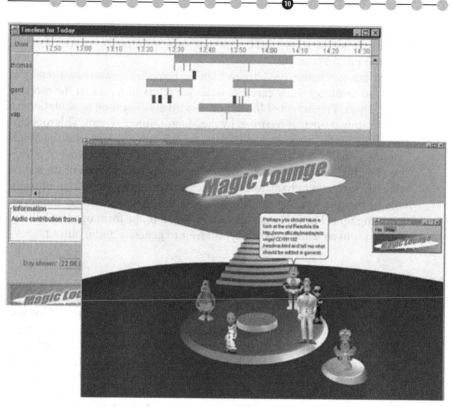

Figure 10.5 Two further tools to access the Magic Lounge meeting memory using a PC. Screenshot in the back: *The Temporal Meeting Browser*. Front screenshot: *The Activity Monitor* for the replay of conversations with animated avatars.

on the main frame and view a message in detail by clicking on its header, causing its full content to be displayed on the bottom frame.

In addition, the Magic Lounge memory tools offer a search function for memory inspection. Search operations can filter the memory to show all messages sent by a specific user, using a particular communicative act, having a particular recipient or belonging to a particular conversation. For instance, a user might be interested in seeing all contributions made by a certain communication partner.

For a more graphical access to the meeting memory the user can choose the *Temporal Meeting Browser* (cf. screenshot of Figure 10.5). This viewer organises all audio and text contributions on a timeline, also called a track, according to their temporal order of occurrence. There is a separate track for each participant and each track is divided into three sub-tracks for the different kinds of events (login intervals, text chat messages, spoken contributions). The track representation is similar to musical score notation. The Temporal Meeting Browser allows the user to navigate back and forth through recorded meetings and inspect individual contributions in a non-linear manner. Messages can be selected for inspection by clicking on them with the mouse pointer.

Last but not least, the *Activity Monitor* provides an animated replay of recorded meetings using cartoon-style characters to represent the meeting participants. The Activity Monitor offers visually appealing presentations of conversational turns as recorded by the Magic Lounge system. This tool employs a dedicated animated commentator character to represent the Magic Lounge system itself. In the example shown in the front screenshot of Figure 10.5, the "Magic Agent" is located in the lower right area of the screen while the five other characters in the central area represent different users. One of them is uttering the text of a message that was sent by the corresponding user. The message is not only shown in the form of a text balloon but also spoken using a speech synthesiser to generate audio output.

10.5 Getting Mobile Users into the Magic Lounge

Unlike many other approaches, such as text chat rooms on the Web or systems for CSCW, we do not assume that users necessarily enter the Magic Lounge via their PCs. Rather, we envisage scenarios in which users may use quite different devices to access the virtual meeting space (cf. Rist, 2000, 2001). In the scenario illustrated in Figure 10.2, two users are at home accessing the Magic Lounge through their PCs while a third user connects through a mobile phone which, in addition to its audio channel, has a tiny LCD display for showing short text messages and minimalist image graphics. This scenario takes into account that people spend considerable amounts of time travelling and commuting, as well as current rapid progress in the development of new portable computing and communication devices, including PDAs, palm computers and mobile phones with built-in microcomputers. With a mobile communication device, community members could use the time spent on travelling to engage in community activities in the Magic Lounge.

In extending Magic Lounge capabilities to mobile devices, it is essential to bear in mind that these devices are very different from each other and from the PC in terms of output and input capabilities. Limited screen space, lack of high resolution and colour, no support for audio and video are among the typical restrictions on the output site, whereas input restrictions may result from miniaturised keyboards and graphical user interface widgets, tiny physical control elements and sparse capabilities for the capture of voice and video input.

Currently, mobile access to the Magic Lounge is through mobile phone based on WAP (wireless application protocol). Our commitment to WAP technology has meant severe limitations with respect to system functionality and user interface design. On the other hand, the choice of WAP technology

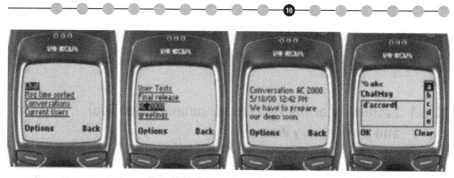

Figure 10.6 Magic Lounge mobile phone WAP interface.

has made it possible to do system evaluation using real devices during the lifetime of the project, as WAP-enabled mobile phones have become available from several manufacturers, including Nokia, Siemens, Ericsson, Alcatel, Motorola and others.

Figure 10.6 shows interaction with the WAP browser of a mobile phone. When entering the Magic Lounge, the services available, such as querying the meeting memory or writing and receiving text chat messages, are listed and can be started by selecting them (first and second screen of Figure 10.6). Magic Lounge memory contents can be presented as text (third screen of Figure 10.6) or via the audio channel. Text chat messages may also be dispatched (fourth screen of Figure 10.6).

10.6 Feedback from the Users

Following a participatory design approach (Bernsen and Dybkjær, 1998a–c; Masoodian and Cleal, 1999; Henry et al., 2000), representatives of potential user groups were recruited and involved in Magic Lounge development from its very beginning. When a sufficiently stable prototype was released after the second year, the users from the Danish isles started testing the system and continued to do so for each new prototype release. To observe and video record the users while using the system, and to systematically collect user feedback through questionnaires and interviews, the islanders attended a series of trial workshops in which they tested the system's technical, functional and usability aspects whilst performing various collaborative tasks. In addition, the evaluation studies carried out with the islanders helped to investigate how the users would choose text chat, spoken interaction and combinations of both modalities when solving everyday tasks, such as organising a summer party or solving Web search tasks. User studies were also conducted with the French user group. These studies had a special focus on evaluating the text chat tool and the memory viewers. Yet another user study was carried out in order to evaluate the usability and user acceptance of mobile access to the Magic Lounge system. In the sequel we report on the outcome of theses evaluation studies.

10.6.1 Evaluation of the Text Chat Communication Tool and the Memory Viewers

As system developers we were strongly interested in studying how our users would work with the text chat communication tool and the various memory viewers when collaborating on a common task. Would they be able to take advantage of the communicative acts labels and would they mind the extra labelling effort? Feedback on this issue was received from tests done at the French and the Danish partner sites.

The French user group was formed by a small team of young women and men from the Youth Service of Villejuif association. As perhaps typical for many interest groups within associations, working together is based on joint interests only and the relationships among the group members are not constrained by an organisational hierarchy. Another characteristic of this particular group was their limited background in computer technology, the group members being occasional users of computers for writing text documents and sending e-mails.

Two sessions were held in the laboratory at separate dates, a discovery session followed by a working session 2 weeks later. In the discovery session, the users received an introduction to the system and had a first opportunity to familiarise themselves with the communication tools and the memory viewers. The purpose of the working session was to observe the users while solving several specific collaborative tasks, such as joint information search on the Web, the organisation of a cultural festival and a travel-planning task. In addition to this task variation, we also varied the communicative setting. In one condition, all users were connected to the system at the same time; in another, some users missed part of a meeting and only joined an ongoing session when it had been underway for some time.

The outcome of the test sessions includes feedback gathered from the participants by means of questionnaires and interviews. In addition, the rich log-files of message exchanges gathered by the Magic Lounge system formed the basis for a statistical analysis. For instance, from the log-file one could see how often a certain message label was used, or in which sequence messages of a certain type occurred most frequently.

As expected, the ability to obtain a record of a past meeting or meeting part was greatly appreciated, especially by users who had to join ongoing meetings. In the test session with the Villejuif user group, the Topic Viewer was among the preferred means of accessing the memory record. An explanation for this preference may be the observation that conversations in

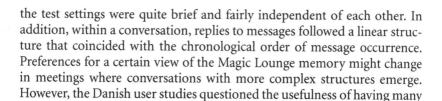
the test settings were quite brief and fairly independent of each other. In addition, within a conversation, replies to messages followed a linear structure that coincided with the chronological order of message occurrence. Preferences for a certain view of the Magic Lounge memory might change in meetings where conversations with more complex structures emerge. However, the Danish user studies questioned the usefulness of having many different view options available, as users have difficulty relocating a message when switching from one view to another.

Several users suggested that the meeting memory should provide a means to inform about the agreements and decisions made in a meeting. While in business meetings it is common to take minutes and explicitly protocol decisions, the situation is quite different in informal sessions. Thus, it is often not clear at the start of the session that any decision making will be done at all, and in the absence of organisational meeting role assignments, people may be reluctant to volunteer as minute takers. At this point it is interesting to ask whether it is possible to infer from the structure of a conversation whether or not agreements were made. In fact, the test session with the Villejuif user group showed strong correlation between the occurrence of a *suggest* message followed by an *accept* message, and the achievement of an agreement between the senders of the corresponding messages. That is, the users were consistent in using the *suggest* and *accept* labels for sending their messages, and a *suggest* message followed by an *accept* message appeared to reliably indicate that an agreement had been reached.

Being forced to label a message with a communicative acts label in order to be able to send the message required some familiarisation effort. One of the users who joined a meeting halfway very much appreciated the labelled information when browsing the meeting memory: "It's a guide, it's a help for structuring thinking before typing." Another user, who was reluctant to use the communicative acts in the discovery session, came to appreciate the labelling of messages during the working session where the usefulness of having a structured view of the memory seemed more obvious to him. Nevertheless, declaring a communicative act for each message remained a heavy constraint for him. Similar problems with distinguishing communicative acts were reported from the tests carried out with the Danish isles user group. In fact, the Danish trials and their analysis have produced a devastating criticism of the use of communicative acts labelling in the Magic Lounge. Thus, users have no way of distinguishing, e.g., *inform* from *report*, or *negotiate* from *suggest*. The set of communicative acts labels offered by Magic Lounge is theoretically arbitrary and has no foundation in the scientific analysis of discourse. Thus, it is impossible to estimate today if general-purpose communication would need 5, 15, 50 or even more labels. The implication is that not even experts will be able to use the Magic Lounge labels in a way which is both systematic and non-arbitrary. No known labelling schemes for communicative acts are general-purpose.

Those schemes that are in actual use are deeply task-dependent, and none of them can be used with anything resembling 100% consistency. And, in Magic Lounge as well as more generally, communication participants often contribute more than one communicative act in a single contribution to conversation. However, Magic Lounge offers one and only one label per text chat contribution. The following results of the French study illustrate these conclusions well.

Thus, in a post-session interview two of the Villejuif users stressed the difficulty of being obliged to select a single communicative act for a long message to which several communicative acts could actually be related. In fact, assuming a one-to-one relationship between a message and a communicative act appears to be an impossible restriction, as users would either have to ignore the labelling system and re-express their thoughts or even split them into a sequence of labelled messages just for the sake of complying with an impossible labelling scheme.

Analysing the log-file containing all messages from the working meeting, we observed great diversity of communicative acts labels during the user test. Yet the labels *reject* and *promise* had not been used at all. The omission of the *reject* label may be related to standards of politeness in the user group. To avoid expressing direct opposition to a proposal, a user would rather *suggest* an alternative or just *inform* that other solutions might exist. On the other hand, the users were reluctant to make strong commitments. This is concluded from the fact that the *promise* label did not occur in the log-file while the *accept* label was used frequently to underscore a commitment.

Our guess is that the use of communicative acts labels depends on a number of factors, including, among others, the personalities of the participating users, the social structure of the group, the tasks to be solved and the topics addressed in conversation. For instance, analysis of the log-files of meetings held among the Magic Lounge developers showed that all labels were used albeit with varying frequency. The great majority (83%) of messages were labelled *accept, inform, request* or *suggest*, while the labels *negotiate, offer, promise, reject* and *report* were used quite sparsely. However, the goal of our evaluation work was not to identify the ultimate set of communicative acts that would benefit all potential user groups solving arbitrary tasks in diverse situational contexts. In fact, apart from introducing the labels we did not constrain their usage, e.g. by having the system perform consistency checks and the like. Thus, users might view the different labels simply as devices allowing them to classify messages according to any group-specific agreement on how to semantically interpret the labels. While this possibility was not considered by any of the user groups, some sessions did include messages on making consultations about the "proper" usage of the labels when sending messages. Although from a linguistic point of view many labels may not well match the actual message content, the user groups did try to use them in a consistent way.

10.6.2 Evaluation of the Use of Speech Versus the Use of Text Chat

A series of trial sessions were conducted with users from the smaller Danish isles user community. The trials involved three different user populations in terms of computer literacy, gender and familiarity with the Magic Lounge software: the Magic Lounge user group from the smaller Danish isles, NIS-Lab administrative staff and the NISLab team of Magic Lounge developers. The trial sessions address different tasks, ranging from website design and web browsing tasks to party planning and Magic Lounge assessments using different methodologies. All tests involved a setup in which users used workstations or similarly equipped portables, thus making available the text chat tool, the audio communication tool and the memory viewers. The audio tracks (a total of about 5 hours of audio data) and text chat records (a total of 205 messages) of all meetings were recorded and stored in the meeting memory. Parts of most sessions were recorded on video as well. The data were subsequently put on CD-ROMs to facilitate analysis. Analysis reveals interesting observations on the use of text chat, speech and spoken dialogue, or both in combination during small-group virtual meetings.

The data consistently show that it is impossible from the chat track alone to obtain a full overview of the topics discussed on the audio track and hence in the meeting as a whole. The audio record generally contains substantial discussions of topics that are completely absent from the text chat record, thus providing a very different perspective on the meeting from the one shown in the text chat record. If a topic raised in the spoken dialogue lies outside the meeting task proper, it is often not reflected in the text chat record at all, no matter whether the topic is the making of a joke, negotiation of how to organise the meeting, the occurrence of a technical problem or usability problem with the Magic Lounge, a meta-comment on a text chat message, comments made during joint web browsing, new ideas to be explored after the session or an important but inconclusive discussion of the Magic Lounge software. Also, if a spoken remark or exchange adds a finer point to the discussion, such points are often not found in the text chat record, presumably because of the difficulties of keeping up with the pace of the spoken conversation when doing on-line text chat meeting minutes. Even major points concerning the task are sometimes absent from the text chat record for the same reason. Finally, lengthy audio discussions often only generate, if anything, a single brief text chat message summarising the conclusion.

Conversely, substantial points may be found in the text chat record although hardly mentioned in the audio record, perhaps because everybody saw that particular text chat message and simply agreed with it. The result sometimes is sophisticated "dual-tasking". For instance, when participants were in agreement with what was in the growing text chat record, they did not comment on it but chose to discuss other issues instead.

Thus, both the audio and the text chat records tend to be used economically, just in different ways. The difference is that, in task-oriented dialogues, the text chat record tends to be goal-oriented and parsimonious: if an audio exchange is not within the agreed scope of meeting minuting (or meeting purpose), it does not get reflected in the text chat record. The audio record, on the other hand, is where to find the unplanned contributions and topics that may sometimes be just as important as the planned ones.

The consistently different roles of text chat and speech found in the Danish data would seem to explain why the audio tracks in our data invariably contain vastly more turns and words than the corresponding text chat records. The following list of observations from the Danish evaluation sessions supports the conclusions made in this section:

- People active on the text chat record are often less active on the audio record, and vice versa.
- Audio helps to quickly unravel the confusion generated by cross-purpose chat messages.
- Users may ignore text chat to save screen real estate for other applications. For instance, text chat is in trouble during joint web browsing.
- Text chat is used for exchange of information which has to be correct and which should not be misunderstood. Speech, on the other hand, was the leading element of communication, making it "live".

The conclusion seems to be that a proper meeting record needs to include both the audio and the text chat records. The text chat record tends to be narrow and focused on the essentials of establishing contact and creating meeting minutes. The audio track tends to be where most of the activity takes place, including many activities that are absent from the chat record. For this reason, the chat record is a poor indicator of the activity of individual participants. To gauge that, both records are needed.

Text chat and speech have different and complementary roles in the kinds of small-group meetings studied in the Danish Magic Lounge trials. Speech is used the most by far and generally appears to be the preferred modality of communication between the meeting participants. In a combined audio/text context in which users have to solve a common task, chat is being used for

- exchanging initial greetings and sending test messages to make sure that the system works and/or that one has understood how to use the system properly;
- keeping a record of task decision points and other important issues.

On the other hand, chat is not being used for

- discussing problems of how to operate the software;
- guiding a user who has problems in, e.g., finding a particular web page;

- discussing sub-task contributions made in text chat;
- discussions in general;
- joking together about the task.

Speech was mainly used for the following purposes:

- to check the audio connection and adjust the volume;
- to exchange initial greetings and farewells at the end of the session;
- to discuss technical problems, usability problems and Magic Lounge functionality issues, including advice on when to call in an assistant, guidance of another participant and discussion of where to view messages (at the very beginning);
- to agree on how to approach the task;
- to task-related discussions, such as websites to visit and which summer house to select;
- to discuss the (seriously meant) sub-task contributions distributed as text messages and
- to make jokes and frivolous remarks about the task domain and the task options.

10.6.3 Evaluation of Mobile Access to the Magic Lounge

Additional evaluation work aimed at gathering feedback on a series of questions and hypotheses related to mobile phone client access to the Magic Lounge. More specifically, we started off with the following questions and hypotheses in mind:

- How difficult would it be for users to access the Magic Lounge via mobile phone, especially if they have not used a WAP-enabled phone before? How much training would be necessary before the mobile client could be used as an everyday communication tool?
- Based on our own experience with mobile phones in general and WAP-enabled phones in particular, we hypothesised that people will have difficulties handling WAP phones regardless of the particular application. Therefore, one of our concerns was that people might get frustrated and blame the application (i.e. the mobile Magic Lounge client).
- Based on their experience with the mobile client during the tests, would the users be able to imagine scenarios in which the mobile client would be useful to them?

The tests were made on a small scale using the devices that had been purchased for system development. Therefore, the studies should be considered as initial evaluation work that provides the necessary input for the formulation of more precise hypotheses to be investigated in a broader field test.

To assess task performance in an objective manner, we identified a number of metrics, such as the total time the user needed to complete a task, the amount of assistance needed and the user's actual sequence of physical device manipulations compared to the minimal sequence of device manipulations required for successful task completion. In addition, we prepared a small questionnaire to collect subjective feedback from the users. For instance, we wanted them to assess how difficult it was to connect to the Magic Lounge, inspect the meeting memory and write a text chat message. We also asked them for which purpose, if any, they would use their mobile device to access a virtual meeting, taking into account their personal, both professional and private, situation.

The evaluation was done with a group of nine people who participated in one or two specific test sessions. While the small group size does not permit the drawing of conclusions of any statistical significance, the gained observations helped to identify those parts of the tasks that were the most difficult to perform. Also, the comments made during the interviews as well as the observations made during the tests provide valuable input for the design of larger field tests aimed at evaluating new WAP services.

In summary, our major observations were

Using the mobile Magic Lounge client seems to be a relatively easy task. We expected inexperienced users to have severe difficulties handling the mobile devices. This seems to be the case. However, we were positively surprised that only little training appears necessary to drastically improve task performance in terms of time required, need for assistance and deviations from the minimal interaction sequence. In addition, the test subjects experienced the exercise as being a bit laborious but otherwise relatively easy to perform. Our concern that users would blame the mobile client because of inconveniences in the overall handling of the mobile device was not supported at all. On the contrary, most test candidates indicated interest in using such a system to, e.g., brief themselves before joining a meeting.

The current interaction design of the mobile client is not optimal. The records of the test sessions as well as the subjective responses from the subjects revealed several questionable aspects of the current interaction structure. Unfortunately, most of the problems encountered are due to the particular mobile phone implementations. For instance, the quality of the displays will certainly improve very fast, and download time will decrease as well. Another serious problem is that we are currently very much restricted in having to over-define the assignments of function keys. As a consequence, any interaction design for an application will be constrained by already assigned functions. This can be very confusing as it is not always obvious whether a certain option is part of the specific application or whether it is related to some other phone functionality, such as the phone's address

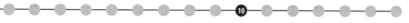
book. Fortunately, some of these problems will certainly become obsolete when switching to the next generation of mobile devices.

If text input is required, it must become more comfortable to use. As expected, text editing on a tiny number keyboard requires some training and is experienced as being quite inconvenient. In addition, most test subjects experienced text editing as being unnecessarily complicated. Unfortunately, the implementations of current WAP-browsers do not allow the introduction of a shortcut so that text entries can be dispatched by pressing a single button when editing has been completed. Although only three text chat sessions were observed, it seems obvious that the current implementation of the text chat function is quite uncomfortable to use. In the design of a revised version of the text chat function, the following issues need to be addressed:

- The additional effort in having to assign conversation names to text chat messages and declare a message's communicative act as required by the current Magic Lounge version is too inconvenient and should be avoided.
- Rather than forcing the user to periodically inspect the message list, there should be a notification (e.g. a beep) when new messages arrive.
- To reply to a message can become difficult since after switching to the editing mode one can no longer see the message to which one wants to reply. At least, it should be possible to switch more easily back and forth between the reading mode and the editing mode.

Finally, we would like to note that a number of factors need to be considered when aiming at broader use of the Magic Lounge mobile access client. Some of these factors are technical in nature, such as inconvenient delays in user–system interaction because of network delays, or the lack of options for designing appealing information presentations for display on the mobile device. Also, whether a system like the Magic Lounge mobile client will be used extensively will depend on the service fees demanded. Even if the Magic Lounge server is made available as a free service for members of a community, the users of the mobile client still have to pay fees to telecoms for using the WAP service, and such fees were relatively high at the time when our tests were made. However, rapid progress in the mobile telephone and computing market raises the hope that mobile access to virtual meeting places will become affordable to the general public soon.

10.7 Acknowledgment

We would like to thank all former members of the Magic Lounge project team: Patrick Brandmeier, Dominique Boullier, Christophe Collet, Laila Dybkjær, Christian Hauck, Claude Henry, Gerd Herzog, Saturnino Luz, Masood Masoodian, Françoise Néel, Peter Rist, David Roy, Jean Schweitzer,

Jérôme Vapillon. We gratefully acknowledge the grant from the EC's former Esprit Long-Term Research division that made our work on the Magic Lounge possible.

10.8 Further Information

For further information, please contact the authors or visit us on the World Wide Web at http://www.dfki.de/imedia/mlounge/

10.9 References

Bernsen, N.O. (2001) Multimodality in language and speech systems – from theory to design support tool. In Granström, B. (ed.), *Multimodality Language and Speech Systems*. Kluwer Academic Publishers, Dordrecht, pp. 93–149.

Bernsen, N.O. and Dybkjær, L. (1998a) Dimensions of virtual co-presence. In Darses, F. and Zaraté, P. (eds), *Proceedings of COOP'98, the Third International Conference on the Design of Cooperative Systems*, Cannes, 1998. Institut National de Recherche en Informatique et en Automatique, Sophia Antipolis, France, pp. 103–106.

Bernsen, N.O. and Dybkjær, L. (1998b) Participatory specification and design ideas from the Danish isles. In *Esprit Long-Term Research Project Magic Lounge Deliverable D4.1a*. Odense University, Denmark, pp. 4–28.

Bernsen, N.O. and Dybkjær, L. (1998c) Bernsen, N.O.: The users' Magic Lounge. In *Esprit Long-Term Research Project Magic Lounge Deliverable D4.1a*. Odense University, Denmark, 1998, pp. 29–45.

Bernsen, N.O., Dybkjær, H. and Dybkjær, L. (1998) *Designing Interactive Speech Systems. From First Ideas to User Testing*. Springer-Verlag, Berlin.

Henry, C., Vapillon, J., Collet, C. and Martin, J.C. (2000) Supporting virtual meetings for collective activities: user-centred design in the Magic Lounge project. In *Proceedings of the i3 Annual Conference*, Jönköping, Sweden, 2000.

Masoodian, M. and Cleal, B. (1999) User-centred design of a virtual meeting environment for ordinary people. In *Proceedings of HCI International'99, Eighth International Conference on Human–Computer Interaction*, Munich, Germany, 1999, Vol. 2, pp. 528–532.

Rist, T. (2000) Using mobile communication devices to access virtual meeting spaces. *Journal of Personal Technologies*, 4(2, special issue), 182–190.

Rist, T. (2001) Towards services that enable ubiquitous access to virtual communication spaces. In Stephanidis, C. (ed.), *Universal Access*. Lawrence Erlbaum, London, pp. 105–108.

10.10 Other Selected Publications of the Magic Lounge Project

Bernsen, N.O. and Dybkjær, L. (November 1997) Questionnaire 1. Eliciting participatory specification and design ideas from the Danish isles. *Magic Lounge Working Paper*, WP41-2-2.html. [Please contact the authors.]

Bernsen, N.O. and Dybkjær, L. (1998) Towards a general characterisation scheme for inhabited information spaces. In *Esprit Long-Term Research Project Magic Lounge Deliverable D1.1*. Odense University, Denmark, pp. 37–43. [Please contact the authors.]

Bernsen, N.O., Rist, T., Martin, J.C., Hauck, C., Boullier, D., Briffault, X., Dybkjaer, L., Henry, C., Masoodian, M., Néel, F., Profitlich, H.J., André, E., Schweitzer, J. and Vapillon, J. (1998) Magic Lounge: a thematic inhabited information space with "intelligent" communication services. In Rault, J.-C. (ed.), *La Lettre de l'Intelligence Artificielle, Proceedings*

of the International Conference on Complex Systems, Intelligent Systems, and Interfaces (*NIMES'98*), Nimes, France, pp. 188–192.

Cleal, B., Bernsen, N.O., Dybkjær, L. and Masoodian, M. (1998) Combined virtual and local meetings in Magic Lounge. In Bernsen, N.O. and Bertelsen, M. (eds), *Abstracts from the First i3 Annual Conference*, Nyborg, Denmark. Odense University, Denmark: The European Network for Intelligent Information Interfaces, p. 7.

Cleal, B., Masoodian, M., Bernsen, N.O. and Dybkjær, L. (1999) Meeting in the Magic Lounge. In Brewster, S., Cawsey, A. and Cockton, G. (eds), *Proceedings of the Seventh International Federation for Information Processing (IFIP) TC13 International Conference on Human-Computer Interaction (INTERACT'99)*, British Computer Society Press, Edinburgh, Scotland, Vol. 2, pp. 73–74 [plus a 10 minute video].

Dybkjær, L. and Bernsen, N.O. (1997) Specification of the Magic Lounge November demonstrator 1997. Odense University, Denmark. [Please contact the authors.]

Dybkjær, L. and Bernsen, N.O. (1998a) Magic Lounge architecture. A first outline. In *Esprit Long-Term Research Project Magic Lounge Year 1 Deliverable D1.1*. Odense University, Odense, Denmark, pp. 15–23.

Dybkjær, L. and Bernsen, N.O. (1998b) Magic Lounge architecture. Scenarios and use cases. In *Esprit Long-Term Research Project Magic Lounge Deliverable D1.1*. Odense University, Denmark, pp. 4–14.

Luz, S.F. and Roy, D.M. (1999) Meeting browser: a system for visualising and accessing audio in multicast meetings. In *Proceedings of the International Workshop on Multimedia Signal Processing*. IEEE Signal Processing Society.

Martin, J.C. and Néel, F. (1998) Speech and gesture interaction for graphical representations: theoretical and software aspects. In *Proceedings of the ECAI'98 Workshop on Combining AI and Graphics for the Interface of the Future*, Brighton, UK.

Martin, J.C., Néel, F., Vapillon, J., Henry, C., Bernsen, N.O. and Rist, T. (1999) Redefinition of the Magic Lounge concept with regard to exploitation possibilities and study of related work. In *Esprit Long-Term Research Project Magic Lounge Year 2 Deliverable D2*. 18 pp.

Masoodian, M. and Bernsen, N.O. (1998) User-centred design of the Magic Lounge. In Greenberg, S. and Neuwirth, C.M. (eds), *Proceedings of the ACM Conference on Computer Supported Cooperative Work, CSCW'98*, Workshop on User-Centred Design in Practice – Problems and Possibilities, Seattle, 1998, The Association of Computing Machinery, New York, p. 417.

Masoodian, M., Bernsen, N.O., Dybkjær, L. (1998) Issues in Magic Lounge software development. In Bernsen, N.O. and Bertelsen, M. (eds), *Abstracts from the First i3 Annual Conference*, Nyborg, Denmark. The European Network for Intelligent Information Interfaces, Odense University, Denmark, p. 7.

Rist, T., Bernsen, N.O., Boullier, D., Briffault, X., Dybkjær, L., Hauck, C., Henry, C., Luz, S., Martin, J.C., Masoodian, M., Néel, F., Profitlich, H.J. and Vapillon, J. (1999) Magic Lounge: a virtual communication space with a structured memory. In Caenepeel, M., Benyon, D. and Smith, D. (eds), *Proceedings of Community of the Future, Second i3 Annual Conference*, Siena, 1999. The Human Communication Research Centre, Edinburgh, Scotland, pp. 157–159.

Rist, T., Martin, J.C., Néel, F. and Vapillon, J. (2000) On the design of intelligent memory functions for virtual meeting places: examining potential benefits and requirements. *Le Travail Humain*, **63**(3, special issue), 203–225.

11

The Digital Hug: Enhancing Emotional Communication by Creative Scenarios

Verena Seibert-Giller, Manfred Tscheligi, Reinhard Sefelin and Anu Kankainen

11.1 Preamble

Informal communication – sharing jokes, expressing feelings and catching up what has happened during a day – is an important part of family life. The MAYPOLE project explored this kind of family communication and developed new product concepts and devices to facilitate such communications. The projects highlights were the developed concepts on the one hand, but equally innovative and worth reporting were the methods that were developed to identify and evaluate the product requirements and specifications. Also worth mentioning is the multidimensional project team, consisting of sociologists, psychologists, designers and engineers from both the academic and industrial world.

In this chapter we will tell the story of the project. But we will not go into detail about product concepts; we will rather tell you about the approach we chose, the puzzle of (game-like) methods that we developed and applied in order to find out more about kids' communications in their social network. Get the impression of our innovative, multidisciplinary journey!

11.2 What It Was All About . . .

Or: why did we do it?

Technology is becoming smaller, cheaper and more widespread rapidly. Devices, which once were perceived as expensive and just for the technically adept, can nowadays be found in most children's rucksacks and pockets. Their usage has changed completely, even though the functions and features of the devices have not changed to that extent. Users, peers or groups adapt the usage of such devices to their needs, not always following the developers or designers ideas of how to use a product.

The special "communication requirements" of children and their social environment were of particular interest for us. Whereby it should be mentioned at this point that talking about "requirements" might be misleading, because most of the communication we studied was related to informal chatting and grooming rather than functional, target-oriented information exchange.

The overall idea and focus of the project was to study communications, as described above, and develop innovative device concepts and/or prototypes, which support them, and carry out intense tests and field studies to verify the concepts and ideas. The targeted device concepts were highly innovative, especially because they targeted communication, which had so far not really been supported by any electronic means. Therefore, we also focused on the development of innovative methods for both studying and testing them.

11.3 Learning Your Users Ways

Or: how do we know what users really want?

The first thing we had to do in order to get started was to get to know our users, the children and their ways. Knowing "their ways" seems to be just the right expression because we were not targeting any specific characteristics or requirements; we expected to learn what we needed to know while carrying out various analysis activities. This was just as challenging as it sounds! We were leaping into an enormous pool of potentially necessary data, knowledge, information and impressions. And especially with children this pool is unfathomably deep and you can find interesting pieces of information everywhere.

So we started activities to collect information about children and their social networks, focusing on the children and people who play a major role in their daily lives such as family, friends or teachers. Our main interest in these networks was to learn more about communication patterns and behaviours, especially regarding informal communications such as chatting and grooming. Also we investigated the motivation for such communications, e.g. "Why do children communicate and what are the contents and patterns of their communicational activities?" or "How do the communications of children vary from adults?" So we were confronted with a long list of interesting questions and challenges. The overall challenge was to identify or develop ways and means to collect valid and durable answers.

So we developed a mosaic of methods and applied them in two different communities in Austria and Finland. The community in Austria included a group of kids, their families and their school environment (teachers, pals, etc.). The community in Finland was built around a scouts group. The kids in both communities were about 10–14 years of age, both boys and girls.

The mosaic of innovative methods turned out to be vital for the study of heterogeneous user groups, especially since single methods were not able to cover the complexity and breadth of the project. We have chosen a couple of methods that we will describe in a bit more detail, so that you can get an impression of what kind of methods we are talking about. Each method should be seen as a piece in a huge puzzle: only when you have all the pieces, you can get the picture right! So in case you wonder about some details or questions, please be aware that you miss out some pieces.

11.3.1 Photo-Story Competition

The idea behind this method was to learn about the kids' values and interests (which would very likely also be part of their daily communication). The method described below, which turned out to be a tremendous success amongst the kids, provided us lots of fun and proved to be highly valuable regarding the insights that we won.

We distributed 31 disposable cameras (sponsored by Kodak) to a school class of 12-year olds. The kids were told to take pictures of whatever they liked, and they could keep a copy of each picture for themselves. The only requirement we had for them was that they should use these pictures for the "photo-story competition" (Figure 11.1). In this competition, their peers (four selected kids) would rate the photo stories and the winner would get a price. We also told them that we would not provide guidelines or criteria on which the rating should take place. The kids would rate the photo stories by whatever criteria they found appropriate. These rules ensured that children would do their best to impress their classmates – not the researchers. This was a very important aspect of the competition, because when we afterwards discussed the results with the kids (why did you find this one best?) we could identify their emotional values and motivations.

So a week after distributing the cameras (and having the pictures developed), we gathered all the kids in a classroom, gave them lots of handcraft stuff (paper, pens, scissors, stickers, etc.) and asked them each to create a photo story with their photos and those materials. Then we carried out the described rating and last but not least we discussed the results with all the kids ("jurors" as well as "creators"). The winner got cinema tickets.

Analysing these photo stories, the photo-story images (Figure 11.2) and the discussion we gathered lots of insights about the artefacts and subjects that mattered to the kids and scenes or happenings that were interesting and special from their perspective. For example one kid had a picture of a boy kissing a girl. This picture was treated treasure-like and no kid would have ever told us about that scene if we had "just asked." The interesting fact about that particular picture was to analyse what made

© CURE

Figure 11.1 The photo-story competition.

it special. In fact it was the mystery about it that made it special, that all kids wanted to see the picture and the owner was therefore "in." As soon as the picture would have been copied numerous times, its "value" for the owner(s) would have decreased massively. So we found an interesting input for a respective electronic device: should the device allow photos (which people would take with it) to be copied? Even if peopled wished that, would it possibly decrease the emotional value of the device (photos?). And how does this perception vary between kids and adults?

We did not really solve these questions, but the photo story identified them as being important. We would not have even thought about them otherwise.

© CURE

Figure 11.2 Photo-story images.

11.3.2 May Market

As another informal way of gathering information, we set up, what they called, a "May Market" (Figure 11.3). The challenge of this method was to learn about the ideas kids had for product concepts and functions.

Three pairs of children, from the ages of 7–9, were invited to define an imaginary product of their own taste and to draw up an advertising slogan for it. First, the children chose their preferred shape from a number of computer toy mock-ups. We then gave them some play money and invited them to "buy" certain functions. The options were displayed on picture cards pinned on a board. They included things like a virtual pet, taking a photo and a romantic fortune-teller.

Once the children had spent all their money on functions, the researchers asked them to think up ways to advertise the product. The girls found it difficult to think abstractly about their imaginary product. In each case they ended up with a product bearing a close resemblance to a mobile phone. The boys (Riku and Joona) found the situation more inspiring and let their imagination rip.

Riku: And you can send notes to your friends.
Joona: That'd be a laugh. Suppose we sent a message to the teacher telling him, "You're fired! Have a nice day. Signed: the school principal."
Riku: You can use it to make videos. I'd put it in Mom's pocket or hide it in the kitchen and spy on her.
Joona: You can play games like Ice Hockey on it, too.

(The boys hold their mock-ups with both hands and pressed imaginary buttons with both thumbs.)

Prior to conducting the method it was difficult to predict any outcome. And also it has to be considered that by offering certain forms and functions

© HIIT

Figure 11.3 "The May Market."

we have already preselected a range of possibilities. Kids might have created totally different things without that selection. But other methods filled this gap.

One important outcome was still that kids love the idea of sending digital images, and even more if they can edit or annotate them.

11.3.3 Checking Out Existing Products

Existing products were also evaluated with children in order to get a better understanding on what kind of features they liked or disliked in a hand-held device targeted specially for them. For example, two groups of boys and girls (9–13 years) were given Game Boy Cameras and Printers that were just released on the market.

The Game Boy Camera was an accessory for the Nintendo Game Boy. It could be used for shooting and saving black-and-white digital images (Figure 11.4) and panorama pictures. Stamps, frames and text as well as animations could be attached to them. The self-made images could be printed out as stickers through the Game Boy Printer. The children participating in the evaluation were given the devices, a brief tutoring and a translation of

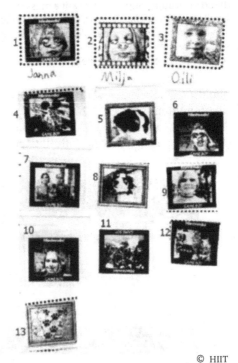

© HIIT

Figure 11.4 Digital images printed out as stickers by children who participated in Game Boy Camera and Printer evaluation.

the basic features of the manual. A week later, they were invited to return the devices and tell about their experiences and to present the images they had created.

All children who participated in the evaluation enjoyed editing the pictures with silly stamps, such as adding ugly monsters or beautiful princess eyes. Moreover, the boys enjoyed the activity of taking pictures by spying on people, play-acting stories and staging silly pictures. They gave away most of their printed pictures. They also spent a lot of time together exploring the product features and possibilities. The girls preferred to take pictures of their family, friends and pets, and they traded pictures with each other. The best pictures were printed out and kept as treasures, glued on a notebook or pencil case.

11.3.4 Scenario Validation – Role Playing

Right at the beginning of the project, we also developed first scenarios of typical communication situations or communication problems, as we thought they might be "typical" for kids. These scenarios were visualised through storyboards Figure 11.5 and were presented to the different groups of kids. On the one hand these scenarios pictured typical situations of children's real life, and on the other hand they also presented sketches of devices that solved communication problems or that enhanced play-like situations.

The idea for doing this was to check whether our ideas of kids' communications were realistic and where they failed. We therefore confronted the kids with these scenarios in various ways, one of them being a role-playing game.

In these role-playing sessions we asked a couple of kids to play scenarios. They were given descriptions only of their specific role and also some different possibilities to (re)act, following respective sketches. In the playing, kids switched roles (e.g. from parent to child) and lived scenarios from various perspectives. In principle that worked very well, although there was a tendency of getting trivial solutions, or solutions that were too "wild."

© HIIT

Figure 11.5   One picture of a storyboard.

One exemplary learning from these role-playing games was about the acceptance of communicational means; e.g., for parents a direct link that forwards information directly to their kid is regarded as highly valuable, whereas all the other included user groups would not accept such a direct link at all.

11.4 What We Had Learned So Far

Or: endless data, valid conclusions?

With the variety of activities that we carried out, we had gathered a huge amount of data. We had lots of scenarios about kids' daily lives, their wishes for devices as well as ideas that could be considered to be "fun" and what was solely "functional." The challenge therefore was to identify combinations of such characteristics, which would serve various purposes within the group of users (kids and their social network).

From the data we had collected, we could not identify one concept or device that would be "it." We had found numerous aspects, functions and characteristics, which were interesting and conceivable. See the mosaic below, which visualises just a selection of the potentials we identified, and learn about the challenge we were focusing on at that point. And consider that we did not want to develop a straightforward functional tool, we had the target to really be innovative. And also consider the time when the project took place – 1996 when mobile phones were still rather rare or used for business.

Play music	Send text messages to friends	Talk to many friends at once	Draw pictures	Make photos	Find my way	Tell parents I am ok	Learn who has same interests	See next scout meetings dates
Tell grandma a story	Show mom a picture of the shoes I want to buy	Tell mom where I am	Make new friends	Spy on friends	Check whether kids are ok	Nobody shall be able to see where I am if I don't want that	Send photos	Share picture only with best friend
See when next bus goes	Check where I am	Share secrets	Learn who is near me	Play computer games	Check appointment	Follow class when I am ill	Create funny pictures	Talk to a friend

So out of these various requirements and ideas we started developing product concepts. Some of these concepts we disregarded rather rapidly, and others were prototyped as paper or physical mock-ups. Finally, one was prototyped as a functional, electronic prototype (hardware and software). Many concepts were tested in various ways (laboratory, focus groups, free confrontations, etc.), and the final functional prototype was tested intensely in two (longitudinal) field trials.

11.5 Device Concepts and Ideas

The variety of concepts and ideas was broad, our fantasies and ideas nearly endless. In many creative sessions, teams of all project partners developed conceptual devices. To give you an impression of the variety of the concepts, let us describe some very diverse ideas.

11.5.1 Atoon

Atoon is a "product" that has no direct communicational functionality that can be used or applied. It was developed by IDEO Europe within the project. It is a product that reacts (in a predefined way) to certain contextual aspects. Atoon can be worn around one's neck, can be put in a pocket or elsewhere near its "owner." It plays music/rhythms. Depending on a variety of contextual aspects (see below) the tune and rhythm changes. The person wearing it can thereby get contextual information. Possible contextual aspects were height, temperature, time (e.g. be home in 15, 10...minutes), closeness to another Atoon (e.g. friend), closeness to another physical object/place (e.g. cinema). Also a set of "tools" was provided, which enabled the users to change the rhythm or the tune of the music. One would only have to touch ones Atoon with the respective tool, and the settings would change. This concept was prototyped as hardware mock-up and checked out with kids. And they loved it!

11.5.2 PIX

PIX is the concept that was concretely prototyped (hardware and software) by NOKIA for the field trials. It is a device that solely allowed taking, editing and sending pictures.
PIX facilitated the users in

- taking and saving still digital images (jpeg-files),
- diting images,
- making series of images,
- blending two images into one,
- attaching a melody to an image and
- sending and receiving images over wireless network (GSM) between the prototypes that were not interoperable with other devices.

The sender could choose either one or more receivers.
The prototypes of PIX were actually functioning, in that the users could take pictures, edit and send them to other PIXs. But to make them work

(and develop the prototypes within a respective budget), we had to put some technology in a rucksack that the user had to carry.

The functions were limited to those operations in order to ensure that the research focus would be limited to communication with digital images. The possibility to use text or audio would have biased the results. We wanted to know more about the advantages and disadvantages of picture communication compared to the communication afforded by phones or pagers. Therefore, we needed a device whose functions would be reduced to picture taking, editing, sending and receiving.

11.6 The Field Trials

Or: pinning down the concept of PIX.

The field trials of the PIX prototypes took place in Vienna and Helsinki. Two different groups of people were invited to participate in the trials: a group of four Finnish boys (12 years) and a family in Vienna. The Finnish boys were all neighbours. Moreover, they were classmates and had similar hobbies. The Viennese family included mother, father, and four children (two boys and two girls, 8–15-years old) and their grandmother. The grandma lived in Vienna, but not with the family as such.

The prototypes were given to the trial user groups for 4 weeks each. We instructed the participants about the usage, the technical backbone, etc., to the extent to which they needed to use the PIXs. Also they were given telephone numbers that they could always call for help or assistance. And finally – but very important – we told them what kind of data our technical logging would provide us, so that they knew how to protect their privacy.

In order to learn as much as possible about how, when and why the participants used PIX, we visited them once a week in their home and interviewed them. Interviews were carried out partially in groups, and partially with individuals. As we had logged all the pictures they had taken and sent (to the PIX server), we also discussed those with them.

In order to give you an impression of how the different persons used PIX, we have collected some samples of usage scenarios. They also demonstrate that the participating user groups (grandparents, parents, children) used the device quite differently.

11.6.1 Creating Stories

The children loved to create stories with series of images. The stories were fictional. They were used to joke or illustrate movie-like scenes. Creating a story was interactive in a way that the children set up the scene for a story by themselves. They created the settings (e.g. blood or fried chicken on a table) and acted the situations by themselves. One of the stories was, e.g., a murder scene where one boy acted as the murderer and the other as the victim. Most of the images used in storytelling were not edited afterwards, but the Finnish boys wished that they could have had more editing possibilities, such as changing the background. Manipulation of the background would have allowed them to create easily new contexts for a story.

The image stories required a lot of work, and the children were proud of their masterpieces. They shared the stories with others by showing them directly from the screen or by sending them over the network.

11.6.2 Expressing Spirituality

The possibility of image editing Figure 11.6 was a new and exciting experience for the grandmother. She could create art with the prototype and shared her creations with her grandchildren. "Now I understand my grandchildren playing with computer toys all the time," she said. She wished that the prototype would include more ways of image editing.

11.6.3 Expressing Affection

For the boys in Finland, pets and girls from their school they liked were the main motives. Those pictures are a good example of images that can only be understood in the context of knowledge about their special context. For neutral observers they were just pictures of dogs or girls. For the recipients, however, they were of great importance.

One Finnish boy created a series of images showing a kissing mouth and a girl. It should say that the receiver of the series is in love with this girl.

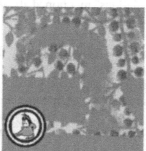

© Maypole Consortium

Figure 11.6   Images created by the grandmother using the implemented modes for picture editing.

11.6.4 Increasing or Maintaining Group Cohesion

One of the purposes of sending images seemed to be maintaining group linkages and friendship between group members. For example, a boy who was not an active member of the Finnish boy group when the trial started sent pictures of himself and of his dog at home to the others. He also received images from other boys. In that way, he got integrated quickly. At the end of the field trial, the boy had become much closer with the other boys.

The grandmother of the family in Vienna also sent images to participate in the family's everyday life remotely. Sending artistic images and images from her dog to her children and grandchildren gave her the possibility to share her life with her family without being obtrusive.

11.6.5 Supporting Conversation

Towards the end of the field trial, the boys in Helsinki additionally invented more utility uses for the device. For example, a boy needed to describe a feature of a computer game to his friend on the phone – which turned out to be difficult. Therefore, he sent a picture of the computer screen and used it as a tool for collaboration.

11.7 What We Learned in the Field

Or: was it worth the effort?

The field trials helped us to understand the possibilities of the PIX product concept better than any other evaluation method used previously. On the basis of the first phase of the project, it became apparent that digital images would possibly support socialisation, and that children would enjoy the editing of images. Little, however, was known about the way children would use images to share experiences or emotions with their relatives and friends. Therefore the "use situations" mentioned above gave an important insight into the actual usage of such a device.

One of the major goals of the field trials was to identify advanced design concepts for visual communication devices. The following list summarises the results of the field trials and should also inspire product designers developing leisure-related devices for children, which use picture-communication as (at least) one of their functions:

- Images are mainly used for joking, expressing emotions and creating art.
- Users' perceptions of images change in character from memory support to the expression of current activities and feelings.
- Users want and use large sets of picture editing possibilities.

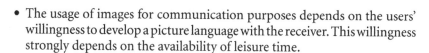
- The usage of images for communication purposes depends on the users' willingness to develop a picture language with the receiver. This willingness strongly depends on the availability of leisure time.
- Images are not enough for functional communication, such as making appointments.
- Therefore, devices should provide the possibility of annotating images with text or audio.

11.8 Looking Back

Or: conclusions on the project.

Looking back, we have to say we had a great time in the project. Especially with all the different disciplines included on the team, extended discussions ensued as to what might be sensible to do and whether some method would provide meaningfully useful results etc. While the technical partners had the prototyping and production aspects in mind, the designers wanted to be highly innovative and creative. The user sites and usability partners tried to focus on what the users would really use/want or go for. At the end we found something that everybody was happy with: not necessarily meaning the PIX device to be the "end product" of the project. But every partner had learned more in their respective domains and skills, about other viewpoints and priorities, and we had all experienced this multidisciplinary and multi-methodologically driven approach to be highly challenging and fruitful.

Summarising the overall content, insights and results of the project, we can state the following: When designing new products, one has to consider the user in all his or her characteristics and contexts and most of all in their daily use. Hence, (prototype) evaluation should be conducted in people's environment in order to gain an understanding of what and how people would use future technologies. For example, the PIX field trials demonstrated that the meaning of the photographic image changed completely as people used products to manipulate and send digital images wirelessly. Digital images were not used by the trial users to memorise past events or relationships – like traditional photographs do – but as tools for creating stories, expressing affection and creating art. Without the experience of the "Digital Hug", one would not have been able to predict that in such a valid way.

11.9 Acknowledgments

We would like to thank all our MAYPOLE partners: IDEO Europe, Meru Research, Netherlands Design Institute and Nokia Research Centre. We are also grateful to Katja Battarbee, Thomas Grill, Ville Haaramo, Juha

Huuhtanen, Kristiina Karvonen, Pia Kurimo, Gerhard Leitner and Aapo Puskala who participated in conducting user research and evaluation.

11.10 For Further Information

Giller, V., Tscheligi, M., Sefelin, R., Mäkelä, A., Puskala, A. and Karvonen, C. (1999) Image makers. *Interactions*, **6**(6), 12–15.

Iacucci, G., Mäkelä, A. and Ranta, M. (2000) Visualizing context, mobility and group interaction: role games to design product concepts for mobile communication. In *Proceedings of Fourth International Conference on the Design of Cooperative Systems (COOP'2000)*, Sophia Ontopolis, France 23–25 May 2000.

Mäkelä, A. and Battarbee, K. (1999a) Applying usability methods to concept development of a wireless communication device – case in Maypole. In *Proceedings of 17th International Symposium on Human Factors in Telecommunication*, Copenhagen, Denmark, 4–7 May 1999, pp. 291–298.

Mäkelä, A. and Battarbee, K. (1999b). It's fun to do things together: two cases of explorative user studies. *Personal Technologies*, **3**, 137–140.

Mäkelä, A., Giller, V., Tscheligi, M. and Sefelin, R. (2000) Joking, storytelling, artsharing, expressing affection: a field trial of how children and their social network communicate with digital images in leisure time. In *Proceedings of CHI'2000*. ACM Press, New York, pp. 548–555.

Maypole CD-ROM (1999) Maypole concept documentation v2.0. Maypole consortium/Meru Research B.V.

12

Ambient Intelligence: Human–Agent Interactions in a Networked Community

Kostas Stathis, Robert Spence, Oscar de Bruijn and Patrick Purcell

12.1 Introduction

It is a weekend morning and you are having coffee and a chat with some friends in your local café. However, this is no ordinary café. It is a café equipped to serve the communication needs of a new type of social context – the digitally connected community. This particular locale is a digital milieu in which information is both ubiquitous and location-independent and it is displayed in advanced communication devices that are embedded in the physical ambience of the neighbourhood. A feature of these communication devices is that they are densely populated with agents of a special type. In this instance these are software agents, performing autonomously a variety of support roles for the local people. In the example of the café in question, these agents may inhabit, for instance, the coffee tables, the customers' mobile phones and their personal electronic key ring tokens.

So, to continue with our narrative, you are mid sentence in an inconsequential social exchange in this café, when suddenly your eye catches some of the information that comes flowing past you along the edge of the screen that forms part of the interactive coffee table at which you are sitting (Figure 12.1). It is an announcement of a chess tournament organised by the local chess club, and you wonder if this is an event in which you might like to participate. You place your finger on top of the moving image on the interactive touch-sensitive screen and by sliding your finger across its surface you drag the item into the middle of the table where the full text of the announcement becomes visible.

After reading the screen text you also notice a hyperlink on the screen, which indicates an electronic entry form for the chess tournament. You proceed to drag and drop this entry form on top of the screen icon representing the Coffee Table Agent – a piece of smart software that autonomously retrieves all the necessary information to enter and take part in the chess

© Philips Design

Figure 12.1   The café's coffee table with an embedded screen showing an announcement of a chess tournament.

tournament. After some time has elapsed, during which you talked with your friends about the film you saw last night, you notice another interesting icon, which is also flowing by on the coffee table screen. However, you do not want to interrupt the conversation with your friends, so with a single hand gesture, you simply capture and transfer this hyperlink icon onto your mobile phone automatically, which you have purposely placed in the middle of the electronic coffee table, so that you can have a look at this new information later on when you get home. The smart software agent resident in your mobile phone soon realises that your regular chess partner might also be interested in taking part in the tournament, so it suggests forwarding the entry form to your friend. On consideration of this suggestion, you agree and it is done.

The electronic coffee table is still active and the Coffee Table Agent subsequently retrieves all the information it can find, which relates to the chess tournament, such as the history of the local chess club and some pictures of the pub where the tournament will be held, and sends this information to your television set at home where you can browse through it later.

The mise-en-scene for the preceding exchange in the café might be considered as utterly conventional for a social encounter in a leafy Scottish suburb, but your interaction with the coffee table and the interactions between the coffee table and your mobile phone are examples of the type of innovative computer-mediated communication that may be commonplace in the not too distant future. Indeed, many of the interactions described in the scenario above are already supported by a prototype system implemented as part of the "Living Memory" project (LiMe, 1997), a long-term research and development programme, that started in the late nineties and was primarily funded by the European Commission. In particular, the above scenario illustrates how current developments in the provision of pervasive information may constructively impact the daily routines of people who live

and work in a specific locality – in effect, generating a new type of digitally *connected community*.

In the Living Memory project, the concept of "being connected" is interpreted as involving social communication and the sharing of information between the members of a community, while "being part of the community" assumes a shared communal memory. The specific aim of the Living Memory project, therefore, has been to stimulate the creation and sharing of grass-roots community incidents and events that reflect day-to-day lives in the neighbourhood. The contents of such shared memory stored in various digital media may include local history and news, individual residents' personal memories, advice based on local communal experiences, event announcements, invitations to such events and so on. These various types of communal experiences form the substrate of the community's collective memory. Opportunities to share and interact with this collective memory can work as a social catalyst for enhancing community contact and neighbourliness generally, as illustrated by the futuristic café scenario described above, for example.

It is a commonplace feature of today's communication climate to create virtual communities from people who are separated by great distances, but who are united by common interests, hobbies or cultures (Rheingold, 2000). By contrast the Living Memory project was conceived to be involved with the physical, co-located community. The aim of the project was to employ advanced information technology to facilitate people who share a physical locality (e.g. an urban neighbourhood or a rural village; Noack, 1994) to help record, interpret and preserve the richness and complexity of their special *local* culture. Thus, the Living Memory community is defined by a shared geographic location, rather than the common practices and shared interests that often define virtual communities. By increasing accessibility to the record of a local communal memory, the Living Memory system aspires to enhance social cohesion, a cohesion that may have been adversely affected by the current dynamic of social and technological change.

An important goal of the Living Memory project has been to empower lay people by enabling them to interact naturally (Bernsen, 2000) with a range of different devices (Norman, 1998), providing easy-to-navigate (Fleming, 1998) personal interfaces (Spence, 1999) based on agents (Stathis et al., 2002) with which to create, assess and annotate the record of their local communal memory. The café scene described above illustrates how news of events may be accessed on an interactive coffee table while people are engaged in social discourse. Many other devices can conceivably be used to afford the members of the community effortless access to community content and easy ways of communicating their experiences with each other. The Living Memory project has explored a range of concepts for interactive devices, such as interactive bus stops and large-screen public displays, together with their contexts of use, as part of an information and communication infrastructure.

These devices can all be connected together to form a local network that serves a neighbourhood's specific information needs. In the design of such a multi-device networked system, one is obliged to anticipate the complexity of the users' needs and the social settings peculiar to a lay-user community. Meanwhile, the designers of the system must realise the overriding need to mask the resultant complexity of the system at the point of use by the community member.

Therefore, to achieve the ubiquity, intelligence and distribution of knowledge required to support facile social interactions by lay people, we advocate a type of system whose main characteristic is based on the concept of agency. In the context of human affairs, both the agent and the concept of agency have long played crucial and defining roles in defining human-to-human relationships. Agency in the human context can take many forms and assume many attributes. Agency may connote the assignment of responsibility (Norman and Reed, 2002a), or the complement of responsibility, and the exercise of delegated authority (Norman and Reed, 2002b). The performance of an agent may imply qualities of trust (Falcone and Castelfranchi, 2001), aptitudes for negotiation (Stathis and Toni, 2005), a capacity to take initiatives proactively and to operate either autonomously or semi-autonomously (Witkowski and Stathis, 2003). Fundamentally, the value of the agent is predicated on notions of integrity and the performance of a professional service to the individual or organisational patron commissioning the service. So much for the performance of the roles and functions of human agency in a day-to-day social or business context; in the alternate digital world of information and communication technologies, it is pertinent to observe that the recent emergence and development of the software agent assumes many of the social traits and behaviours of its human exemplar.

In the digital world of computation an agent is a software entity capable of autonomous and flexible action, having the ability and intelligence to choose an action without the direct intervention of humans or other related software agents (Jennings et al., 1998). In other words, the agent has control over its actions and internal states. Flexible action means simply that the agent is *responsive* to changes in its environment, *proactive* in the pursuit of its goals and *social* in its interactions with other agents and the people it deems relevant in its immediate operating domain. In the context of the Living Memory project, software agents aim to have the capacity to relieve the members of a connected community from having to explicitly make their connections and associations with the mundane details of communal memory content such as local information, specific physical locations and particular fellow community members.

Once we have accepted the concept of an individual software agent, we can think about how such agents could work together in a multi-agent system. A multi-agent system is a complex software system that is built from many agents interacting both with other agents and with the people (Singh, 2000) using the system. Each agent in a multi-agent system has

Figure 12.2 A human community enhanced by a society of agents.

both incomplete information and a limited viewpoint as to the extent of its operational environment. There is, therefore, no global control in the operation of the community-based system (Mamdani et al., 1999). As a consequence, information in the system is decentralised, and computation occurs asynchronously.

More intuitively, one might say that agents work in an individual island of information, able to communicate only with other agents and humans that are within their information horizon. In this respect, a multi-agent system is much like a society of people in which there is also a limited amount of central control, and in which people operate according to their individual time schedules. Moreover, as in human societies, organisation in multi-agent systems is achieved primarily through the communication between agents. Thus, we can build multi-agent systems that operate like societies of agents (Pitt et al., 2001), which mirror much of the dynamics of the interactions that occur in the human communities and organisations in which these agents operate (see Figure 12.2).

The question that begs for an answer then is how to design a system in which agents need to anticipate and satisfy the needs of lay people, who may use a range of different devices and objects to access and share an evolving and situated collective memory pertaining to their neighbourhood. Indeed, this design effort does need to take into account not only the interactions that take place between individual persons but also those between people and the various devices/objects that they use (possibly mediated through the roles of software agents). Furthermore, the interactions between the various kinds of software agents that reside in different devices, whose effective interactions are necessary to achieve the desired flow of information and accessibility to local community information, are equally important. All these interactions need to seamlessly complement each other in order to achieve the enhancement of social interaction between people within the connected community that the Living Memory project ultimately aims for. Therefore, the task of designing information systems for people, in which software agents are intrinsically embedded, is a very special challenge indeed.

In summary, therefore, a community system relying on intuitive user interfaces (de Bruijn et al., 2002) and software agents (Stathis and Toni, 2004) may create an intelligent ambience (Markopoulos et al., 2004) for the community and make it responsive to people's needs. The prevailing requirement in this context is the quality of the enhanced interactive experience on the part of the people who use the system from different devices and objects available in the environment of the community. The enhancement of the interaction is determined primarily by the effectiveness of the underlying organisation of the system and the interaction design process. In such a process we have to carefully consider not only the interactions between humans and other humans in the community, but also the organisation of interactions between humans and the various kinds of software agents that inhabit the connected community. The autonomous behaviour of such agents has important implications in the way people communicate via the system. This extended model of interaction design, which includes interactions between individual software agents and interactions between these agents and users, forms the topic of Section 12.3. In the next section (Section 12.2) we start by briefly introducing an approach to the design of interactive systems that takes into account the expertise offered by a range of disciplines. Indeed, when it is necessary to support many of the social interactions between people in a community, it is of paramount importance that we start the design of the system from a thorough understanding of social interactions. Section 12.4 summarises the work presented and outlines possible directions that we believe are worth pursuing as future work.

12.2 Designing for Social Interactions

In this section we argue that the challenge of designing ambient intelligence for supporting community interactions can be met only by the application of an appropriate design methodology. This methodology should be capable of capturing into a community system all the relevant types of social interactions taking place within a connected community. In achieving this goal we introduce into the development process a stage in which we focus on capturing these community interactions. A model of these community interactions forms the basis of any system development. The nature of such a model and the process by which it is derived are described in Section 12.2.1. Then, in an effort to bootstrap the design process, we present a description of the basic elements that have to be included in any model of community interactions, namely the people and places that make up the community,

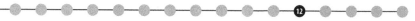

and the events that take place as a result of these people interacting with each other.

12.2.1 Capturing Community Interactions

Computers are no longer used exclusively in work environments by people who are specially trained in their use. Instead, many computer systems are currently being designed for people without special training in the execution of their common everyday tasks. This expansion of the domains in which interactive information systems are being applied has created a need for an interdisciplinary design approach that can cope with the multitude of factors thrown up by the explosion of the numbers of lay users (Pitkin, 2001). Today, the design of interactive systems faces the challenge of facilitating the successful adoption of new information technology by these lay users.

In order to design systems that can support people in a wide range of different social contexts, we need to broaden the scope of the kinds of interactions that need to be considered in the design process. In particular, it is important to realise that one has to consider all aspects of a system meant to be used by ordinary people in order to facilitate and possibly augment a broad spectrum of everyday activities. Within such a wide context of use, the designer can no longer assume that the system will typically be used by a single person, interacting with a single computer, in order to accomplish a clearly defined task within a well-defined social environment. Many computer systems are now designed to be used by lots of people, each of whom may use a range of devices in the pursuit of ill-defined and often ambiguous goals. To complicate matters even further, people may often want to collaborate in the achievement of a common goal, such as when two or more people are playing a game together like chess, which creates many inter-dependencies between the activities of these people. Therefore, the design process needs to cater for all the interactions that are relevant to the experience of all those users interacting with the system.

In the design of systems for application in local communities, we have to draw on the experience of a range of seemingly diverse disciplines such as those of social scientists, social psychologists, ergonomists, computer scientists and social planners (Bannon, 1997). The process of design starts with a description of the day-to-day activities in which the people who will use the system are typically involved. The resulting technology should be capable of serving real human needs in areas where information and communication technology has not, until recently, been applied such as in connected local communities that are the subject of this book. In this particular context, this descriptive process must provide information about people's activities in the context of their families and community. Furthermore, because we are particularly interested in capturing the dynamics of *local* communities, knowledge about people's activities and perceptions in the context vis-à-vis their neighbourhood can additionally provide valuable contributions.

The results of this multidisciplinary effort in the study of local communities or organisations are subsequently used in the development of a narrative description of the relevant social interactions (Bødker, 2000). That is, from the experiences and observations of the researchers studying people's interaction habits we create a set of use-cases, personal histories and scenarios (Carroll, 2000) that illustrate the many idiosyncratic ways in which people interact with each other and their environment. We refer to such a narrative description of the system as the *interaction model.* It is important to note that at this stage it is important that the interaction model illustrates how interactions between people are mediated by electronic communication devices such as mobile phones and computers (e.g. through e-mail). Indeed, any new system development needs to take into account people's experience with existing technology. These experiences may determine to a large extent the expectations that people have about new technologies, which may play an important role in deciding whether newly introduced technology will be adopted or rejected by the target groups of users.

In the remainder of this section, we describe the major building blocks of a model of community interactions, namely the people, places and events that form the constituencies of the community. Like the work of Kindberg et al. (2002), we are interested in people and places within the community. Unlike Kindberg et al., however, here we shift our attention from objects to the narratives that make up the interaction model of a community, on how people interact with each other and the nature of the events that result from these interactions. In the case of local communities such as neighbourhoods, we need to describe in detail how the character of these interactions is mediated by the locations that make up the territory of the community. In short, local people interact with each other as part of their day-to-day lives, initiating and taking part in events that take place in community locations such as the local school, the pub and the many other places that play a significant role in community life.

The process of building an interaction model is delicate, because the success of introducing new technology may depend on it. It is also complex, because of the many factors that need to be taken into account. It is very important therefore that the knowledge gathered during the design of community systems, in terms of both their successes and failures, is taken into account in future projects. In that way, previous successes may be replicated while previous failures may be avoided. Therefore, we conclude this section with a description of an abstract framework for community interactions in which we explain how people in a community are connected to each other and to the places.

12.2.2 People

A physical community (as distinct from the virtual kind) consists of those people who are the members of a given local precinct, or neighbourhood. Membership of such a group normally requires that a person, for the most part, resides, works or socialises within that community.

Individuals within this type of community (say "Our Town") may organise themselves into interest groups, which act as a collective unit, like the chess club referred to in the introduction. Although groups are often emergent structures whose members are normally members of Our Town, there could be cases where a group may have members that belong to another co-located community – say "Next Town". Such individuals would be implicit members of Our Town in that they are members of a group, which forms a constituent part of Our Town, either through working or socialising there.

More complex and formal groups of people organised for a particular end can be members of a community. Typically, these may include businesses that support the economic life of the community, and other institutions such as schools, hospitals, churches and local government, if they are located in the physical locality of that community. As with informal groups, the members of such organisations would be implicit members of a community such as Our Town.

When information and communication technology infrastructures enhance community interactions by using special purpose devices connected via a community network, membership (Agostini et al., 2002) is thereby extended to include

- *remote* membership – where a person's participation in community life is facilitated without that person being necessarily present in the locale of the community. This would include, for example, locals who are students in distant colleges.
- *temporary* membership – where a person's participation in community life is facilitated because that person is temporarily within the locale of the community. This would include, for example, tourists or workers on temporary assignments.

Remote membership presupposes an already existing relation between the remote member and the local members or the locality of the community; for example, consider a community member leaving a community to attend university or work assignments. Temporary membership, on the other hand, facilitates people that are within the locale of the community for a limited amount of time to access the community services available, for instance tourists and other transient residents.

Orthogonal to the types of community membership are the different roles that members play within the community. Within a community we can typically identify a number of different roles that people can assume while taking part in community activities. For instance, peoples' roles can be defined by their official capacity, such as their professions or other non-professional capacities, for example as chairperson of the local chess club. In addition, people can assume roles that are linked to their social position, such as magistrates. Other peoples' roles may be based on personality traits, such as that of activists and gossipers. In turn, some of the roles community members play will be delegated to their agents, which in turn will use these roles to access resources in the artificial world of the community (Toni and Stathis, 2003), thus mapping parts of the community interaction in the artificial environment of the connected community.

12.2.3 Places

What distinguishes local neighbourhood communities from other types of communities is that interactions between members are located within the physical territory that the community occupies (Harrison and Dourish, 1996). For this purpose we principally identify a local community by its physical territory. One way to define the physical space of a local community is by a set of significant places that have some important meaning within the community and helps to define its unique *genus loci*. The motivation behind this is that a location determines the context of communication and social interaction, and in certain cases has important implications for people's needs, in particular their need to share experiences with others or to remain private. Based on this observation, we rely upon the notion of *accessibility* to distinguish between two different kinds of locations:

- *Public* places are locations that all people in the community have the right to access, such as a shopping mall, a public park and a coffee bar;
- *Restricted* places are locations that only certain community members have the right to access, such as people's homes, the offices of a local business and a private club.

The notion of restricted places is abstract enough to allow for a range of such places, ranging from those with privileged access as a result of people belonging to groups (e.g. being member of a club or a business) to those with private access as a result of ownership (e.g. the privacy of one's home).

12.2.4 Events

We assume that interactions between community members relate to specific events and, as a result, these events can be used to characterise the way in which members of a community are connected to each other and to the specific locations within the community. Consider again the example described earlier in the introduction of this chapter, where it is suggested that you may play chess with a regular chess partner. The event of playing a game of chess connects both you and your chess partner together with the location in which you play. This social connection is reinforced every time you play chess together at that particular location.

Memories of events involving people and places in the community create the shared interpretation of experiences from past events. Such shared interpretation often results in the formulation of a community's collective memory, which together with the present activities and future events in the community form the basis for its members to form close social bonds that keep the community together. Indeed, the existence of a collective memory provides the context for meaningful interactions between the members of the community, allowing for a shared understanding of those interactions. In fact, information is exchanged constantly in any given community – not just within a network of family, friends and neighbours, but also between members of a community who do not personally know each other. Often, this information is picked up incidentally, almost subconsciously. Whether gazing out of the window in a bus, or eavesdropping in a café, we continuously build a picture of our local environment. These cumulative and random happenstances and experiences define to a significant degree how we may feel about being part of a given community.

The occurrence of events is governed by the norms that are prevalent within a particular community. The intended function of norms is to guard the way in which the collective memory develops and grows. Alternatively, we can think of norms as a way of specifying what kind of events ought to take place between people and places in the community. Locations are important because they impose constraints on the way people behave or interact with each other. Locations are also important because they may determine various forms of social constraints in a community context. For example, the behaviour of people is normally different in a public place, such as a coffee bar, than in a private place such as their homes. In this context (Laurier et al., 2002), sociology plays an important role too, to form a basis for observing various norms of behaviour from one community location to another.

12.2.5 Connective Tissue

12.2.5.1 Smoke Signals

We investigate the basic structure of the narrative description of a connected community system through the definition of an abstract framework that we call *connective tissue*. Intuitively, the connective tissue refers to the social, organisational, cultural and technical fabric that connects the members of a community to each other and, in the case of a local community, to the significant physical locations that make up the territory of the community. Our hypothesis here is that every community has a connective tissue that gives to this community a recognisable structure that is distinct from other communities. The framework presented here identifies the basic abstract components of the connective tissue, namely people, places and events. Given a particular target community, these components must be specified further to define the concrete interaction model to be supported by the community system.

We refer to this framework as a community's *connective tissue*, a term we apply to the various forms of social fabric that may constitute the basis of a community's social interactions. This connective tissue has been derived from a study of a leafy suburb of Edinburgh in Scotland called Corstorphine that served as the case study focus of the Living Memory project's efforts to understand this community's social interactions. Firstly, a series of semi-structured interviews were carried out with a representative selection of community members (Laurier et al., 2001). These interviews were subsequently encoded to identify the types of interactions that featured in the interviewees' answers. These codes were then elaborated and refined by asking further questions in order to reveal more detailed information about these interactions. The themes in the analysis were finally integrated into a narrative by selecting those that had been found valid and significant by the citizens who took part in the field study.

12.3 Agent-Based Connective Tissue

The interaction model that results from attempts at capturing the interactions between community members forms the basis from which new community systems can be developed. The next step in the design of the system is therefore to augment the interaction model with examples of how newly developed systems would fit into the existing social interaction patterns. This stage in the development mainly serves two purposes. Firstly, it gives social scientists and technologists a common frame of reference within which to work. Secondly, the augmented interaction model, which describes the functionality of the new system, can be validated by asking potential users to comment on it.

In the technical scenarios we have been setting out for tomorrow's connected community, it was felt that a distinctive term was needed to capture the special character of the growing interlocking web of interpersonal, familial and social interactions and communications. Given its physiological resonance, the term "connective tissue" was coined to refer to this social web of ever-increasing density, power, speed and technical sophistication.

To show how to build a system based on the connective tissue we describe here the Living Memory system. In the Living Memory system people carry

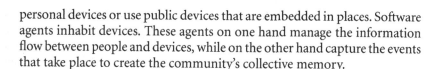
personal devices or use public devices that are embedded in places. Software agents inhabit devices. These agents on one hand manage the information flow between people and devices, while on the other hand capture the events that take place to create the community's collective memory.

12.3.1 Designing Agents

12.3.1.1 Agents Per Se

Models of social interactions can be quite complex, especially when they attempt to capture the activities of lay people in environments such as local communities. Complexity arises here from the sheer variety of possible human-to-human interactions and of the contexts in which these interactions can take place. As some of these interactions will be carried out with computing devices (e.g. portable, mobile or fixed), the software capabilities in these devices must be able to support lay people in performing their social activities and, ideally, enhance community spirit. The assumption here is that a device should be capable of supporting one or more users to increase the opportunity for computer-mediated communication between members of the community without, as a consequence, increasing the users' information burden due to excessive interaction with different devices and other people.

In order to deal with the complexity of interactions between lay people in a connected community, we investigate the construction of interaction models that have as a feature computing devices containing software agents. In this context the issue then becomes how to design the functionality of agents to manage the complexity of the interaction that is facilitated by a device. For an agent to decide how to behave appropriately in particular circumstances of the interaction, we propose that each agent is designed to play a particular *role*; see, for example Zambonelli et al. (2003). Informally, we can think of a role as the set of tasks and activities that are *assigned to* or *required* or *expected from* an agent. We assume that the role that an agent plays in its capacity of acting on behalf of people in a connected community should be designed in such a way that it lifts the information-processing burden off people interacting with devices and likewise with other people via these devices.

By embedding agent-supported interactions in people's social activities we aim to increase the fluidity and ubiquity of community interactions without increasing the overhead – of keeping track of these interactions on the part of the human – In particular, suitable specification of roles for agents running on electronic devices can make such devices aware of other devices, agents and people within an electronically connected community. The characteristic and expected behaviour of an agent through roles facilitate the creation of interfaces for the devices that people use that are persistent, active and adaptable to the highly dynamic requirements of interactivity in social settings. What is special about the interaction design of a multi-agent system in this way is that the designer's task is to decide which interactive processes should be the reserve of humans, and which can be delegated to be carried out by agents.

To integrate interaction design and agents, the interaction model should specify the agent roles in terms of the tasks and activities that agents have to carry out on behalf of people. In this description, the designer has to specify how a community member interacts with an agent, confirm that a role is active, indicate how it should be performed and check whether the tasks that the agent performs as part of its role have been accomplished as required. For example, suppose you have a software agent residing on your mobile phone and you receive some information about a chess tournament, as with the example in the introduction. As you play chess you might ask your agent that any time you receive a message about chess the agent should play the role of *disseminator* (Stathis et al., 2002) so that the people you play chess with from the chess club you are a member of receive this information too. At the end of all this you should be in a position to verify that your agent indeed played the role properly, to ensure that all your friends that play chess with you indeed received the message.

The example above indicates that the interaction designer also needs to specify the nature of the communication that takes place between agents acting in particular roles. For instance, in the example of the chess tournament, the agent of one of your chess partners is required to reply to your agent when it receives the information that your agent sent to it. Your chess partner's agent may reply by informing your agent that if your agent in the role of disseminator sends an item of information, such as the announcement of the chess tournament, then it has to acknowledge receipt of this message, by perhaps further acknowledging that the message has been brought to the attention of your chess partner.

In other words, the interaction process should specify how agent functionality in terms of roles would characterise the communication protocols that agents use to communicate, including how these protocols ensure computational feasibility. This includes knowing not only how it is possible, for example, to decide when information in the system is no longer relevant, but also what representations of interactions are computationally desirable. It is important that we avoid an explosion of communications, which may be caused by agents forwarding information to other agents without taking into consideration the possibility that some of these agents may have already received that particular information from another source. For instance, when your agent forwards the announcement of the chess tournament to your chess partner's agent, his or her agent may, in turn, forward it to the agent of someone you both occasionally play chess with, and to which you have already sent the announcement. In that case, it is important that agents have cognitive abilities that allow them to remember when to stop, thus

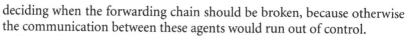
deciding when the forwarding chain should be broken, because otherwise the communication between these agents would run out of control.

To rectify possible naïve computational behaviour it is paramount that the interaction design team devises communication mechanisms between agents in their roles where each agent can draw appropriate conclusions from the structure of a message and adjust its behaviour accordingly. For example, a simple solution to avoid the problems that uncontrolled information forwarding could lead to is to include in the design of an agent communication protocol a list of recipient agents as a characteristic feature in the forwarding message.

The result of embedding agent roles and communication protocols in the interaction design process is the creation of a library of roles and protocols that are based on ethnographic data describing the communication requirements of people living in a particular community. In other words, the approach is motivated by what people do and not necessarily by what the designers think, while at the same time the functionality of agents in the community-based system can be constructed and reused flexibly, when it is required.

12.3.2 My Agents

Mobile devices are typically carried around by the members of the community, who use these devices to interact with their environment and to capture the connective tissue that is created as a result of these interactions. Interacting with the environment means that mobile devices interact both with other mobile devices and with the embedded static devices. For these interactions to take place it is necessary that devices are visible to each other and that links can be established between them so that the agents within the devices can communicate. In our work on the Living Memory project we have introduced a distinction between mobile devices that are personal and ones that can be shared between a number of people. These two types of mobile devices are described next.

Personal mobile devices are devices that are owned and used privately by a single person. Information stored in a mobile device has a personal and often private character. This information includes the owner's personal history – a record of the owner's interactions with people and devices, personal profile – a record of the owner's interests and needs, and acquaintance model – a record of the owner's contacts.

Personal mobile devices are already in common use and include mobile phones, personal digital assistants and pocket personal computers. However, to enable these devices to function as part of a connected community system, these standard devices need to be augmented in two important ways: one is the inclusion of personal service agents, and the other is that these devices must be registered with the connected community in order to receive the services provided by the system. Registering with the connected community system is very similar to registering a mobile phone with a network service provider (such as Vodafone in the UK) in order to receive telephone services.

Personal devices are inhabited by at least one *personal service agent* (Mamdani et al., 1999), an agent type designed to manage the storage of community information and its access for an individual member based on a model of the device's user. We allow either the user to update parts of the user model explicitly via the device interface or the personal service agent to infer parts of the model implicitly according to what the user has done. In this context we identify a number of roles that mainly manage the interests of the user, the social network of the user as it is captured by the user's acquaintance model and the history of the interactions the user performs via the device.

We also allow for certain mobile devices to be shared. Shared mobile devices, such as tokens, are devices that are used in the community to store information that is not private and which people may therefore want to share with others by handing these devices over. Shared mobile devices provide a way of capturing and physically carrying community content.

12.3.3 Ambient Intelligences

To allow accessibility of the community system by everybody we introduce static devices. A static device is a device that is typically embedded into people's everyday environment, in such a way that when these people interact with their environment they are generating the electronic events that are necessary to capture the connective tissue. In order to augment and strengthen the community's connective tissue, the Living Memory system has been designed to have a ubiquitous presence in the environment of the community. To reflect this, the interfaces to the Living Memory devices have been designed to be informal and accessible, blending with the immediate physical local environment and integrating with tangible community life. Natural crossing and meeting points in the community, such as tables in a local café, can be used as the nodes in the electronic connective tissue that are inhabited by agents.

The interactive coffee table is just one example of a class of information artefacts that can be embedded in the local community environment to display locally relevant information. In this section we explain how ordinary people interact with the coffee table, a large screen embedded in the environment of a shopping mall and an information display that forms part of an interactive bus stop. Although these artefacts represent only a small fraction of the possible devices that can be embedded into the environment of a local community, they are representative of the type of situations in which embedded electronic interfaces supported by agents can be used to establish or enhance the community's connective tissue.

© Jonathan Olley

Figure 12.3   Interacting with the coffee table in a community location.

Interactive coffee tables represent devices that occur in locations, such as pubs and coffee bars, where people meet informally and spend some time in social interaction. The coffee table and a close-up of its interface are shown in Figure 12.1. The coffee table is a device that facilitates information dissemination in a way that is unobtrusive, but also very accessible. Sometimes, information can be retrieved on the coffee table while talking to other people, as illustrated in the scenario presented at the beginning of this chapter. At other times, the coffee table may become the focus of interaction between the people sitting around it.

Information items presented on the coffee table's display move slowly from left to right around its perimeter (Figure 12.3). Since the embedded display is clearly within reach, the coffee table is specifically designed to respond to touch (finger placement) in order to identify an item of interest. Once an item has been identified, a finger movement can then lead to amplification of the information in an identified item and, if desired, its subsequent movement towards a mobile devices or electronic token placed upon the table. This incorporation of information in a mobile phone or electronic token facilitates later examination, an examination that might be entirely inappropriate – indeed, probably intrusive – within the social environment in which the coffee table is located. Such direct engagements are natural to the user.

In the local shopping mall, interactive large display screens can be mounted to show information that may be relevant to the shopping public. As people wander around the mall, they may notice any information items that attract their attention. On the mall display information does not scroll;

Figure 12.4 Information display embedded in a bus stop.

rather it is presented as a slideshow of images (Spence, 2002). The sole interaction with the mall display is that of brushing a special device mounted underneath the display with a personal token (or other mobile personal device). The token then stores a pointer to the item of community content such that further examination of the item can take place at a later time in the user's home or on the coffee table when more focused interaction with content is acceptable.

The bus stop (Figure 12.4) represents an obvious place to display information, as witnessed by the advertising already present in these locations. People waiting for the bus have a variable (but usually brief) period of time in which they are often not engaged in any particular activity. During this time they may want to interact with an embedded information display when an item of information attracts their interest. The bus-stop display may present, at its lower level, a scrolling sequence of information items in a manner similar to the coffee table. In addition, some information may be permanently visible, such as bus-route information and the expected arrival time of the next bus. Because this artefact does not allow interaction by touch very easily, a remote pointing device allows users to "point" at an item of potential interest, whereupon it is displayed in more detail in a large, central area of the display. By brushing the screen with an information token, items of interest can be stored and carried along for later examination when more time is available.

We started this chapter by giving an example of how information and communication technology, with the support of software agents, is used to bring to our attention information that is relevant to our everyday activities (the announcement of a chess tournament), and to redistribute this information to the people we know for whom this information might also be relevant (our regular chess partner). For one thing, we did not have to search for this information actively. Instead it came to our attention because we were in a certain location (i.e. the local café in which people sometimes play chess). In addition, we did not have to inform other people, as this was taken care of by the system in the form of a software agent acting on our behalf. The result of all these electronic interactions is that the connections between these people may be strengthened through possible face-to-face interactions (taking part in the chess tournament).

As a consequence of being tied to a particular location, shared static devices provide a view of the community that is characteristic of its location. In order to provide such location sensitivity, shared static devices are home to one or more *location service agents* (Stathis et al., 2002) that manage models of the locations in which these devices are situated. To allow a location service agent to construct a location model we readjust the user model used to cater for locations. One important role that we identify for the location service agent is one that allows the agent to create a location profile. This location profile will allow the agent to stereotype the kind of device usage by people who visit this location, and, as a result, make device useful to people who are likely to visit again this location in the future. In addition, the location service agent also manages a model of the neighbourhood devices for a location and the history of the interactions people perform via the device.

12.3.4 Servicing the Memory

Typically, members of a connected community use the system in order to contribute documents that are of interest to others or to access existing documents that are of interest to them. These documents, together with records of their history, form part of what is referred to as a collective memory of the community. Documents have their own life cycle, which starts at the time documents are created and submitted to the system. For each document a record is made of when members of the community access and annotate it. The process of creation and subsequent annotation creates a history from which we can derive additional properties such as how popular a piece of information is, or how a piece of information relates to other pieces of information at any time.

The amount of information and communication technologies embedded in the physical environment of a community will determine the extent to which we can capture some of the community's connective tissue electronically. We capture the community's connective tissue electronically through descriptions of electronic events. We assume that agents that are resident in devices are able to observe the occurrence of electronic events through the interaction between people and devices. An electronic event occurs whenever people interact with electronic devices such as the coffee table mentioned in the introduction. For example, when you stored a link to the announcement

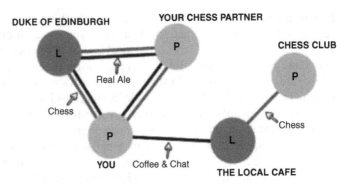

Figure 12.5 A link–node model representing the connective tissue between you, your regular chess partner and the Duke of Edinburgh pub. The model also shows how the announcement of the chess tournament that originated at the chess club may have reached you through links with the local café where you have lunch every day and the chess club meets once a week.

of the chess tournament on your mobile phone, an electronic event took place that can be recorded and described in a structured way. From these structured descriptions of electronic events, it should be possible for agents to extract relevant information that they can consequently use to initiate actions, which might involve interactions with other agents and people. Thus, using the structured description of your interaction with the coffee table, your agent was able to decide upon forwarding the announcement to your chess partner.

The way in which we structure the description of electronic events in the various community locations determines what the computation flow in the system will look like. We use electronic events to capture the electronic connective tissue as a *node–link model* (Toni, 2001), where *nodes* represent members and locations in the community, while *links* represent ways in which they are connected. Figure 12.5 depicts how you, your chess partner and the café are connected according to the scenario that was presented in the introduction.

In order to capture the complexity of the connected tissue, we further assume that each link consists of a number of *fibres*, where each fibre is created as a result of events taking place jointly between nodes (see Figure 12.6). Each fibre is associated with a *topic*. For example, playing a game of *chess* is the topic of the *fibres* that connect you with your regular chess partner. Additional fibres may describe the link between two nodes. For example, you may also be linked with your chess partner through a fondness for Real Ale that you might have in common and that you like to talk about during your games of chess.

An important parameter of a fibre is its *strength*, determined over time by events concerning a particular topic that links the two end nodes. The more of these events occur, the stronger is the fibre associated by the topic of

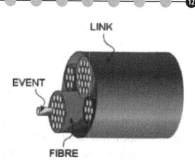

Figure 12.6 Anatomy of a link.

those events, but with a built in decay to cater for dynamic situations. Thus, the chess-fibre is strengthened every time you and you chess partner play a game of chess, but it weakens when you stop playing. Together, all the links between nodes, including their fibres, make up the electronic connective tissue of the community.

What the discussion above suggests is that the representation and management of community information to produce a "living" community memory is a very complex task. To address this issue we introduce a type of agent that we call *memory service agent*, whose functionality is designed to capture the events that take place in the community and produce an information model based on these events. To make the information that flows in the system active a memory service agent is capable of disseminating community information. Dissemination here is understood on one hand as *filtering unwanted* information, while on the other *retrieving only relevant* information from the system according to the interests of people stored in profiles stored in personal or static devices. In addition, to make the information stored in the system more like a memory, information that is not used by people "sediments" and becomes difficult to access, while information that is used often "stays on the surface" and is more easily accessible. It is here that we integrate the link–node model we described earlier with the functionality of memory service agents whose role is to manage implicitly the full complexity of the resulting connective tissue for a given community.

12.4 Conclusions

The purpose of this chapter has been to study the provision of ambient intelligence by developing a community system with the aim to support the citizens of a local community in performing their everyday social activities. The viewpoint taken has been that a system of this kind is for the people who use it and not in any sense for the benefit of the developers or social policy makers who will eventually introduce it and maintain it.

The focus of our investigation has been to empower the people of a local community with enhanced interactions that are accepted by all as part of the community environment. Acceptance of this kind can be achieved on the one hand by making people participate in development of the interactive system at all stages, from design to implementation. In this way, people feel they own the system by responsibly and accountably participating in deciding what the system does or the information it is likely to contain. On the other hand the system development must take the people's needs and preferences not only at design time but also while it is operational. In summary, we have tried to build a community system that is accessible by and inclusive to different social strata of the local community, trying to make it part of the people rather than an alienating artefact produced by a group of technical experts.

Our system development, which was based on ethnographic data, suggests that empowerment of people through a community system can be construed as if the community has a "connective tissue", a link–node model that we have used to abstract how people are connected in the community through interaction and access to the community's collective memory.

What appears to create (or strengthen) the links of the tissue are the events that happen between community members at different places of the community. The proposed community system we have experimented with captures a large collection of these events electronically and organises them in a complex but meaningful and easily accessible collective memory. However, as the captured interactions accumulate through system use, the connective tissue becomes difficult to manage.

To deal with the complexity of the connective tissue resulting from a large size of community interactions, in our approach we have advocated a particular software development approach based on software agents. Agents in our study were used as a suitable technology to support proactive and intelligent information management for the connective tissue; such information could be about a single community member, a community group or the community as a whole.

For our experiments we deployed multiple agents in a prototype to support the connective tissue within a specific local community. Our focus was more on demonstrating proof of the connective tissue concepts rather than delivering a fully functional system. Still, our experimental prototype has implemented most of the interactions presented in this chapter, including the way agents capture events in the connective tissue from a set of different networked devices: the coffee table, the large screens in the bus stop or in the shopping centre. The interactions we have implemented are characteristic of proactive information management for people that use the community system, including what information is communicated from an agent to another. The only interaction that we have not implemented was that between the mobile phone and the coffee table we presented in the introduction. The reason for this was that agent platforms running on mobile devices were not available at the time we performed our experimental work. The implemented version of our prototype, however, did support the use of tokens, electronic objects used to capture information in one device and explore this piece of information on another.

From a development perspective, we found multi-agent systems technology mature enough to support the kind of interactions required within community systems, especially information management. Agents not only provide a useful development paradigm, but also an intuitive metaphor for people interacting with the system. However, this last claim we never evaluated properly in our prototype. We believe that the notion of agency, in particular the concept of people interacting with autonomous software components, raises a number of important issues.

One of them is about the kinds of socio-cognitive mechanisms that people and agents would need to share in order to naturally interact between them. Even if we suppose that such human–agent interaction is possible, we also need to cater for frameworks that would make people trust agents and/or multi-agent systems to manage their community interactions, especially those that are private to them. In other words, we would require designing interactive systems whose agents are adaptable and accountable to people's needs. We believe that resolving some of these issues will certainly determine the long-term acceptance of ambient intelligence and, in particular, the kind required for connected community applications such as the one discussed in this chapter.

12.5 Acknowledgements

We acknowledge support of the LiMe (Living Memory) project, part of the EU Esprit I-cubed LTR Programme, Project No. 25621. We especially thank our partners on LiMe: Queen Margaret University College Edinburgh, Philips Design and the Domus Academy. Research on software agents continues by the first author, thanks to partial support by the IST programme of the European Commission, Future and Emerging Technologies under the IST-2001-32530 SOCS project, within the Global Computing proactive initiative. Research on tabletop interactions in the context of communities is continued by the second and third authors, thanks to generous support from Mitsubishi Research Laboratories.

12.6 References

Agostini, A., De Michelis, G., Divitini, M., Grasso, M.A. and Snowdon, D. (2002) Community interactions and memories: the role of community systems. *Interacting with Computers,* **14**(6), 689–712.

Bannon, L. (1997) Dwelling in the "Great Divide": the case of HCI and CSCW. In Geoffrey, C.B., Susan, L.S., William, T. and Gasser, L. (eds), *Social Science, Technical Systems and Cooperative Work: Beyond the Great Divide*. Erlbaum, New Jersey.

Bernsen, N.O. (November 2000) Natural interactivity, *I3 Magazine*, no. 9.

Bødker, S. (2000) Scenarios in user-centred design – setting the stage for reflection and action. *Interacting with Computers*, **13**, 61–75.

Carroll, J.M. (2000) Five reasons for scenario-based design. *Interacting with Computers*, **13**, 43–60.

de Bruijn, O., Spence, R. and Chong, M.Y. (2002) RSVP browser: Web browsing on small screen devices. *Personal and Ubiquitous Computing*, **6**(4): 245–252.

Falcone, R. and Castelfranchi, C. (2001) The human in the loop of a delegated agent: the theory of adjustable social autonomy. *IEEE Transactions on Systems, Man and Cybernetics, Part A: Systems and Humans*, **31**(5), 406–418.

Fleming, J. (1998) *Web Navigation: Designing the User Experience*. O'Reilly, Sebastopol CA, USA.

Harrison, S. and Dourish, P. (1996) Re-place-ing space: the roles of space and place in collaborative systems. In *Proceedings of the ACM Conference on Computer Supported Cooperative Work (CSCW'96)*, Boston MA, USA, 16–20 Nov.

Jennings, N.R., Sycara, K. and Wooldridge, M. (1998) A roadmap of agent research and development. *Autonomous Agents and Multi-agent Systems*, **1**, 7–38.

Kindberg, T., Barton, J., Morgan, J., Becker, V., Caswell, D., Debaty, P., Gopal, G., Frid, M., Krishnan, V., Morris, H., Schettino, J., Serra, B. and Spasojevic, M. (2002) People, places, things: Web-presence for the real world. *ACM Journal Mobile Networks and Applications*, **7**(5): 365–376.

Laurier, E., Whyte, A. and Buckner, K. (2001) An ethnography of a cafe. *Journal of Mundane Behaviour*, **2**(2), 195–232.

Laurier, E., Whyte, A. and Buckner, K. (2002) Neighbouring as an occasioned activity. *Space and Culture*, **5**(4), 346–367.

LiMe (1997) Living Memory Technical Annex. ESPRIT PROJECT 25621. Philips Design, Eindhoven.

Mamdani, E., Pitt, J.V. and Stathis K. (1999) Connected communities from the standpoint of multi-agent systems. *Journal of New Generation Computing*, **17**(2), 381–393.

Markopoulos, P., Eggen, B., Aarts, E.H.L. and Crowley, J.L. (eds) (2004) *Proceedings of the Second European Symposium on Ambient Intelligence (EUSAI'04)*. Lecture Notes in Computer Science, Vol. 3295, Springer, Berlin.

Noack, D. (1994) Blacksburg electronic village: testbed for the future. *Government Technology*, **7**(5), 20–76. (See also http://www.bev.net/)

Norman, D.A. (1998) *The Invisible Computer*. MIT Press, Cambridge, MA.

Norman, T.J. and Reed, C.A. (2002a) Group delegation and responsibility. In *Proceedings of the First International Joint Conference on Autonomous Agents and Multi-Agent Systems*, pp. 491–498.

Norman, T.J. and Reed, C.A. (2002b) A model of delegation for multi-agent systems. In d''Inverno, M., Luck, M.M., Fisher, M. and Preist, C. (eds), *Foundations and Applications of Multi-Agent Systems*. Lecture Notes in Artificial Intelligence, Vol. 2403. Springer-Verlag, Berlin, pp. 185–204.

Pitkin, B. (2001) Community informatics: hope or hype? In *Proceedings of the Hawaii International Conference on Systems Sciences*, Maui, HI, January 2001.

Pitt, J.V., Mamdani E. and Charlton, P. (2001) The open agent society and its enemies: a position statement and research programme. *Journal of Telematics and Informatics*, **18**(1), 67–87.

Rheingold, H. (2000) *The Virtual Community: Homesteading on the Electronic Frontier*. MIT Press, Cambridge, MA.

Singh, M.P. (2000) A social semantics for agent communication languages. In *Proceedings of the IJCAI Workshop on Agent Communication Languages*. Springer-Verlag, Berlin.

Spence, R. (1999) A framework for navigation. *International Journal of Human–Computer Studies*, **51**(5), 919–945.

Spence, R. (2002). Rapid, serial and visual: a presentation technique with potential. *Information Visualization*, **1**(1), 13–19.

Stathis, K., de Bruijn, O. and Macedo, S. (2002) Living memory: agent-based information management for connected local communities. *Interacting with Computers*, **14**(6), 639–642.

Stathis, K. and Toni, F. (2004). Ambient intelligence using KGP agents. In *Proceedings of the European Symposium for Ambient Intelligence*, Eindhoven, November 2004. Lecture Notes in Computer Science. Vol. 3295, Springer-Verlag, Berlin.

Stathis, K. and Toni, F. (2005) The KGP model of agency for decision making in e-negotiation. In *Proceedings of the Joint Workshop on Decision Support Systems, Experimental Economics, and e-Participation*, Graz, Austria, pp. 110–121.

Toni, F. (2001) Automated information management via abductive logic agents. *Journal of Telematics and Informatics*, **18**(1), 89–104.

Toni, F. and Stathis, K. (2003) *Access-as-You-Need: A Computational Logic Framework for Resource Access in Artificial Societies, Engineering Societies in the Agent World III*. Lecture Notes in Artificial Intelligence, Vol. 2577. Springer-Verlag, Berlin.

Witkowski, M. and Stathis, K. (2003) A dialectic architecture for computational autonomy. In Nickles, M., Rovatsos, M. and Weiss, M. (eds), *Proceedings of Agents and Computational Autonomy: Potential, Risks, and Solutions*, Melbourne, Australia. Lecture Notes in Computer Science, Vol. 2969. Springer-Verlag, Berlin.

Zambonelli, F., Jennings, N.R. and Wooldridge, M.J. (2003) Developing multiagent systems: the gaia methodology. *ACM Transactions on Software Engineering and Methodology*, **12**(3), 317–370.

Part D
Mediated Human Communication

Photo: Courtesy of Human Connectedness Group/MLE

Beyond Communication: Human Connectedness as a Research Agenda

Stefan Agamanolis

13.1 Introduction

Our interactions and relationships with other people form a network that supports us, makes our lives meaningful, and ultimately enables us to survive. The Human Connectedness research group has been exploring the topic of human relationships and how they are mediated by technology. This chapter presents, together for the first time, several major pieces of work from the 3 ½ years of the group's existence that highlight the increasingly varied and subtle nature of technology-enabled human communication. Discussion on these projects is collected into seven major sub-themes: extended family rooms, keepsakes of the future, intimate interactive spaces, slow communication, socially transforming interfaces, sports over a distance, and minimisation of mediation.

13.2 The Importance of Human Relationships

Humans have a fundamental need for contact with other humans. Human psychology is undoubtedly a complex subject of study and our understanding of it is always changing, but there is a wide body of evidence and experimentation to support this claim. Lewis et al. (2000) describe in detail the essential regulating effects that social contact and healthy relationships have on human mental and physical well-being, and the consequences that arise from a lack of these requirements. They also recount how these consequences are often most devastating early in life, as was unwittingly discovered in the 13th century by Frederick II, then the King of Southern Italy.

Frederick II ran an experiment to determine if children had an inborn language – a language that they would simply start uttering if they never learned or heard any other in the early years of their lives. His method was to raise a group of children who would never hear language by forbidding their foster mothers and nurses from speaking to and even holding them. The babies otherwise had all their basic survival needs satisfied – they were kept fed, warm and clean. The unexpected and striking result confounded the early researcher: all of the subjects died while they were still infants.

Similar results have been stumbled upon at later points in history, notably during the Second World War during which orphaned infants were kept in institutions employing practices based on the new germ theory of disease transmission. These infants were again kept well fed, clean and warm, but they were not handled or played with in order to lessen the risk of disease exposure. The result was a weak and unhealthy group of infants who, ironically, were drastically more susceptible to the very diseases that the institutional measures were designed to guard against. Forty percent of the infants who caught the measles died, compared to only 0.5% in the general population.

Thus, humans (and in fact all mammals) are not self-regulating creatures. The latest theories propose that our physiology is at least partly regulated and stabilised by others around us through a variety of channels including facial expression, physical touch, hormonal signals and so on. This mechanism for the mutual exchange and internal adaptation between mammals is known as limbic resonance. Disruptions in these channels have serious consequences, especially during infancy but also throughout life:

> Adults remain social animals: they continue to require a source of stabilisation outside themselves. That open-loop design means that in some important ways, people cannot be stable on their own – not should or shouldn't be, but can't be. (Lewis et al., 2000)

Other authors report findings that put a lack of social contact into perspective with other threats to well-being:

> ... social relationships, or the relative lack thereof, constitute a major risk factor for health – rivalling the effects of well-established health risk factors such as cigarette smoking, blood pressure, blood lipids, obesity, and physical activity. (House et al., 1988)

Our interactions and relationships with other people form a network that supports us, makes our lives meaningful and ultimately enables us to survive. Various things threaten our ability to form and achieve balance in the kinds of relationships that we want and need to have with others. Personal circumstances may play a role, such as a need to travel or live in a different place apart from family and friends in order to fulfil work responsibilities. Trends that exist at a societal level have an impact as well and are a subject of increasing investigation. Putnam (2000) describes how people in American society increasingly lack social interactions, and how this loss of social interconnectedness jeopardises health on both a physical

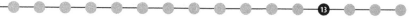
and a civic level. Many of his observations arguably apply to other modern societies as well.

New technologies undoubtedly affect our behaviour in forming and maintaining relationships. The introduction of various forms of transportation has allowed one's circle of friends and family to become increasingly dispersed over great physical distances. At the same time, communication technologies ranging from the postal mail to the telegraph and telephone have allowed people to maintain close relationships that might not have otherwise survived those distances. Internet chat rooms and social networking sites have allowed users to form and carry on friendships and romantic relationships in ways not possible before. Flat-rate home broadband and PC-based video conferencing programs enable distant family members to hold virtual gatherings and "visit" with each other for extended periods of time.

However, the effects of new technologies may not always be supportive of human contact. Automated bank machines, vending machines, self-service gasoline pumps and automobiles reduce opportunities for casual contact with sales clerks, neighbours and other passers-by that we used to have in earlier times. While they can help in maintaining relationships over great distances, anecdotal evidence suggests that telephones and other remote communication devices can also emphasise the existence of that distance and place a strain on a relationship. Some studies, such as one undertaken at Carnegie Mellon University, suggest the use of computers and the Internet may contribute to social isolation and individual stress levels (Kraut et al., 2002). Whether or not a single technology has a positive or negative effect is likely dependent on individual tendencies and the character of the relationship in question.

13.3 The Human Connectedness Research Group

The Human Connectedness research group was established to explore the topic of human relationships and how they are mediated by technology. Its work is grounded by the beliefs that humans have a basic need for social and intimate contact with others and that new technologies can have a positive effect in supporting this contact. The ultimate mission of the group is to conceive a new genre of technologies and experiences that combat some of the effects mentioned earlier and allow us to build, maintain and enhance human relationships in new ways. It also aims to enable new kinds of individual bonds and communities that were not possible before but may be beneficial or fun.

Beyond imagining new forms of communication and social interaction, the group aims to understand how new technologies change the way people can *be related* to each other – in the same way that people are *related to* each other in families or *attached to* each other in close relationships. In addition to applications that address individual relationships, the group has also worked on projects that involve, for example, supporting awareness and collaboration between distant groups of people, and on enabling new forms of cultural exchange.

The group aims to build a technological framework for applications in this domain, taking advantage of the infinite bandwidth and processing-rich computing environments of the future and the opportunity to extend these networked media environments into our architectural surroundings as well as into interfaces that sit close to our bodies and are always with us. Consequently, its innovations often reference the established research domains of ubiquitous computing and wearable computing. The group is equally interested in forming a design framework that includes an understanding of sociological and psychological factors to help shape these systems in a fashion that reflects the needs and sensibilities of the groups within which they operate.

The group gains inspiration for the development of its prototypes from a number of channels that include the results of scientific studies (psychology, sociology), observations of people and how they interact (ethnography) and ongoing dialogues with potential users of new technologies (participatory design). It endeavours to build technologies not just "because we can" but that are a response to human problems and desires. Its guiding philosophy can be summed up in the phrase "When anything is possible, what really matters?"

The Human Connectedness group was formed in September 2001 at Media Lab Europe, the European research partner of the MIT Media Lab. It operated for approximately 3 1/2 years at the lab's premises in Dublin, Ireland. The following sections survey several of the projects undertaken by the group over this period of operation, with additional details and background available in the accompanying references and on the group's website (http://www.medialabeurope.org/hc). These projects are collected into seven major sub-themes:

- Extended family rooms
- Keepsakes of the future
- Intimate interactive spaces
- Slow communication
- Socially transforming interfaces
- Sports over a distance
- Minimisation of mediation

13.3.1 Extended Family Rooms

Modern western homes consist of rooms used for different functions: kitchens for cooking, bathrooms for bathing, bedrooms for sleeping and so on. A typical room in many households is the "family room" – a room in which the family traditionally gathers for various activities: watching television, hosting guests, playing games, etc. Consider a broadening of this concept: the "extended family room", defined as a new kind of room in houses of the future, instrumented with various technologies so that inhabitants can achieve an enhanced sense of connectedness with relatives in different places in space and time.

One of the first prototypes that emerged from our group is a potential fixture of the "extended family room", though we originally developed it with office environments in mind. The system, called *iCom*, connected several laboratory areas at the MIT Media Lab with Media Lab Europe on a 24-hour basis in order to support background awareness, chance encounters and ad hoc audio-visual meetings between remote research colleagues – things we felt were important to build a sense of family and togetherness between the two labs and at the same time were not well supported by other communication media. The notion of a "media space" was pioneered in the late 1980s in research projects like that at Xerox PARC which connected several offices and common areas in multiple geographic settings via continuous audio and video links (Bly et al., 1993). Many variations on the theme of media spaces were developed in a number of research institutes thereafter, such as the VideoWindow system (Fish et al., 1990), Portholes (Dourish and Bly, 1992) and Montage (Tang and Rua, 1994) to name a few.

At each iCom location there was a large-screen projection and seating area visible from and integrated within a larger workspace (Figure 13.1). The characteristics of each connected space were similar and the researchers located in these spaces generally knew each other. A sofa and coffee table in front of the screen emphasised use of the locations as informal socialising areas. The screen displayed several live streams of video captured from two cameras in each location, one mounted above the screen with a view of the surrounding work area and the other situated on the coffee table. A trackball enabled users to rearrange the windows or enable audio connections for meetings or casual interaction with the remote sites. The system addressed some potential privacy issues by synchronising the screen projections at each site. What you saw on the screen is what the other sites saw on their screens, and nothing was recorded or available to view in other places.

We built on prior research in media spaces in a few ways, including adding a community messaging functionality whereby users could send text messages via e-mail to the iCom screen like a bulletin board. These postings could help members of one connected site learn more about what was going on at the other sites beyond what was visible on the screen. The titles of these postings were listed in chronological order with varying size to reflect the age and popularity of a posting, and clicking on a title caused its full text to be displayed (Figure 13.2). The system conserved bandwidth by reducing frame rates where no activity was detected and by adjusting transmitted resolution to reflect the size of each video window. Use of connectionless networking

311

(a)

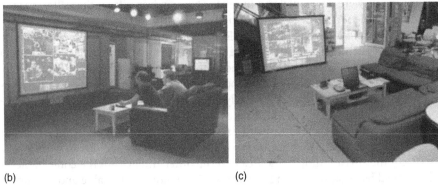

(b)   (c)

Figure 13.1   iCom stations at Media Lab Europe (a) and the MIT Media Lab (b, c).

protocols enabled the system to operate effectively on a commercial (and often problematic) transatlantic Internet connection.

Our iCom system operated more or less continuously for over 3 years and served as a lightweight communication tool for holding project meetings for cross-lab collaborations as well as for informal socialising between remote friends (including occasional flirting). A major aim of the iCom, like some media space efforts predating it, was not to create "yet another video conferencing system", but rather to use audio-visual media in an architectural installation in order to make it feel like the other connected sites were *physically adjacent,* just like an ordinary window is a portal into an adjacent space. This architectural aim was often difficult to explain to visitors who spent only a few minutes in our space. Feedback from those exposed to the

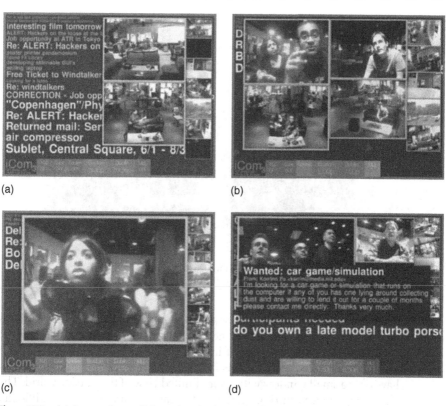

Figure 13.2 (a) Screen shots of iCom showing it in a background state (a), in use for a meeting (b, c) and while reading a posted message (d).

iCom for longer periods suggests that after an initial "novelty" period of a few days, those who frequented the iCom spaces began to think about the installations more in this spirit. Some even reported feelings of isolation when their site was occasionally out of order due to a projector malfunction or network outage.

Feelings of isolation are the focus of another major project of the group called the *Open Window*. Hospital patients often feel isolated from the outside world and disconnected from the people that love them, especially if an illness requires residing within a single room for an extended period. These factors can lead to depression and a reduced potential for healing. The Open Window attempts to counteract these effects by creating an always-on ambient portal from the patient's room to a familiar place or environment to which the patient feels a strong connection.

A collaboration with a haematology ward at a local hospital, the Open Window is particularly targeted to bone marrow transplant patients who must undergo an intense chemotherapy program and are allowed only a

313

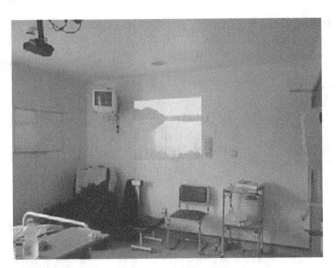

Figure 13.3   Hospital room with an Open Window installation.

limited number of visitors for several weeks while their immune systems recuperate. We interviewed former patients of this ward and studied the characteristics of the ward itself in depth. The rooms the patients inhabit are small and filled with various intimidating medical technologies. Most have only a small window with a very limited view of the outside world. The illness experienced during therapy often causes patients to have difficulty focusing on simple foreground mental tasks like reading a book or watching television. All of these factors contribute additional mental strain and feelings of isolation to an experience that is already very physically challenging.

The Open Window prototype creates a projection on a wall of the patient's room that displays a live yet low frame-rate video stream from a place chosen by the patient, such as a window facing the patient's garden, a room in the patient's house or a favourite hilltop view (Figure 13.3). These video images are captured with Web or mobile phone camera technologies installed in the desired places. The connection is only one-way because, despite their enthusiasm for the overall concept, interviews with patients revealed they generally did not feel comfortable with the idea of camera images from their rooms being transmitted elsewhere, even if it was only to family members or other patients. Given that they have telephones available to them, they speculated just a continuous unidirectional link would be the most helpful.

Like the iCom, the Open Window aims to have an architectural effect – to transform the isolated patient *space* into a *place* that feels physically adjacent to somewhere the patient knows well and considers strengthening. The prototype aims for an ambient design that conveys an impression of the other place while not drawing attention to itself and, most importantly, not overwhelming the patient's senses. The patient sees a single moderately

(a)   (b)

Figure 13.4   The connected café tables from the Habitat project.

static image projection that subtly updates itself once every few seconds or minutes. The hope is that the ongoing presence of this connection will have a positive effect on the patient's mental state and healing potential. This hypothesis is being formally tested as part of the project, though results will take some time to obtain since the prototype is currently installed in only two rooms at the cooperating hospital.

The group has also explored furniture as a platform for awareness applications in the "extended family room", as have a variety of parallel and earlier projects such as Peek-a-Drawer (Siio et al., 2002), Roomware (Streitz et al., 2002), the Water Lamp (Wisneski et al., 1998) and the RemoteHome exhibition (2003). We developed the *Habitat* project, which consists of two networked café or kitchen tables in different locations. For example, one table might be in your own house and the other in that of your elderly parents or grandparents. Each table integrates a computer, an ISO-standard RFID (Radio Frequency Identification) tag reader and a video projector. Unique RFID tags are embedded in objects typically placed on kitchen tables at each site, such as cups, plates, books, games and so on. Placing these items on one table causes messages to be sent to the remote table, which displays a graphical representation of the objects (Figure 13.4a and b). Multiple tagged objects can be sensed simultaneously.

The system operates in both directions, conveying impressions of presence and household activity around the tables at each site. But beyond conveying what objects are on the tables at a single moment, the design of Habitat endeavours to enable a more time-extended awareness of the recent history and rhythms of use of these tables. For example, the longer an item remains

Figure 13.5 Screen shots of Habitat, showing appearance just after a cup is placed on the remote table (a), some time later (b), just after the cup has been removed from the remote table (c) and some time later again (d).

on one table, the larger its image will grow on the remote table. When items are removed, their representations at the far end turn grey and fade away slowly (Figure 13.5). Several other visualisations were developed and testing is planned to investigate which of these are the easiest to understand by different kinds of users. The overall aim of the project is to determine if it is possible to convey awareness of rhythms over a distance, and if doing so can support levels of reassurance and intimacy similar to those possible when living in physical proximity.

13.3.2 Keepsakes of the Future

Keepsakes are an important aspect of human relationships. We often present people who are important to us with objects and things to remember us by and to remind us of each other. How can various technologies enhance the

Figure 13.6   The whiSpiral shawl as worn by a model.

bonds that these keepsakes represent? From a wearable computing point of view, this domain was an interesting one to us because keepsakes are often things that sit close to our bodies and that we carry with us all the time (like pocket watches, wedding rings and so on). Thus in addition to any technical challenge, there is a design challenge of making non-intrusive objects that do not feel like "devices" and that are comfortable to have nearby for long periods.

Inspired by these challenges, we developed the *whiSpiral*, a keepsake that explores how technology can enhance the way garments and accessories evoke memories of relationships. The whiSpiral is a spiral-shaped shawl that consists of several miniature audio recording modules integrated directly in the textile, each capable of storing a short voice message (Figures 13.6 and 13.7). The locations of the modules are made visible by exposing some of their electronic components on the exterior of the shawl, covered by a protective material resembling three white leaves. Microphone connectors are denoted by a yellow leaf.

The whiSpiral could, for example, be given as a going-away present when someone leaves a job or moves to a new place. Friends of the recipient can record messages by removing the yellow leaves one by one, attaching a microphone and speaking into each module while pressing a small button (Figure 13.8a). When finished, the whiSpiral appears completely white and is ready to present as a gift (Figure 13.8b). The whispers are released when sensors located in each audio module detect a soft caress or wrapping

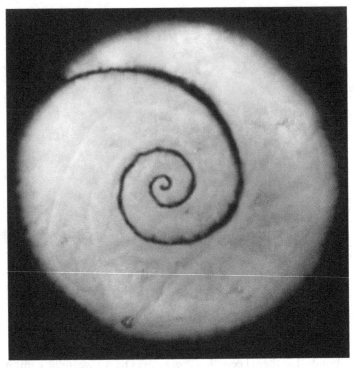

Figure 13.7   The whiSpiral shawl spread on a flat surface to show spiral shape. The audio modules are visible at different points along the spiral.

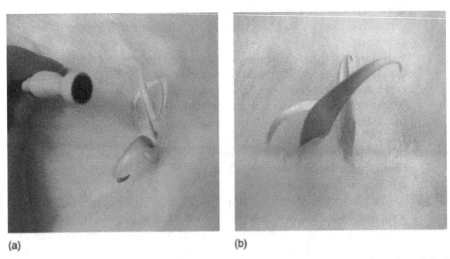

(a)   (b)

Figure 13.8   An audio module integrated in the whiSpiral textile, showing the coloured leaf indicating the location of a microphone connector (a), and appearance after the coloured leaf is removed and a message is recorded (b).

(a)

(b)

Figure 13.9   Appearance of the Floral Display when the flower is closed (a) and open (b).

movement. The messages are stored in persistent memory that will not be erased when the battery occasionally needs to be replaced. The whiSpiral is inspired by the power of a simple human voice to evoke rich memories of a person or relationship, and by the power of a whisper as a medium of intimacy. Similarly, just as they are worn close to our skin, articles of clothing and jewellery presented to us as gifts are reminders of the closeness of our friends and loved ones, especially when we are far away. In this way, the whiSpiral is a keepsake that aims to allow the simple intimacy of a whisper to be carried in a garment that you can wrap around you, take everywhere and even pass down to future generations.

Other projects from the group incorporate network connections between a keepsake and external information about a distant partner, or between two distant keepsakes being held by different people. The *Floral Display*, for example, is a flowerpot with a large pink flower that only blooms when the researcher's girlfriend logs into her computer at her university and closes when she leaves (Figure 13.9).

Figure 13.10 Disassembled Floral Display prototype.

The pot itself is wireless and can be carried from room to room. Inside the pot there is a simple radio transceiver and some motors that allow the flower to be opened and closed on command (Figure 13.10). The pot communicates to a nearby base station computer that checks the girlfriend's login status at regular intervals.

The Floral Display can be thought of as an example of *ambient media,* a topic investigated in many earlier efforts such as the ambientROOM (Ishii et al., 1998) and the AROMA project (Pederson and Sokoler, 1997). Ambient media function in the periphery of sensation without demanding conscious attention, unless perhaps a dramatic condition exists that would warrant interruption. In the process of developing our concept, we realised that it was possible to sense and connect almost any piece of information to any kind of display. In this case, we felt the design problem took priority – that appropriate sensing and display choices based on preferences and the character of the relationship would be more meaningful than the arbitrary mappings enabled by commercial products like the AmbientOrb (Ambient Devices). Thus, the form and function of the Floral Display was created through a dialogue between the researcher and his girlfriend, who said she liked flowers and the colour pink.

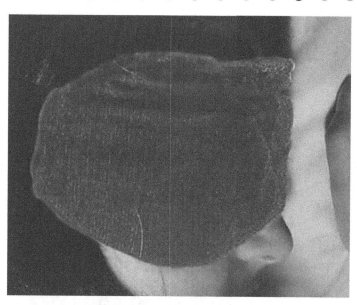

Figure 13.11 The Aura prototype sleeping mask with integrated electro-oculargram.

Partly inspired by the Floral Display, the *Aura* project takes one step deeper in the way it attempts to convey not *what* one is doing but *how* one is doing. Certainly one of the most delicate forms of connection between close partners is rooted in a sense of awareness of each other's emotional state. For example, one partner can often tell if the other is feeling down by interpreting, sometimes unconsciously, a variety of subtle signals to which they have been attuned over a period of time, like body movement, facial expression, voice quality and so on. Physical or temporal separation, consequently, can impede partners from maintaining this kind of intuitive awareness.

Aura investigates the possibility of reinstating this form of subtle awareness regardless of separation. The prototype consists of a sleeping mask with an embedded electro-oculargram that can detect eye movements typical of Rapid Eye Movement (REM) sleep (Figure 13.11).

Data from the mask are used to grossly estimate whether or not the wearer has had a good night's sleep, which is in turn used to infer if he or she is in a good or bad mood the following day. This information is transmitted to the remote location and mapped to music compositions or selections that play inside a precious box recalling a jewellry or music box (Figure 13.12). By opening the box the remote partner can listen to music that was composed from their loved one's previous night of sleep.

Music was chosen as a medium because we felt it was something that could evoke the visceral quality of the emotions inferred from the captured data. Conceptually, Aura aims to enable the user to not only listen to but also *feel*

321

Figure 13.12   The Aura keepsake music box.

their distant loved one's emotional state. The project highlighted a number of difficulties in designing remote awareness systems, especially those that use physiological measurements as a basis for capturing emotion. In the end we felt that a greater understanding of the mechanisms of human emotion was required to produce communication devices capable of abstracting and reconstructing emotional information effectively.

One final "keepsake of the future" to note is the *Portrait of Cati*, a portrait that senses and reacts to the proximity of the spectator. The woman portrayed in the piece, Cati, at first appears neutral and indifferent. When a spectator physically approaches the portrait, her facial expression changes (Figure 13.13). For example, she might smile, frown or look angry or surprised. As the spectator gestures or moves nearer and farther, her expressions become more and less pronounced. When the spectator leaves, she returns to a neutral state. If the spectator returns at a later time, Cati's face may change in a different way. There are a total of 50 different expressions she might make. The proximity of spectators is tracked using a hidden electric field sensing device that can detect extremely small movements, on the order of millimetres. As an experiment in future forms of portraiture, Portrait of Cati suggests ways that technologically enhanced portraits can offer a more dynamic understanding of the identity and personality of the subject and forge a deeper connection between the subject and spectators of the portrait.

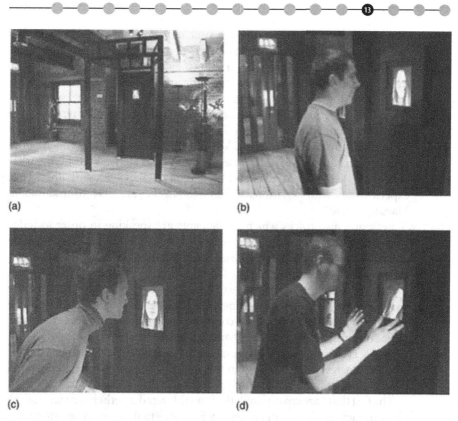

Figure 13.13 Portrait of Cati exhibit, showing the entire structure (a) and three reactions of the subject in the portrait to the presence of a spectator (b–d).

13.3.3 Intimate Interactive Spaces

Interactive communication systems can sometimes feel cold and impersonal, limiting the sense of closeness between distant parties that can be conveyed through them. For example, traditional multi-party video conferencing systems often display participants in separate windows, in scenes that often look like the title sequence from the TV show *The Brady Bunch*. We feel the visual separation characteristic of this kind of design can introduce a confrontational "them versus us" dynamic into a meeting or interaction. Some systems employ audio-based automated camera selection algorithms to switch between views of the active participants. This approach could result in an even greater sense of separation since individual users do not appear on the screen together, and it also limits awareness of the inactive participants.

One way to avoid these potential issues is to get rid of the "window pane", which was the approach we used in the *Reflexion* project. Reflexion is an interpersonal video communication system that operates like a "magic mirror" in which you see a reflection of yourself overlapped with the reflections of other participants in remote locations. Each participant uses a separate station equipped with a camera, monitor and computer connected to the Internet. Using a custom image segmentation algorithm, the system extracts the participants from their backgrounds at full frame rate, and then combines them all together into a single video scene. The participants are composited over a common backdrop, which could be an image, a document or a movie. This interface metaphor was derived from an earlier system created in the Media Lab (Agamanolis et al., 1998), and others have experimented with variations of it as well (Ishii et al., 1994; Morikawa and Maesako, 1998).

Reflexion also tracks which participants are speaking in order to judge who is the centre of attention. Active participants are rendered opaque and in the foreground to emphasise their visual presence, while other less-active participants appear slightly faded in the background in a manner that maintains awareness of their state without drawing undue attention (Figure 13.14). The system smoothly transitions the layering and appearance of the participants as their interactions continue. Every participant sees exactly the same composition in order to enhance a sense of inhabiting a shared space. A central server handles control messages that synchronise the screen compositions at each station. The system uses a multicast peer-to-peer networking strategy for audio and video transmission to achieve low latency.

The fact that participants in Reflexion are layered together and can "touch" and interact with each other directly in the virtual video scene potentially creates a space with a more intimate social dynamic, one that could be more appropriate for many kinds of applications. Feedback from informal user tests suggests that Reflexion might be a useful device for family communication – for example, to connect the households of members of an extended family who want to gather and watch old home movies or perhaps a live soccer match on television. Other scenarios were also suggested, such as distance learning environments and multi-user remote interactive theatre spaces.

We used the same custom image segmentation algorithm in a different type of installation intended to create connections between people separated by *temporal* rather than spatial distance. This installation, called *Palimpsest*, consists of a large rear-projection screen and camera aimed across an interaction area, which could be a hallway or passage inside a building, or a special area dedicated to the installation. A computer digitises camera images and controls the projected video display. Images of passers-by or participants entering the interaction area are extracted from the background and layered into a video loop that repeats itself every few seconds.

Because the video is looped, if a passer-by lingers in the space, she will see a delayed copy of herself entering the space from several seconds ago, and even more layers if she remains longer, together with the layers generated

Figure 13.14   Screen shots of the Reflexion system in use for a meeting. The three participants are located in different places.

by other passers-by from earlier points in time. These layers accrue on the screen over several minutes, hours, even days, creating a visual that collapses time and compresses the recent social goings-on of the given space.

A Palimpsest is a manuscript consisting of a later writing superimposed upon an original writing. This word was borrowed for the title of this project that aims to superimpose layers of recorded social interaction and present them as a single visual. Increasing the persistence of these interactions potentially raises awareness of the social history of a place and allows the

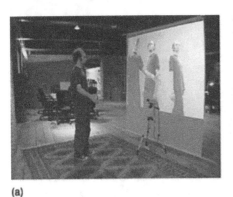

(a)   (b)

Figure 13.15 A passer-by interacts with himself as captured at an earlier point in time in the Palimpsest prototype at Media Lab Europe.

viewer to witness the human crowd that has passed through a seemingly quiet and empty space. Even if totally alone, a passer-by is able to "transcend time" and become a part of this community, and to interact with its members, including oneself (Figure 13.15a and b).

In addition to an ongoing placement in our laboratory, we installed the Palimpsest at a local film festival (Figure 13.16). Once passers-by understood what was happening on the screen, a few actually used it to investigate if

Figure 13.16 A young girl dances with delayed copies of herself and other earlier spectators in an installation of Palimpsest at a local film festival.

certain friends or companions were also present at the venue. Many suggested changes, such as increasing the loop duration or adding sound capabilities. Occasionally a passer-by would make an obscene gesture, but interestingly, after realising such a gesture would repeat itself over and over, many of these perpetrators would return and try to conceal their gestures. Others, however, seemed to gain huge satisfaction from filling the screen with as many looping obscene gestures as possible. The Palimpsest also sparked a new research project investigating new forms of interactive theatre in which a single actor can play all the parts in a short play, layering in each performance after the last in an asynchronous fashion.

However, prototypes like Reflexion and Palimpsest still have their limitations. We felt that with both of these systems, as well as with iCom and other media space efforts described previously, it was still possible for users or passers-by to perceive a sense of separation from each other because they see each other through wide-angle views captured by cameras mounted far away from them. The user always has to maintain a certain distance from these cameras in order to be seen by the remote participants in the space. The intuitive behaviour in real life of physically moving nearer to or farther from people, and any social meaning underlying such movement, does not translate well into these spaces.

In order to allow a greater sense of intimacy, we developed a new computer vision system to enable interaction at a very short distance to the screen surface, to the point that users can actually touch it. The installation, named *Passages*, creates a bidirectional link between distant locations in which passers-by can interact with and touch the projected silhouettes of their remote counterparts as captured by the vision system. This system, which involves a camera mounted behind the user pointing towards the interaction screen, maintains a near-perfect registration of the participants' bodies to their projected silhouettes even when they are standing directly against the surface. The overall aim of the installation is to create the feeling of touching the shadow of someone standing directly behind a translucent screen (Figure 13.17).

13.3.4 Slow Communication

Trends in the design of interpersonal communication technologies in the past several years have been towards efficiency, flexibility and mobility. The mobile telephone, for example, allows you to talk on the telephone anytime, anywhere, no matter what else you might be doing. They also provide numerous additional features not directly related to communication: clocks, calendars, games and so on. Audio quality may not be the best, but it is enough to get a message across. In all these ways, the mobile telephone is a little bit like "fast food".

Figure 13.17   A passer-by interacts with the silhouette of a remote counterpart in an installation of Passages.

The work of our group proposes a converse approach to the design of communication technologies, one that results in the equivalent of a gourmet restaurant experience. Just as the "slow food" movement celebrates fine food and the enjoyment of an entire meal experience, "slow communication" devices can be thought of as celebrating our ability to telecommunicate. Such technologies do not necessarily emphasise mobility and large sets of features, but rather the purity and singularity of the communication experience.

One example of such a technology is the *Iso-phone*, which could be described simply as a telephone combined with a flotation tank. Iso-phone users are submersed completely underwater and wear a special mask that blocks out external visual distraction while providing compressed air for normal breathing (Figure 13.18). An earlier prototype involved a helmet that held the user's head just above the water (Figure 13.19). The water is heated to body temperature, dulling the sense of touch and blurring the physical boundaries of the user's body. The only sensory stimulus presented is a two-way voice connection to another person possibly using the same

Figure 13.18 Iso-phone mask in use completely under water.

apparatus in another location. Underwater headphones deliver high quality sound while insulating the user from external noise pollution.

In some ways an Iso-phone tank could be thought of as an extreme version of the traditional telephone booth or telephone box. Part of our aim in engaging in an unusual experiment like the Iso-phone was to challenge mainstream design practice as well as people's assumptions about what telecommunication can be like. Equally important was our desire to investigate the value of sensory deprivation in communication systems in a way that important user behaviours might be revealed more quickly than, for

Figure 13.19 Iso-phone initial prototype helmet.

Figure 13.20 Iso-phone tanks as exhibited in the Hauptplatz of Linz, Austria, during the 2004 Ars Electronica Festival.

example, if we just made a minor change to a current technology. Feedback from trials of the Iso-phone at exhibitions, such as that at Ars Electronica (Figure 13.20), suggests that the system enables a more relaxed stream-of-consciousness style of conversation that might be well-suited to activities requiring high levels of creativity such as brainstorming meetings. Participants also lose their sense of the passage of time and display unexpected behaviours, such as gesturing not only with the arms but also with the legs and the entire body, which are unencumbered while floating in the tank.

Another project addressing the theme of "slow communication" is *Mutsugoto/Pillow Talk*, an intimate communication device intended for the bedroom environment. Mutsugoto was inspired by the observation that more people now than ever carry on long distance relationships with romantic partners, sometimes for extended periods of time. However, the communication systems used by these partners to stay in touch are almost always impersonal and generic. E-mail, for example, is often read and written on the same computer and at the same desk that one uses for any other kind of communication. Phone calls and SMS messages are sent and received between partners on the same devices used for work and business. The form and function of Mutsugoto is designed to more strongly reflect the character

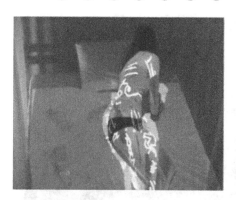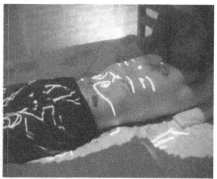

Figure 13.21   Mutsugoto installation showing drawings projected on the bodies of users.

of an intimate bond, specifically by allowing distant partners to communicate through the language of touch as expressed on the canvas of the human body.

Mutsugoto is meant to be installed in the bedrooms of two remote partners. Each partner lies on the bed and wears a special ring that emits an infrared beacon, visible only to a camera mounted above. A computer vision system tracks the movement of the ring finger and projects virtual pen strokes on the user's own body. The silhouette of the user is also captured and serves as the "canvas" for this drawing. The completed drawing is transmitted to the remote site where the same silhouette is projected softly on the bed. After lying in the same position, the distant partner can reveal the drawing by tracing the ring finger around the body. Special bed linens and curtains were crafted to enhance the mood of this romantic communication environment (Figure 13.21).

13.3.5 Socially Transforming Interfaces

A "socially transforming interface" can be defined as an interface that, when wielded, enables its user to become a more sociable entity than he or she would be normally. Such an interface might allow two people to meet and talk to each other who might not have otherwise. The interface or the technology acts as a facilitator, breaking the ice between people and creating conditions that make interaction easier and less intimidating to enter into. Having broken the ice, a basis may be formed for interactions to continue after the interface is left behind and the experience finished.

One possible example of a socially transforming interface is the *WANDerful Alcove*, a playful installation in which participants use magic wands to interact with an enchanted landscape in a large screen projection. Two magic wands are instrumented with hidden three-axis accelerometers

Figure 13.22   Two people wielding wands in the WANDerful Alcove installation.

and connected via a radio link to a base station computer. The computer detects gestures made with each wand and causes magic spells to happen in the onscreen environment (Figure 13.22).

The intention of the installation is for the magic wand interface to have a socially transforming effect on its users, serving as a catalyst for ad hoc interaction and collaboration in the story experience. Two wands are made available in the installation, one that creates lighting and another that creates explosions. With some experimentation, the novice wizard will gain skills in the use of the wand and learn special magic gestures that cause a more controlled reaction in the story scene, either the creation of a rainstorm or the emergence of the sun (Figure 13.23a and b). The participant is challenged to be physically active, focusing not only on his own actions but also on that of the other, to share magic power and create something together. For example, if the two wizards collaborate and perform their special gestures at the same time, a rainbow emerges (Figure 13.23c).

Initial meetings and introductions mark the first moments of building new relationships. Yet these important moments are often awkward or forgotten, sometimes because of the natural failings of human memory (not being able to recall someone's name) or because there is a lack of a catalyst for a richer interaction. Whereas the WANDerful Alcove strove to break the ice through immersion in the role of a sociable character, another project in the group, the *iBand*, tries to address these problems in a more explicit way. The iBand is a wearable bracelet-like device that exchanges information

(a)

(b)

(c)

Figure 13.23   Screen shots of the WANDerful Alcove.

about its users and their relationships during the common greeting gesture of the handshake, which is detected by the device (Figure 13.24).

Earlier products and prototypes, like the Lovegety, Synchro.beat (Swatch) and GroupWear (Borovoy et al., 1998) can help people who have similar interests or profiles meet each other. Others like nTag, CharmBadge and SpotMe are focused on supporting this functionality in conference

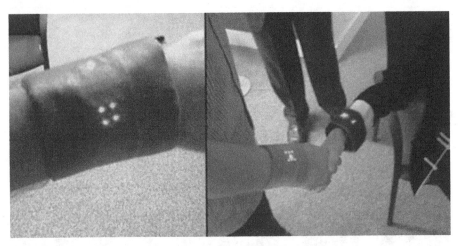

Figure 13.24 The iBand prototype in use during a handshake.

scenarios. Zimmerman (1996) uses touch to transmit information via a weak electrical current running between two people's bodies. The iBand differs from these related efforts in the way it aims to augment gestural language by transmitting information only during a handshake. When worn, the circuit board and battery lay flat under the wrist and an infrared transceiver is positioned near the back of the thumb pointing towards the hand such that it is visible to an IR transceiver on another device when shaking hands. A handshake is detected via infrared transceiver alignment combined with hand/wrist orientation and gesture recognition using a two-axis accelerometer.

In a full experience with the iBand, the users first enter contact/biographical information into a kiosk, which stores it in a database and assigns a unique ID number to their iBand. The users can also create a personal logo that appears on the LED display woven into their device. When the user shakes hands with another iBand user, ID numbers and logos are exchanged and stored. The LED display cycles through the stored logos at a pace reflecting the number of hands that have been shaken. When the user returns to the kiosk, it displays a list of new contacts by looking up the collected ID numbers in the database.

The iBand project seeks to explore potential applications at the intersection of social networking and ubiquitous computing. Social networking sites like *Friendster* and *Orkut* allow people to build relationships in an active social cyberspace, and the iBand attempts to break the concepts behind these popular sites outside of the web browser and into everyday life. For example, when one shakes hands with someone else, their iBands could tell them what friends they have in common, what business colleagues they have both dealt with or what people they have both met recently.

During two evaluations of the iBand at actual social networking events, it became clear that it was impossible to introduce a new parallel or background technology without changing the fundamental nature of a subtle gesture like the handshake in some way. The iBand essentially combines two conventional gestures (the handshake and the business card exchange) into a single new "handshake and business card" gesture and effectively eliminates the availability of either old gesture on its own. Thus, iBand users are handicapped, and at the same time they are enabled in a different way with an *invented* gesture. Consequently, users lack a set of rules and conventions about how to use the new "techno-gesture" and must create these on their own as they use it. This suggests that through careful design, it may be possible for a wearable technology to enable an invented language of social gestures with its own affordances and constraints that have particular effects in certain social situations. Such gestures might better support particular applications for which traditional gestures are felt to be inadequate.

However, before you even get to the point of shaking hands with someone, there is still the problem of finding an interesting person to talk to and the right conditions under which to do so. Take urban environments for example, which can often be unfriendly and isolating places in which to live and work. Nevertheless, sometimes you might find yourself open to a chance encounter with a stranger or curious about someone nearby yet unable to initiate contact for fear of appearing awkward or intrusive in an environment already full of distractions.

Inspired by these observations, we wanted to explore ways that new technologies could facilitate very lightweight contact between people who happen to be nearby. We felt that one way to support such contact would be through the creation of a shared experience that could serve as a starting point or catalyst for an interaction. In the *tunA* project, the main focus of this shared experience is music. tunA is a mobile wireless application that allows users to share their music locally through hand-held devices. Users can either listen to their own music or "tune in" to other nearby tunA music players and listen to what someone else is listening to (Figure 13.25).

The application displays a list of tunA users that are in range, gives access to song information and enables peer-to-peer audio streaming over ad hoc WiFi links (Figure 13.26). The audio stream timing/delay algorithm enables the audio playback to be closely synchronised on the source and any destination devices. This enables someone to, for example, tune into another person sitting across the aisle on a bus or train, and each could be nodding heads, gesturing or dancing in perfect synchrony, just as if he or she were both listening to the same conventional radio station. SoundPryer (Axelsson and Ostergren, 2002), Bubbles (Bach et al., 2003) and Sotto Voce (Aoki et al., 2002) are a few close relatives of tunA but lack this tight synchronisation aspect.

tunA users can "bookmark" songs that they hear while tuned into someone else's player, and later review these bookmarks, or transfer them to a

Figure 13.25 Illustration of the concept underlying the tunA project.

computer where they might purchase the songs for themselves. They can also bookmark another person they have come into contact with through tunA, and, for example, be notified if that person comes into range again. An instant messaging feature allows users to exchange short text messages without necessarily knowing anything about each other except what they are listening to. In user studies, this messaging feature was felt to be an essential component to support the desired ice-breaking effect of the application.

13.3.6 Sports over a Distance

Sports and physical exercise are good ways of introducing people to each other and maintaining relationships. For example, if you move to a new job, you might play a game of tennis or go to the gym with a co-worker to break the ice and get to know each other better. In addition to their physical benefits, various studies have shown that working up a sweat and getting your adrenaline moving while playing sports increases your propensity to bond socially with teammates and competitors (Zahariadis and Biddle, 2000). Inspired by this, the group created *Breakout for Two*, an interactive installation for playing sports over a distance that aims to combine the positive effects of sports described above with the advantages of telecommunication technology in bridging distance.

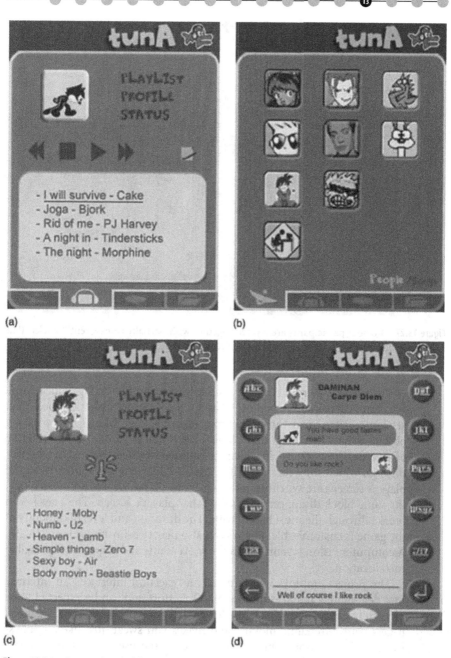

Figure 13.26 Screen shots of the tunA user interface, showing the local user's song playlist (a), a pictorial list of other users in range (b), the playlist of a remote user (c) and the instant messaging interface (d).

Figure 13.27   Remote participants are layered together with partially transparent "blocks" that must be struck during the course of a game of Breakout for Two.

The game is a cross between soccer, tennis and the vintage video game Breakout. Participants in remote locations throw or kick a physical soccer ball at a large wall to break through a projection of virtual "blocks" that partially obscure a live video image of the other player (Figure 13.27). The effect is one of a virtual game "court" in which the competitors are separated by a barrier through which they can interact (Figure 13.28). The game boards on the player's screens are synchronised – when one player breaks through a block, the same block disappears from the other player's screen. The player who breaks through the most blocks wins. The duration and level of difficulty of the game is customisable. Players can also play two-on-two (Figure 13.29). A computer vision system tracks the movements and impacts of the ball at each location.

The project employs what we call an "exertion interface", an interface that deliberately requires intense physical effort and can be expected to be physically exhausting when used for an extended period of time. In short, it gets your adrenaline moving and makes you sweat, just like any physical exercise or sport. Others have explored the use of exertion in computer interfaces (Telephonic Arm Wrestling, Virtual Tug-of-War), particularly in commercial game machines (Dance Dance Revolution, Kick and Kick), but none that we found have addressed exertion and broadband audiovisual telecommunication used together to create an online sport environment.

Figure 13.28   A Breakout for Two installation, illustrating the two-sided virtual "playing court" effect created by the large screen display.

Our hypothesis was that augmenting an online sport or gaming environment with exertion would enhance the potential for social bonding. The heightened state of arousal induced by the exertion also potentially makes the interaction more memorable. We conducted a study to test these hypotheses and to evaluate the effects of exertion interfaces. The results were encouraging: players of Breakout for Two said they got to know each other

Figure 13.29   Scenes from a two-on-two game of Breakout for Two.

better, felt the other player was more talkative and were happier with the audio and video quality in comparison to a control group playing an analogous game using a traditional non-exertion keyboard interface. Further studies are needed to investigate these effects more thoroughly.

13.3.7 Minimisation of Mediation

For a variety of reasons, we do not always have a good sense of what "everyday life" is like in other places in the world, and having this sense might be helpful in improving relationships between people in different cultures. Our impressions of distant societies and their inhabitants are often mediated by the biases of numerous "third parties" – researchers, directors, producers, camera people, distributors, censorship organisations and so on. Additional constraints arise from inherent restrictions of popular forms of media: at the very least, experiences must fit a certain time slice or page count to be considered palatable to a mass audience, and therefore editing must occur. We feel these kinds of factors degrade the full sense of awareness and appreciation that people can achieve of other peoples and places, above cultural stereotypes and clichés. Inspired by these observations, we developed *RAW*, a new kind of audiovisual recording tool together with a method for processing and presenting the material captured with the tool, that aims to enable a more direct, minimally mediated relationship between its user and a later audience, possibly in a far away place or time.

The RAW tool consists of a digital still camera and a high-quality digital stereo audio recorder that captures 60 seconds of sound *before* and *after* a picture is taken. Audio is recorded binaurally using high-quality miniature microphones that are placed in the user's ears, just like "earbud" headphones (Figure 13.30). This design was chosen in an attempt to enable the later audience to immerse themselves "into the shoes" of the person who originated the content they are experiencing, and to place greater emphasis on the subjective point of view of this original source. Material captured with the RAW tool is archived in a raw form, with no deletion or modification allowed by the user or any third party acting between the user and the audience.

In the domain of video-making we felt that audio was typically considered secondary – that audio was seen as supplementing the visuals, not the other way around. At the same time we believed audio, and especially ambient audio, held great potential for conveying certain kinds of impressions of everyday life with a richness not possible with visual media. We decided against using video as a recording medium because we wanted to innovate in the field of audiovisual expression so that sound and image could have a different and more equal dialogue. The design of the RAW system aims to consider both sound and image of the same importance. The audio provides context to the image and the image provides a context for the audio.

Figure 13.30   Depiction of the RAW audiovisual recording tool, consisting of a digital still camera and ear-microphones for binaural audio recording.

A handful of other projects have experimented in the domain of "audiophotography" (Frohlich et al., 2000; Martin and Gaver, 2000), though none have addressed the capture of audio before a photograph is taken.

Since a tool like RAW would potentially be utilised in diverse areas of the world, we felt it was critical that it be tested and considered relevant and valuable within a plurality of cultures. We first conducted workshops in Dublin and Paris which helped us refine a method of presenting RAW records in a way that maintains the "minimal mediation" ideal of the project. An interactive installation with a timeline allows audience members to jump to desired points in each record, or just to sit back and let the moments come and go at the same pace they were captured (Figure 13.31). Each photograph reveals itself slowly while the pre-photo sound plays, and then fades away slowly during the post-photo sound segment (Figure 13.32). This cycle progresses more rapidly if several photographs were taken in close succession.

We also mounted a large-scale study over 3 weeks in the African country of Mali during which 25 people used the RAW tool to capture aspects of their daily life. The records created by our participants were inspirational to us in their originality and immediacy. Through this and the earlier experiences, we were also able to uncover several distinct categories describing the ways participants would make use of the tool for different storytelling and documentation purposes. These span a range from social modes of

341

Figure 13.31 Screen shot of the RAW exhibition program showing the timeline interface.

engagement to more journalistic styles and inwardly reflective commentaries, and they emerged despite the care we took to not suggest any particular styles or themes to our participants in our introductory discussions with them. Overall, the feedback from both participants and viewers of the resulting exhibitions reaffirmed our belief that introducing a new technology does not always have to increase a sense of something "standing in between" two parties in a relationship.

13.4 Directions for the Future

After operating in Dublin, Ireland for over 4 years, Media Lab Europe closed its doors in January 2005 due to unfortunate political circumstances stemming from a lack of an agreement between its major stakeholders on how its activities should be funded and managed in the future. As a result, the seven research groups based at the lab at the time ceased their operations, and about 50 talented researchers were forced to start looking for new opportunities.

Figure 13.32 Storyboard depicting the way an image in the RAW exhibition is gradually revealed and then faded to black during the playback of the audio recorded before and after the photo was taken.

Nevertheless, the vision of the Human Connectedness group is still alive and some of the projects discussed above are already moving forward in new locations on independent tracks of commercialisation or further research. The author hopes that the group will be able to resume operation in some form in the future.

13.5 Acknowledgments

Thanks to all the members of the Human Connectedness group and its collaborators for their great ideas and hard work in developing these and other projects: James Auger, Arianna Bassoli, Mike Bennett, Joëlle Bitton, Jonah Brucker-Cohen, Alice Buckee, Elena Corchero, Céline Coutrix, Cian Cullinan, Anna Gavin, Fran Hegarty, James Harris, Tomoko Hayashi, Marije Kanis, Matthew Karau, Jimmy Loizeau, Ciaran McGrath, Julian Moore, Florian Mueller, Aoife Ní Mhóráin, Dipak Patel, Ben Piper, Denis Roche and Niall Winters. Please see http://www.medialabeurope.org/hc for detailed project credits. Thanks to Bakhtiar Mikhak for the "extended family room" metaphor.

13.6 References

Agamanolis, S., Westner, A. and Bove, V.M., Jr. (1998) Reflection of presence: toward more natural and responsive telecollaboration. *Proceedings of SPIE Multimedia Networks*, **3228A**, 174–182.

Aoki, P.M., Grinter, R.E., Hurst, A., Szymanski, M.H., Thornton, J.D. and Woodruff, A. (2002). Sotto Voce: exploring the interplay of conversation and mobile audio spaces. *Proc. CHI 02*, ACM Press, New York, pp. 431–438.

Axelsson, F. and Östergren, M. (2002) SoundPryer: joint music listening on the road. In *Ubicomp 2002 Adjunct Proceedings*, Viktoria Institute, Göteborg, Sweden, pp. 39–40.

Bach, E., Bygdås, S.S., Flydal-Blichfeldt, M., Mlonyeni, A., Myhre, Ø., Nyhus, S.I., Urnes, T., Weltzien, Å. and Zanussi, A. (2003). Bubbles: navigating multimedia content in mobile ad-hoc networks. *Proc. 2nd International Conference on Mobile and Ubiquitous Multimedia*, Linköping University Electronic Press, Linköping, Sweden.

Bly, S.A., Harrison, S.R. and Irwin, S. (1993) Media spaces: bringing people together in a video, audio, and computing environment. *Communications of the ACM*, **36**(1), 28–47.

Borovoy, R., Martin, F., Resnick, M. and Silverman, B. (1998) Groupwear: nametags that tell about relationships. In *CHI 98 Late-Breaking Results*, ACM Press, pp. 329–330.

Dourish, P. and Bly, S. (1992) Portholes: supporting awareness in a distributed work group. In *Proceedings of CHI 92*, ACM Press, pp. 541–547.

Fish, R., Kraut, R. and Chalfonte, B. (1990) The videowindow system in informal communications. In *Proceedings of CSCW 90*, ACM Press, pp. 1–11.

Frohlich, D., Adams, G. and Tallyn, E. (2000) Augmenting photographs with audio. *Personal and Ubiquitous Computing*, **4**(4), 205–208.

House, J.S., Landis, K.R. and Umberson, D. (1988) Social relationships and health. *Science*, **241**(4865), 540–545.

Ishii, H., Wisneski, C., Brave, S., Dahley, A., Gorbet, M., Ullmer, B. and Yarin, P. (1998). ambientROOM: integrating ambient media with architectural space (video). *CHI 98 Conference Summary*, ACM Press, New York, pp. 173–174.

Ishii, H., Kobayashi, M. and Arita, K. (1994) Iterative design of seamless collaboration media. *Communications of the ACM*, **37**(8), 83–97.

Kraut, R., Kiesler, S., Boneva, B., Cummings, J., Helgeson, V. and Crawford, A. (2002). Internet paradox revisited. *Journal of Social Issues*, 58(1), pp. 49–74.

Lewis, T., Amini, F. and Landon, R. (2000) *A General Theory of Love*. Random House, New York.

Martin, H. and Gaver, B. (2000) Beyond the snapshot from speculation to prototypes in audiophotography. In *Proceedings of DIS 2000*, ACM Press, pp. 55–65.

Morikawa, O. and Maesako, T. (1998) HyperMirror: toward pleasant-to-use video mediated communication system. In *Proceedings of CSCW 98*, ACM Press, pp. 149–158.

Pederson, E.R. and Sokoler, T. (1997) AROMA: abstract representation of presence supporting mutual awareness. In *Proceedings of CHI 97*, ACM Press, pp. 51–58.

Putnam, R. (2000) *Bowling Alone: The Collapse and Revival of American Community*. Simon & Schuster, New York.

Siio, I., Rowan, J. and Mynatt, E. (2002) Peek-a-drawer: communication by furniture. In *CHI 02 Extended Abstracts*, ACM Press, pp. 582–583.

Streitz, N., Prante, T., Müller-Tomfelde, C., Tandler, P. and Magerkurth, C. (2002). Roomware: the second generation. *CHI 02 Extended Abstracts*. ACM Press, New York, pp. 506–507.

Tang, J. and Rua, M. (1994) Montage: providing teleproximity for distributed groups. In *Proceedings of CHI 94*, ACM Press, pp. 37–43.

Wisneski, C., Ishii, H., Dahley, A., Gorbet, M., Brave, S., Ullmer, B. and Yarin, P. (1998). Ambient displays: turning architectural space into an interface between people and digital information. *Proc. CoBuild 98 First International Workshop on Cooperative Buildings*, Lecture Notes in Computer Science, vol. 1370, Springer, Heidelberg, pp. 22–32.

Zahariadis, P. and Biddle, S. (2000) Goal orientations and participation motives in physical education and sport: their relationships in English schoolchildren. *Athletic Insight – The Online Journal of Sport Psychology*, **2**(1).

Zimmerman, T.G. (1996) Personal area networks: near-field intrabody communication. *IBM Systems Journal*, **35**(3–4), 609–617.

13.7 Websites of Other Referenced Projects and Products (All Last Visited 12 April 2005)

Media Lab Europe, http://www.medialabeurope.org

Human Connectedness, http://www.medialabeurope.org/hc

Slow Food, http://www.slowfood.com

Charm Tech Badge, http://www.charmed.com

nTag, http://www.ntag.com

SpotMe, http://www.spotme.ch

Synchro.beat, Swatch, http://www.swatch.com/synchro/index2.html

Telephonic Arm Wrestling, http://www.normill.ca/artpage.html

Virtual Tug-of-War, Ars Electronica Futurelab, http://www.aec.at/en/futurelab

Dance Dance Revolution, Konami, http://www.konami.com

Kick and Kick, Konami, http://www.konami.com

Ambient Devices, http://www.ambientdevices.com

The RemoteHome, http://www.remotehome.org

The Presence Project: Helping Older People Engage with Their Local Communities

14

William Gaver and Jacob Beaver

> 'You are old, Father William,' the young man said,
> 'And your hair has become very white;
> 'And yet you incessantly stand on your head—
> Do you think, at your age, it is right?
>
> Lewis Carroll, *Alice in Wonderland*

14.1 Introduction

Anybody who has witnessed the explosive growth of mobile telephony, i-Pod fetishism and online grocery deliveries knows that digital and electronic technologies can work profound changes in our personal, social and cultural lives. We might hope, however, that new technologies will go beyond commercial value to provide societal ones as well, helping address some of the significant societal problems of our time. This might mean rethinking our approach to technologies – what they can do, how we interact with them and the methods we use to develop them.

Presence was a 2-year project that investigated ways that technology be used to increase the presence of older people in their local communities. At the outset, our ambition was to discover whether digital technology could be used to alleviate the risks of increasing isolation and alienation amongst older members of an ageing population. As the project progressed, however, it became clear that even this daunting issue was only the tip of an iceberg. For to understand whether we could increase the presence of older people, we had to address a host of related, equally thorny, issues, issues that are fundamental to understanding how technology should be designed more generally. How should we think about older people and what approaches might help us learn about them? What functions might technology offer them, and what forms might it take? Finally, what methods can we use to design technologies that integrate deeply and well with everyday lives and local communities?

In this chapter we describe how we addressed these issues in the course of designing to increase the presence of older people. To start with, we discuss how we came to understand older people and the approach we took to the project. Our answers to these two issues went hand in hand and led us in unexpected directions. For as we developed clarity about how we wanted to address the older people, our methods became increasingly idiosyncratic. Better to call it methodically idiosyncratic, since our approach was far from frivolous, although it was often playful.

14.2 Methodical Idiosyncrasy

To understand this seeming contradiction, it might be helpful to start by considering a comment by Kenneth Koch, a poet who spent some time teaching poetry in a nursing home in New York City. In an article for the *New York Review of Books* (20 January 1977), Koch described the problems and pleasures of his experience, which were two sides of the same coin.

> There is a hush-hush feeling about old people, the feeling that what they really want and need is quiet and peace ... [and that challenging] subjects, all having to do with power, energy, and passion, will cause them pain and make them feel empty, because they feel unconnected to them, being themselves weak and infirm. Both ideas are wrong. Strong feelings do not vanish. Passion and energy are what life is all about ... (p. 42)

It seems that little has changed in 25 years, except that Koch says "old people" and we now say "older people". Koch's problem as a teacher of poetry was our problem as designers: on our side, we had to overcome the stereotype of older people as frail, needy and "nice"; on their side, they had to overcome the stereotype of modern technologies as complicated, screen-based and involving global information.

How does one overcome stereotypes? The short answer must be: by focusing on the particular and eschewing generalisations, which can tend towards the stereotypical. A longer answer, which we can only sketch here (see Gaver and Hooker, 2001, for a fuller account), lies in a questioning of the scientific method – the traditional model of research in new uses of technology. For science tends to privilege the general over the particular, blurring minor differences in order to focus on the bigger picture. This is a powerful strategy for reaching fundamental understandings, but it has its costs. The problem is not that science itself produces stereotypes, since over-generalisations are avoided by delineating the circumstances in which findings are expected to hold. The problem is that this way of working enforces a kind of prejudice against the one-off, the subjective, the interesting but unjustifiable – all of which can serve as a basis for the designer. In other words, science's bias towards essential principles distances the scientist from the details of the field of inquiry. For the scientist, this is both necessary and valid. But for the scientist-designer, such distance has to be dangerous. It carries the risk of stereotyping, among other things, because it fosters intellectual and emotional immunity. Whatever the scientist-designer might learn about what Koch calls the "strong feelings" of specific older people, they would not be able to acknowledge these ambiguities because they cannot express them, at least not in objective terms. In this respect our scientist-designer would be, as it were, ontologically mute.

Three key decisions arose from this rationale.

First, to overcome stereotypes on both sides, we needed to reveal the older people, and ourselves, as individuals. We needed to see them as particular people in a particular place, with particular desires and frustrations. And they needed to see us not as experts who flew in with esoteric agendas, but as ordinary people, with our own interests and foibles. Otherwise, why should they venture to expose themselves to us? And how could we presume to design anything of use to them?

Second, the design process was as important as the result. In a word, it had to be engaging, both for us and for the older people. If a person is not engaged by something, they would not really exert themselves on its account – and exertion, or passion, discloses character. Or to put it another way, we were willing to abandon "truth" (of the empirical order) if we thereby encouraged people to share their personal truths, which often do have universality, of the kind studied by psychologists and sought by artists.

Third, we needed to find ways of communicating with the older people, of opening and pursuing a dialogue that allowed subjectivity to flourish. How could we fire their imaginations, and in turn fire our own? What opening gambit would sidestep the deadening hand of the "rational" and unlock the aesthetic or emotional response? Our solutions were inspired partly by the contemporary arts and partly by psychoanalytic theories, but both these lines of inquiry resulted from our chief decision – to use the language of design as our primary medium of communication. Designers, like scientists, know much that they cannot say; unlike scientists, their goal is to express their knowledge through design itself.

* * *

Individuality, engagement, subjectivity – with such tenets, we knew, even before we visited the three sites of our research, that the Presence project would be a kind of adventure. After all, our aim was to *seek* the unpredictable, to create new situations for communication and insight, both during the design process and as a result of it. We also knew that whatever innovative interaction techniques we might create, they would probably be

site-specific – perhaps relevant only to those who had helped in their gene-
sis. But this did not concern us. Although it may sound perverse, being short
sighted was key to our approach. Our whole effort was to focus attention on
what was in front of our noses: to heighten awareness, through interaction
design, of the particularity of a place, and thus also of the particular role
technology could play for the people in that place, for the spirit of the enterprise was not to prescribe, but to sensitise. The true measure of our success
lay less in the general applicability of a specific interactive system – though
we believed that general results might come from an intense preoccupation
with the particular – than in the attempt to change people's ideas about
technology itself. First and foremost, we wanted to help the older people see
that technology could reflect richer versions of their lives, that it could tell
more rewarding stories about all of us.

The Presence project itself is a story, beginning with design-led user studies that probed older people's attitudes, proceeding to conceptual designs
for new services, and finally testing working prototypes in the communities
themselves. To do justice to the nature of the story, this account focuses
on its unfolding, rather than on generalisations abstracted from the narrative. We try to present, in words and pictures, what five researchers at the
Royal College of Art, five partner groups, and 58 older people in three local
communities in Europe achieved over the course of 2 years – and also what
we decided not to attempt. When the means are as important as the end,
the crucial question for the designer is not whether something is possible,
but whether it is interestingly difficult.

14.3 Who Are They? And Who Are We?

In 1997, when we started work on the project, our research partners were
already involved with three communities. Our initial knowledge of the sites
was scant, but our partners informed us that they complemented one another in terms of geographical position, of urban density, education, wealth
and beauty. Our brief initial visits confirmed these contrasts and strengthened our resolve to treat the older people in each site as individuals living
in unique cultures.

Peccioli, near Pisa in Italy, is a small hilltop village dating from medieval
times (Figure 14.1a). Narrow streets wind among stucco buildings, with
panoramic views of the Tuscan countryside. Modern technologies are most
evident in the cables draped over peeling walls and in the clutter of television
antennae, all angled towards an unseen target. By and large, the older people
seemed happy, although more and more younger people have been leaving
Peccioli for cities like Pisa, where both our partner groups, Innovative Devices and Engineering for Automation and Scuola Superiore S'Anna, were
based.

Figure 14.1 Presence worked with older people from three sites: (a) Peccioli, Italy; (b) The Majorstua, part of Oslo, Norway; (c) The Bijlmer, near Amsterdam in the Netherlands.

The Majorstua is an affluent district of Oslo, Norway (Figure 14.1b). It strikes one as a comfortable place to live, with clean streets and cosy shops. The older people, all educated and well-heeled, had formed an Internet club in the library with the help from our partner groups, Human Factor Solutions and Telenor. Our partners told us that the local government had tried using the Web to further local democracy, but their initiatives had not aroused much enthusiasm.

The Bijlmer, in the Netherlands, is mammoth and stark (Figure 14.1c). A planned housing community south of Amsterdam, it was built in the 1970s in an attempt to lure inner-city workers into the countryside, but failed due to poor transport and financial difficulties. Nowadays the Bijlmer is the first home for many immigrants to the Netherlands. It is notorious for its high levels of unemployment, drug abuse and crime, as local researchers from the Netherlands Design Institute had noted, and as we soon sensed.

So much for demographic, ethnographic and historical data. Although our partners told us many interesting facts, and we saw many interesting things, to repeat them here would be essentially touristic. Casual impressions and pocket histories tell you as much about the visitor as the visited, and all of it is peripheral. Our challenge was to get *inside* these people – to discover what animated them, what angered them, what gave them hope.

* * *

It is worth noting that, just as we were observing the older people, they were observing us. What did they see? They saw four men and one woman, between the ages of about 25 and 40. We were dressed casually, so we thought, but it is possible we seemed outlandish in some respects. The style of a shoe, perhaps. The design of an earring. Or simply the pallor of our English complexions. But one thing is certain: we did not look or act like anybody's idea of scientists. We joked, and talked about the weather. We tried to be ourselves.

14.4 Cultural Probes

Traditional methods of user research set up a sort of game, with implicit rules limiting the relationship between researcher and researched. Content areas are controlled and the participants constrained. Thus enquiries intended to inform technology development tend to produce answers based on what people already understand about technology. This is inevitable, whether you employ theory-based methods that encourage a doctor – patient relationship or participatory design methods where the researchers behave as servants offering technological know-how without examining the assumptions behind users' wants. The problem is in the paradigm of question-and-answer. Although they manifest themselves in different ways, the "rules" of each method allow both sides to present themselves as they wish to be seen. As Kenneth Koch said of his poetry assignments in the nursing home, direct questions tended to elicit "conventional and general statements which will be the 'right answer'" (*New York Review of Books*, p. 41).

To put it very simply, a certain type of question implies a certain type of answer. Since we could not dispense with the question-and-answer paradigm, we had to find a way of subverting the "rules". Inspiration came from those infamous rule-breakers, the Surrealists and Situationists, who developed tactics that exploited artistic and socio-political conventions in order to expose the superficiality of those very conventions (see Plant, 1992; Levy, 1995; Andreotti and Costa, 1996). Like us, their aim was to put people into a kind of rational free fall – spin them around a bit and see what happened.

Of course, to do this would mean spinning ourselves around too, that is, revealing ourselves. To put it very simply again, an interesting question says a lot about the person asking it. There is a useful analogy here to the Rorschach inkblot test, the best known of the projective tests used by psychoanalysts

Figure 14.2   A Cultural Probes pack.

(see Gregory, 1987; Eysenck and Kean, 1995). Created by pouring ink onto paper and then folding it, the inkblots are complex symmetrical shapes with no inherent representational meaning. Subjects are presented with these patterns and asked to describe what they see in them, like children finding animals or people in clouds. Now, within the psychoanalytic framework it is considered crucial to hide the therapist's own beliefs and concerns, and the inkblot does seem to shield them. (In fact, use of the Rorschach test is fraught with theoretical difficulties.) But what if the inkblot were not used as a shield? What if it were used, not for diagnosis, but as a springboard for communication? The ambiguity of projective tests seems ideally suited to engender fascinating dialogues. After all, children who are cloud-watching do it as a shared activity, for mutual enjoyment. And in just this spirit we started to assemble our Cultural Probes.

The Probes took the form of activity packs that we distributed to older people in each community, with the invitation to respond over time as they pleased (see Gaver et al., 1999). The core materials were developed with Elena Pacenti of the Domus Academy, and consisted of postcards, maps, a camera and a photograph album (Figure 14.2).

Postcards combined open-ended questions with evocative images to provoke discussion. A mundane medium for the dissemination of artistic imagery, postcards demand only the most casual of writing. They allowed us to ask a wide range of questions in a relaxed way, hinting at topics we found interesting but also providing opportunities to tell us almost anything. Three of our questions were "What advice or insight has influenced you?" "What is your favourite device?" "What use are politicians?"

Our maps were derived from the Situationists' psychogeographical maps, and required the older people to indicate the emotional topography of their communities: local landmarks, where friends lived, places they found intriguing or dangerous, and so on. The maps were printed in the form of envelopes, and stickers were included for ease of annotation.

The cameras were disposable, but repackaged to efface their commercial origins. A variety of picture requests were printed on the backs, such as "the view from your window", "the clothes you'll wear today", "something beautiful", "something ugly", "something boring". Remaining film was to be used in any way that might help us to know the person better – the choice was entirely theirs.

The photograph album came with the instruction to "tell us your story in six to ten pictures". This encouraged the older people to send us their own photographs or clippings, which we copied and sent back to them. When questioned what we meant by "your story", we were deliberately vague, preferring to see how the person would judge the relative importance of, say, their personal history, their families or their current activities.

The older people were able to question us about the photograph album, and other materials, because we presented all the Probes in person. Although we had not originally planned to do this, in hindsight it was a crucial part of the process, for it allowed us to hand out the Probes as if they were gifts, to stress that they were experimental and indeed might not work and to assure the older people that, far from being in control, we were as curious as they were about what would ensue. The Probes thus helped us by acting as catalysts for fruitful conversations, while the conversations helped the older people to understand that what we wanted was, on the face of it, the most obvious thing in the world – to know more about them.

And then we went off. We left the older people to engage with the Probes at their own pace, sending us occasional clues from afar, like the data sent back from an unmanned spacecraft, which offers tantalising glimpses that are open to diverse interpretations.

14.5 Data

To disseminate the Cultural Probes, we visited each site in turn over several months. Returned items appeared in our post for about 6 weeks after each visit, which meant that new material flowed in continuously for about 5 months. However, the return rates varied, which was a telling indication of differences between the groups. The Oslo group diligently returned almost

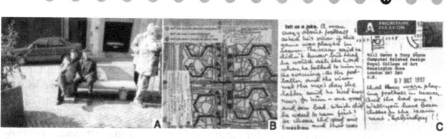

Figure 14.3   Sample Probe returns: (a) a photograph from Peccioli; (b) a map from the Bijlmer; (c) a postcard from the Majorstua.

all the material. The Bijlmer group returned slightly over half, their engagement seeming to depend on whether the tasks stimulated them. Finally, the Peccioli group returned less than half, despite their initial enthusiasm. We took this as a sign that older people in Peccioli were well-meaning but happily distracted by their daily lives – an important consideration for our designs.

As we sorted through the mass of returned material, further differences began to emerge, some of them epitomised by particularly striking items that then acted as beacons for us: a photograph of friends in an Italian piazza, a map of the Bijlmer with extensive notes about the "junkies and thieves" in the area, a joke about death from Oslo (Figure 14.3). The more we pored over such fragmentary, elliptical tokens, the more we saw in them.

Indeed, the Cultural Probes were so successful that it might be tempting to develop a formal methodology for them. But we remain sceptical. The results might be beautiful, but as heartless and superficial as an advertising brochure. We prefer to view the Probes as embodying an attitude towards research. It was their experimental, even risky, nature that contributed as much to their success as the particular materials we used.

* * *

As we began to prepare our "replies" to the returned Probes, we decided to take another risk. We had noticed that many of the most suggestive items were the returned photographs. Just as our picture requests had, at first sight, a surreal irrationality, the resulting snapshots often had something of the impact of surreal images. Why was this?

The photographs, often compelling in their very ordinariness, suggested narratives, some desolately humdrum, others more quirky. Of course, every figurative image suggests a narrative of some sort, because of our notions of cause and effect. But when you give an image prominence, as when send it to a team of researchers, in effect you are saying, "There is more to this than meets the eye". The observers then find themselves, almost automatically, constructing narratives to account for such unapparent importance. Thus the image becomes a "moment", imbued with hidden drama. It acquires a new, temporal aesthetic, the impact of which is entirely dependent on what *we* imagine. Nowadays one tends to call such an effect "surreal", but the Surrealists were only one group of experimental film-makers, photographers and painters who have played on our assumptions about narrative in various ways (strange juxtapositions are another favourite) in order to create liberating or disturbing effects (see Mamet, 1991).

We decided to attempt something similar in creating our design proposals. Instead of presenting one or two neat scenarios for design concepts – the classic storyboard method – we would shower the older people with minimalist and impressionistic proposals, like a handful of old frames picked up from a cutting room floor.

This approach gave us a way of replying in kind to the returned Probes, for it would allow the older people to participate imaginatively in the development of the proposals. They would have to put in effort, to work out what we meant, just as we had to work with their photographs in order to envision their personal narratives. But we soon discovered that the approach was itself liberating for us, as designers. It allowed us to work fast and freely, unconstrained by the usual demands of rationalising every angle of a single concept and thereby becoming increasingly wedded to what is, after all, merely a proposal.

* * *

This stage took about 6 months (including the assimilation of the returned Probes) and culminated in the creation of two workbooks, both containing a section on each of the three sites. One workbook led to the other. The first presented preliminary ideas and the second more developed concepts, but both used a variety of imagery to create fragmented narratives for the systems we proposed, and also contained descriptive text to provide a framework.

We discussed the first workbook with our partners, and the second with the older people, who were particularly forthcoming. This feedback prompted our design proposals to crystallise rapidly. What had started as loose constellations of ideas, jostling in our minds for many weeks, now cohered as integrated systems, and 3 months later we were able to present our refined proposals in the form of a multimedia presentation. Separate screens described each community to set the context, then showed animated diagrams of the systems we suggested, and offered a "catalogue of parts" that indicated the kinds of physical artefact that might comprise the systems.

That, then, is what we did. But it begs the larger question of *how* we did it. How did we extrapolate from the returned Probes, and then elaborate the results and amalgamate them? What "tools" did we employ in the process? Unfortunately – or perhaps fortunately – there is no helpful answer, except to suggest looking at the workbooks themselves. The evolution of the

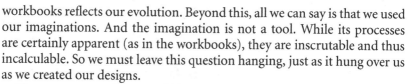
workbooks reflects our evolution. Beyond this, all we can say is that we used our imaginations. And the imagination is not a tool. While its processes are certainly apparent (as in the workbooks), they are inscrutable and thus incalculable. So we must leave this question hanging, just as it hung over us as we created our designs.

Now we come to the most dramatic part of our narrative: what we proposed, and how it did – and did not – lead to working prototypes.

14.6 The Proposals: Why and How

From the first, Peccioli had suggested to us the use of radio to interconnect the village in new ways, and to extend it beyond its walls. As a technology that points simultaneously towards the traditional and the high-tech, radio seemed to resonate with Peccioli's juxtaposition of rustic architecture with exposed wiring, television aerials and satellite dishes. The returned Probes encouraged this notion, and some of our proposals explored how radio could amplify the sociability of this close-knit community. We imagined older people communicating with two-way radios, and the idea developed into a proposal for transmitter–receiver units, each devoted to a single wavelength. Separating the transmitter from the receiver would allow people to exchange them, setting up flexible communication networks. People could either talk in pairs or, by passing along their transmitters, join in with others, forming loops and chains of communication. If a transmitter was left somewhere – in a field, say – then everyone in the chain would hear the noises it picked up (Figure 14.4).

Other proposals followed from this idea of transporting sounds from the countryside to the village. We could plant transmitters in woods, with built-in microphones, to carry birdsong into people's homes. We could even attach wind vanes, to vary the direction of the microphone. Or we could harness the transmitters to cows, to witness the ambience of their journeys. Maybe we could try all these ideas, and others, to build up a rich pastoral radioscape.

Tourists could engage with Peccioli's radioscape, too. We could equip roadside lay-bys with low-power transmitters, playing sounds picked up from a visible landmark, or stories and songs collected by the villagers. We might also leave devices in guest houses so that tourists could participate, thereby providing fresh stimuli both for them and for the locals.

At heart, our ideas envisioned an alternative to commercial radio – an alternative that was there right at the beginning, a 100 years ago, before the wavebands were swamped with generic music, advertising and news. Ever since the mid-1890s, when radio tests and informal communications started to interfere with naval communications, amateur broadcasts have been under attack, with the result that nowadays regional differences tend to be discounted instead of appreciated (see Hill, 1978; White, 2003). Indeed,

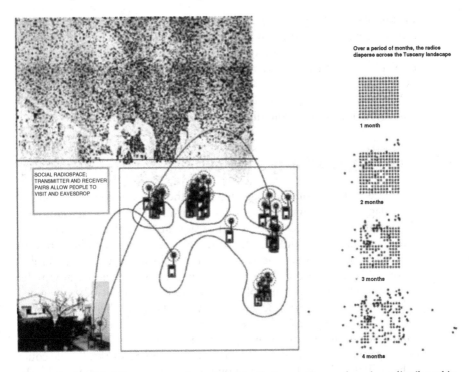

Over a period of months, the radios disperse across the Tuscany landscape

1 month

2 months

3 months

4 months

SOCIAL RADIOSPACE;
TRANSMITTER AND RECEIVER
PAIRS ALLOW PEOPLE TO
VISIT AND EAVESDROP

Figure 14.4 For Peccioli, we proposed a variety of radio transmitters and receivers distributed in the village and surrounding countryside.

rather than contest radio's cultural uses, some people choose to forgo all human civilisation, preferring to travel to remote deserts and mountains in order to record the clicks, tones and textures caused by high-level atmospheric phenomena (see Strauss & Mandl, 1993; McGreevy, 1996). To listen to these sounds is to experience radio as a wilderness. In their delicate and elusive beauty, they remind us how thoroughly the electronic spectrum has been settled, zoned and appropriated.

Our proposals for Peccioli were an attempt to reclaim radio as an unofficial, and unofficious, medium of communication – promoting casual conversation, linking strangers with locals and bringing the long vistas of the Tuscan countryside up close into people's homes.

* * *

For the Majorstua, we looked at ways of provoking the older people to engage more with local issues and events. It seemed they had little sense of a community identity, partly because the district merged with other areas of Oslo and partly because opportunities for local involvement, like the library's Internet club, tended to look outward to the wider world. Could

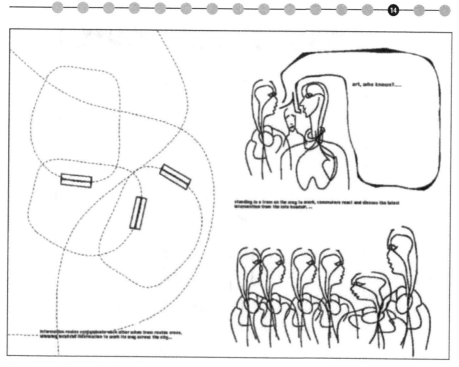

Figure 14.5 The Digital Boudoir proposal suggested that older people could circulate questions to which local citizens could reply.

we find a happy medium, a communicative space that lay within the city limits yet was beyond the older people's immediate socio-economic and age boundaries?

Our first idea was to exaggerate the privileged position of these older Internet users. We envisaged transforming the functional room where they met into a "Digital Boudoir". Filled with luxurious furniture-technology hybrids, the boudoir would allow the group to lounge in comfort while exploring the diverse pleasures of the Web. From this point, the notion of the boudoir expanded to include its function as a place for discussion, the centre of a community-wide conversation about local issues. We saw it as a kind of headquarters (HQ) for the older people, generating viewpoints on controversial issues, which would then be displayed on billboards and printed on tickets and shopping receipts. Perhaps these viewpoints could literally *be* viewpoints: we could install viewfinders around the area, like those found at tourist spots, except ours would overlay the city with social and political information, creating an interactive psycho-geographical map.

Over time, these ideas gelled into our proposed system (Figure 14.5). Special furniture in the library would allow the older people to sort through information found on the Web, or provided by local politicians, and extract interesting topics. Other furniture would mediate their discussions, helping

357

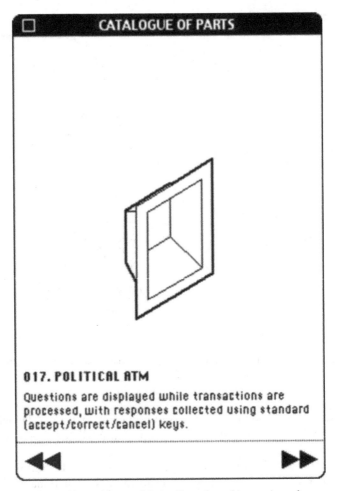

Figure 14.6 Questions from older residents might be disseminated in a variety of ways, presented as a "catalogue of parts" in the proposal.

to distil them into questions for public release. These questions would be relayed to the local tram station and then transmitted to passing trams, displayed inside, or at following tram stops, or in nearby cafes. People could respond by pressing "yes" and "no" buttons on associated devices.

Other possibilities for disseminating questions included the use of automated teller machines (ATMs) and periodic calls to public telephone boxes. Whatever the method, the purpose would be to collect a certain quota of answers that the older people could analyse, reporting the results to the community and its politicians (Figure 14.6). Meanwhile unpopular questions, which did not meet their quota, would be sent to isolated areas to languish.

This proposal, more than the others, sought to emphasise the responsibilities as well as privileges of being present in a local community – present

in the sense of being noticeable, and also present in the sense of being aware of one's surroundings. By casting the older people as curators, or perhaps hosts, of the local political culture, we hoped to encourage them both to express their views publicly and to pay more attention to other inhabitants of the Majorstua.

* * *

The Bijlmer was the most complex, and contradictory, of the three communities. Although many of the Probe returns highlighted the sinister aspects of the area – racism, threatening behaviour, the police's reluctance to interfere, others evinced a defiant pride in the place. They pointed out the richness of its cultural mix, the opportunity for social gatherings in the market beneath a main road and the custom of bringing caged birds outdoors to sing in the sunshine.

Nonetheless, our initial design ideas were overwhelmed by the prevailing perception that fear is the Bijlmer's defining characteristic. We speculated about the use of laser pointers to draw attention to miscreants, and toyed with the notion of projecting warnings on the outside of their flats. These thoughts led us to an even more drastic proposal: municipal cages, into which older inhabitants could lock themselves and enjoy the outdoors without trepidation, just like their songbirds (Figure 14.7). Seemingly a tasteless joke, this concept dramatised the problem of designing with an obsession for security. What starts out as a protection system ends up as a cage.

Working through such uneasy preoccupations was perhaps necessary, and certainly useful, for it enabled us to see that our problem with the Bijlmer was not ours alone. We were simply reflecting the Bijlmer's problem with itself, and to address this we had to address the *idea* of the Biljmer. Our thoughts turned towards ways of prompting, gathering and communicating the residents' collective self-image. Early ideas included home-made soap operas projected onto billboards along nearby roads; linked security cameras that create a network of "virtual neighbourhoods"; a "mains radio system" broadcasting voice and data over the existing power network within the buildings; robots called "vent crawlers" that travel the ventilation ducts, recording and broadcasting fragments of conversation; and lastly, electronic or paper maps that indicate the area's emotional topography, assembled by older people using pagers to signal where they felt safe or anxious (Figure 14.8).

Even our most radical proposals met with serious consideration from the older people. Using their feedback as guidance, we started integrating our sketches into a single system called Projected Realities. This would project people's beliefs and attitudes from their homes, through their neighbourhoods, to the outskirts of the Bijlmer (Figure 14.9). The idea was to commingle individual elements of expression to create a kind of public face for the area.

Figure 14.7   Early proposals for the Bijlmer explored how an obsession with safety might end by caging older people.

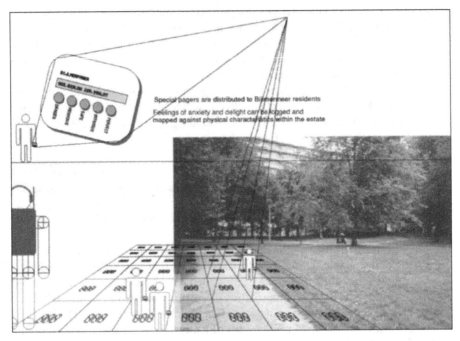

Figure 14.8   "Psychogeographic pagers" in the Bijlmer might allow older people to tag locations as threatening or desirable.

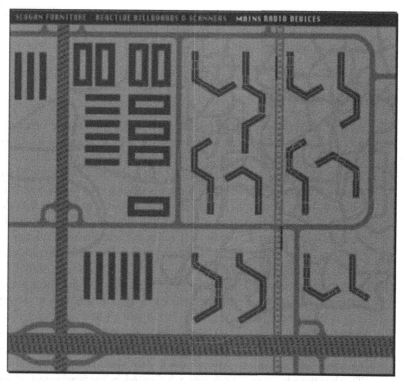

Figure 14.9 The Projected Realities proposal for the Bijlmer suggested propagating expression from homes to neighbourhoods to the outskirts of the area, as shown in an interactive sketch of the system.

We foresaw three main stages. First, the older people would collect images and slogans about life in the Bijlmer. Other residents would access these via a mains radio system, and select images they felt were representative of them at that moment. Second, the choices from each housing block would be amalgamated, and the local mood reflected in "slogan furniture" in the immediate vicinity. Third, the slogans would be processed periodically by a neural network (see Rumelhart et al., 1986). This would select images that best reflected all the slogans, and these images would then be displayed on large "image boards" near the surrounding commuter roads and railways.

The virtue of neural networks is that, unlike more traditional forms of data processing, no element in the system depends exclusively on another. Instead, structure at each level depends on patterns over the whole of the preceding one. It is a wonderfully subtle process, given that you design input conditions that create effective pattern matching. Our system would have to be elementary. Although it could only suggest the power of a sophisticated neural network, we felt this would suffice for the Projected Realities proposal.

361

Each person's image would be distributed over the entire network, with the end result – the images representing the Bijlmer to outsiders – emerging from the layers behind it.

14.7 Experiments

After the first year of the project, we tested approximate versions of our proposed systems in each site. About these experiments we shall have to be brief. Much of the interest lies in the details, and this is not the place go into details. Moreover, it was already clear that due to constraints of time and money, we would be able to implement working prototypes in only one site. The tests were, in effect, a knockout competition. Our criteria for taking one proposal forward were that it be technically possible in a short time, and more importantly, that it offer a stimulating intellectual challenge.

Unfortunately, the main challenge to the Peccioli radioscape was not intellectual but political. Many major radio stations have signal repeaters on a hillside only 10 kilometres away from the village. These saturate the airwaves. Almost the entire spectrum is in corporate hands, and trespassing might have brought legal action. Our 20-watt FM transmitter (capable of being received for over 16 kilometres) was just too powerful to use. Of course, the radioscape could be transmitted over higher frequencies set aside for amateur use, and the older people provided with specialised receivers. This has the advantage of creating a dedicated radioscape, foreshadowing the possibilities opened by digital radio. But the sound quality would be reduced. It goes without saying that commercial interests only relinquish second-rate electromagnetic real estate to the public. However, we were convinced that the public – at least, the older public – would enjoy a pastoral radioscape. Many of the older people were enthusiastic about the idea of recording the sounds of their landscape. They met us to discuss likely locations, creating a festive atmosphere, more like an impromptu party than a briefing. Later we all tramped the surrounding countryside, holding out microphones to catch the bleating of sheep and the soft plink-plonk of a nearby stream (Figure 14.10).

Our proposal for the Majorstua was both a technical and an intellectual challenge. And it remains so, although our experiment there was a surprising failure. We had decided to concentrate on finding a way to simulate the socio-political effects of the system, rather than its technical underpinnings. Armed with advice from Design Age Network about common age issues, as well as topics more specific to Norway and Oslo, we prompted the older people to think of contentious local questions. Then we recorded them talking into a PC-based telemarketing system. We fed the numbers of all the local telephone boxes into the system and set it to dial each one systematically, ready to pursue a limited, pre-scripted interaction with passers-by (Figure 14.11). But no one answered. We had already put up signs in the

Figure 14.10   Testing the Pastoral radioscape concept in Peccioli.

telephone boxes, so now we adjusted the system to make it easier to use. Yet we received only two meaningful answers out of the hundreds of calls we placed. Finally it dawned on us that when a public telephone rings, the one certainty is that it is not for you. Our signs were informative, but it seems that information was not enough. People did not participate in the experiment simply because they were unaware that an experiment was going on. Despite this setback, we were paradoxically enthused. Several new avenues beckoned, such as putting more official notices in the telephone boxes, or even painting and labelling some, to mark them out as polling centres. On reflection, it was too late to try new ideas. However, we had learnt an important lesson about the necessity for eye-catching street furniture, which we resolved to bring to our work in the Bijlmer.

We travelled to the Netherlands with 100 sheets of card stock, cut to 2 metres by half a metre, and with digital scanners and a high-powered data projector. In preparation for our arrival, the local researchers had issued each older person with a disposable camera to take pictures around the Bijlmer. We scanned these shots, and also used the scanner as a kind of horizontal camera, to capture images of the participants hands, spectacles, jewellery – anything intimate. Then we passed around small booklets printed with slogans derived from the Probe returns, as well as from other sources. After heated discussions about conditions in the area and the misunderstandings of outsiders, the older people agreed on a set of nine slogans,

Figure 14.11   Test of the Digital Boudoir concept in the Majorstua.

some based on comments in the booklets but most produced by them during the meeting. We printed the slogans and mounted them on cards, in sets of 10. Now all that remained to do was to hire 10 clothes stands, and hang up our slogan boards (Figure 14.12).

Passers-by all reacted with interest, and we soon realised that the most beguiling slogans were slightly ambiguous or detached in tone ("I am from another country"), or were particular and personal statements ("I like a few drinks once in a while"). By escaping classification into known forms of public display, such as the demands of political protest and the imprecations of religious fanaticism, such comments invoked curiosity rather than instinctive dismissal.

That evening we undertook the second part of the test in the local tram station: back-projecting our scanned images onto the window of a small booth, for people to see as they disembarked from trains or waited on the platform. As with the slogans, we learnt that ambiguity and intimacy (of

Figure 14.12   Test of the Projected Realities concept in the Bijlmer.

content and style) were most effective. The images of hands seemed to press against the display, as if from another space, and were particularly disquieting.

Again, passers-by stopped to talk and tell us about their lives in the Bijlmer, and soon our biggest decision had made itself. We would test our working prototypes here. It was not simply that our experiments had been successful in technical terms, or that the Bijlmer presented us with richer social and cultural issues than the other sites. What convinced us was the fact that our technical achievements had enabled us to raise these issues. The local inhabitants were more than responsive. Many seemed eager to pursue the project further.

14.8 Projected Realities

Although we had a fair idea of how the Projected Realities system might work, we had made almost no decisions about the forms that particular elements should take. The final version of the system coalesced slowly, during a process that may appear more logical in the telling than it did as we experienced it. Linear refinement and problem solving tend to emerge as such in hindsight, just as the shape of a narrative becomes apparent when you reach the end.

Our first critical decision was to abandon the idea of using the mains radio, as it was too complicated for a short test. Instead, we decided to

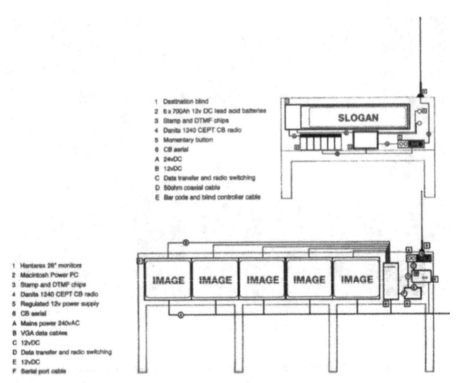

1 Destination blind
2 6 x 700Ah 12v DC lead acid batteries
3 Stamp and DTMF chips
4 Danita 1240 CEPT CB radio
5 Momentary button
6 CB aerial
A 24vDC
B 12vDC
C Data transfer and radio switching
D 50ohm coaxial cable
E Bar code and blind controller cable

1 Hantarex 28" monitors
2 Macintosh Power PC
3 Stamp and DTMF chips
4 Danita 1240 CEPT CB radio
5 Regulated 12v power supply
6 CB aerial
A Mains power 240vAC
B VGA data cables
C 12vDC
D Data transfer and radio switching
E 12vDC
F Serial port cable

Figure 14.13 Schematics for the Sloganbenches and Imagebank components of the Bijlmer Projected Realities system.

reuse our telemarketing system, which would prompt volunteers to select images from a printed "menu" kept near their telephones. Prior to this, the older people would have rated these images along three dimensions – such as "positive–negative" – inspired by semantic differentials (a standard approach to measuring meaning in psycholinguistics). The ratings would provide numerical representations for the images, which could then be mapped together using simple algorithms based on neural net processing.

The functions of the Sloganbenches (previously "slogan furniture") and Imagebank (previously "image boards") were also kept simple, as were their forms, indicating that they were components of a system that was still incompletely developed (Figure 14.13). We had rejected LED displays on the benches, and LCD displays or video walls for the images, because of their associations with commercial interests (who can afford such technology) and resultant impersonality. Instead, we decided to insert fabric scrolls in the benches. Scrolls seemed most approachable as they were reassuringly mechanical, they even allowed us to incorporate handwriting, and yet they were electronically controllable. As for the images, these would flow over a bank of large monitors, segment at a time, compelling passers-by to piece

Figure 14.14 A Sloganbench on site in the Bijlmer.

them together. We were confident that little more was required in the way of design. Both objects would be intriguing enough by virtue of their conceptual uniqueness. However, as a final touch, we used wood-grain laminate for the finish. This emphasised the association with domestic furniture, and suggested our intention of giving access to the Bijlmer's interior spaces.

Finally, we considered a variety of ways that the system might communicate: SMS messages, infrared, even wires. Eventually we concluded that dial tones sent over CB radio would be robust and affordable. And so, on a wet spring day, we installed the CB equipment in a borrowed flat high in one of the largest housing blocks, and went to work.

The following days were all productive, although dogged with unforeseen difficulties. For instance, due to the problems of sourcing electricity and of obtaining permission for a good roadside site for the Imagebank, we had to content ourselves with installing it near a shopping arcade. Nevertheless, the system generated enormous enthusiasm, both from the older people participating and from passers-by. The Sloganbenches (Figure 14.14) were probably most effective at catching people's attention, holding it, and then translating it into self-expression: either they would turn the scroll to select their own slogan, or they would start chatting about some aspect of the Bijlmer. Children were particularly drawn to the benches, using them as a

Figure 14.15　The Imagebank on site in the Bijlmer.

cross between playground furniture and an interactive toy. The result of the children's interest was that the benches acted springboards for dialogue between them and the older people. This was as delightful as it was unexpected, and suggested a potential for future development.

The Imagebank (Figure 14.15) was a more passive object than the Slogan-benches, yet it also exerted a noticeable magnetism. The key to the success of both pieces lay, we believe, in their essential ambiguity, in the tension between the familiar and the strange. A few feet off the ground, the Image-bank offered a kind of multiple, public version of a domestic TV, and much of it what it showed was domestic, although not immediately apparent as such. The benches, on the other hand, *were* benches. You could sit on them. However, everything about them – the sketchbook form, the unusual materials, the changing slogans – was mysterious. Such ambiguity of form and function seems to draw people out of themselves at the same time as sending them inwards. It sets them wondering, to themselves and out loud, creating a mood of contemplative amiability. Engendering this mood among inhabitants of the Bijlmer was a positive outcome in itself, quite apart from the conversations that ensued with the older people and members of our team.

The conversations themselves were wide-ranging, and this is another aspect of the system that could be developed. Although the commitment of our

Figure 14.16   Bijlmer inhabitants enjoying a Sloganbench.

team encouraged engagement with the project, it might be stimulating to involve outside facilitators – artists, social commentators, politicians – who are involved in local issues and would welcome the opportunity to explore them via the slogans and images, and then follow this up in conversation – informal conversation, that is. Anything else would smack of "user feed-back", of covert agendas, thus defeating the purpose of the whole enterprise. Let the participants (including the facilitators) say what they want, as they want.

Could the Projected Realities system continue? We can only say that in our talks with the people of the Bijlmer, they were unanimous in their support of it (Figure 14.16). (They protested only when they learnt that the furniture was experimental, and not a permanent addition to the area.) And a museum in the Bijlmer expressed interest, as did a community group in south London. However, in a museum environment it would be difficult to achieve the community ownership we were advocating, and redeploying the system would necessitate a rethink of all its social components: meetings with the older people, input from local participants, appropriate display materials. Therefore we decided that rather than attempting to maintain or develop the system, we would carry the lessons it had taught to new projects and contexts – and also try to share them, as we have done here.

369

14.9 Over and Out

Leaving the Bijlmer reminded us of leaving behind the Cultural Probes – that unmanned spacecraft sending us clues from afar. The difference was that, this time, we were going out of range. Our involvement in the Presence project was over. However, the impetus we helped to create seemed to promise some kind of continuation. When we were packing up in the Bijlmer, a number of the older people were actively engaged in new projects of their own, and it was not clear that they even noticed our departure.

So, although the Presence project did not leave any physical traces, we believe it did leave behind ideas and experiences that were beneficial to the groups in each site. After all, the conditions of one's life are, ultimately, what one makes of them. They depend less on tangible than imaginative power. Like a workshop, or an exhibition, or a theatrical event, the process through which we led the older people changed how they perceived themselves, and encouraged them to reflect and perhaps act on the insights that emerged.

The process also changed us. One of the key lessons we learnt from the project was that older people do not want or need to be cosseted. At each stage of the design, the older people responded to our speculations and provocations with engaging vitality, whether positive or negative. In fact, they were often bolder than we were. No, there is nothing hush-hush about older people, and they should not be treated as if there were. They should be *respected*, but not in the usual, rather condescending sense of the term.

Afterwards, we missed them. We missed their lively pessimism and their wry optimism. And, though it may seem contradictory, we hope that they missed us, the systems we proposed and our mutual engagement in the project. If people were unhappy that the story ended, they might question why technology does not tell a great many more such stories.

14.10 Acknowledgments

The work here was pursued as part of the Presence project, sponsored by the European Union's Future & Emerging Technologies Unit as part of the I3 initiative. The Royal College of Art team were Anthony Dunne and William Gaver (coordinators), Ben Hooker, Shona Kitchen and Brendan Walker. Throughout, "we" refers to these five people. Jacob Beaver is a writer in the department. We worked with seven other research teams in four countries: the Netherlands Design Institute, Design Age Network at the RCA, four local groups, and another design team from the Domus Academy in Milan. We would like to thank all our partners and the participants in the three sites. Particular thanks are due to Jakub Wejchert for his visionary leadership of the I3 initiative.

14.11 References

Andreotti, L. and Costa, X. (eds) (1996) *Situationists: Art, Politics, Urbanism.* Museo d'Art Contemporani de Barcelona. Barcelona, Spain.

Eysenck, M. and Kean, M. (1995) *Cognitive Psychology.* Erlbaum, Hove, UK.

Gaver, W., Dunne, A. and Pacenti, E. (1999) Cultural probes. *Interactions Magazine,* **vi**(1), 21–29.

Gaver, W. and Hooker, B. (2001) *The Presence Project.* RCA CRD Research Publications, London.

Gregory, R.L. (ed.) (1987) *Oxford Companion to the Mind.* Oxford University Press, Oxford.

Hill, J. (1978) *The Cat's Whisker: 50 Years of Wireless Design.* Oresko Books, London.

Levy, J. (1995) *Surrealism.* Da Capo Press, Cambridge, Mass USA. Reprinted by Da Capo Press [first printed by Black Sun Press, 1936].

Mamet, D. (1991) *On Directing Film.* Viking/Penguin, NYC, USA.

McGreevy, S. (1996) *Electric Enigma: The VLF Recordings of Stephen P. McGreevy.* [CD audio]. Irdial Discs, London.

Plant, S. (1992) *The Most Radical Gesture: The Situationist International in a Postmodern Age.* Routledge, London.

Rumelhart, D.L., McClelland, J.L. and the PDP Research Group (1986) *Parallel Distributed Processing. Vol. 1, Foundations.* MIT Press, Cambridge, MA.

Strauss, N. & Mandl, D. (1993) *Radiotext(e).* Autonomedia, New York.

White, T.H. (2003) *United States Early Radio History.* Retrieved from <http.earlyradiohistory. us/index.html>

15

Informing the Community: The Roles of Interactive Public Displays in Comparable Settings

Antonietta Grasso, Frederic Roulland and Dave Snowdon

15.1 Preamble

Effective communication is a fundamental prerequisite of the sustainable community. In the projects reported here, large-scale interactive screen displays have been proposed and evaluated as augmented means to promote communication and information exchange within and across communities, either in public spaces or at work sites.

The authors of this chapter have recently been involved in the design and deployment of two such systems. Campiello is a multimodal interaction system conceived to promote information exchange in cultural heritage sites. MILK (Multimedia Interactions for Learning and Knowing), the second system in this chapter, was subsequently developed to exploit the mapping of Campiello's interaction features onto other types of community structures, such as distributed work organisations.

The authors draw conclusions from these projects, by first presenting what a general architecture and set of functionalities for such systems might look like. On the basis of these common features several different requirements are identified and are presented for the two settings, namely the public precinct and the distributed work organisation.

While requirements obviously touch on issues of content creation and selection, they also touch on issues of reactivity of the system, content layout, degrees of interaction and finally privacy.

As the output of this exercise, the lessons learned from these explorations will therefore focus on the different requirements that emerged in the design of the two systems, which had started out looking very similar in terms of the range of functionality offered.

15.2 Introduction

It was a few years ago that Mark Weiser started to identify a new trend in the development of information technology, which he called *ubiquitous computing* (UC) and identified as the third computing wave after mainframe and personal computing.

In his own words:

> the third wave of computing is that of ubiquitous computing, whose crossover point with personal computing will be around 2005–2020. The "UC" era will have lots of computers sharing each of us. Some of these computers will be the hundreds we may access in the course of a few minutes of Internet browsing. Others will be embedded in walls, chairs, clothing, light switches, cars, in everything. (Weiser and Brown, 1995)

UC is, of course, not only about having computational power embedded in objects and devices ranging in size, but about having a multitude of information sources connected with a multitude of delivery systems. The multitude and interconnection aspect is what eventually fully characterises this new paradigm and opens up new possibilities, which in turn set new challenges on the side of the user interaction. In fact we can imagine that such an interconnected world of computation has the potential to increase the level of attention and complexity people have to deal with. For this reason Weiser also outlined an interaction paradigm to which the UC technology should comply: the one of calm computing. The core idea of calm computing is that if there is such a large potential for information delivery in the environment, the UC technology has to be designed in such a way as to support peripheral awareness and easy transitions from the background to the foreground and vice versa. Fully following this paradigm, Ishii et al. (1998) have started to look at architectural space as a suitable interface between people and digital information. They have named this enlargement of the traditional graphical user interface to the whole space surrounding the user as *ambient displays*. Of particular relevance to this concept is the feature that "Ambient Displays are well suited as a means to keep users aware of people or general states of large systems" and "can be used to create a persistent, yet non-intrusive connection between loved ones, bringing people a sense of community through shared awareness" (Ishii et al., 1998). Moreover in this original definition the ways the information is conveyed by the ambient display are not confined to being only textual and visual: the possibilities are deliberately kept open to all the human senses, in an attempt to explore how best to exploit the affordances of the space for supporting a peripheral provision of information in the sense proposed by Weiser.

In Figure 15.1, a common scheme for describing ambient displays in a conceptual way is presented. In this scheme the following modules are presented:

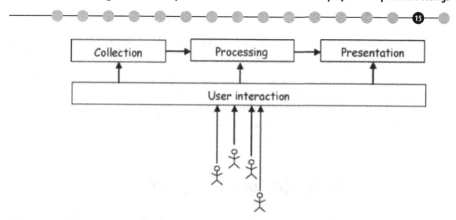

Figure 15.1 Overview of an ambient display.

- *Collection*: This is the module that monitors one or more information sources.
- *Processing*: This is the module that decides what information passed by the collection module to process and how. This is where specific ambient information is taken into account for prioritisation decisions.
- *Presentation*: This is the module that maps the information to the possibilities of the chosen output.
- *User interaction*: Not all ambient displays are interactive; however, interactivity could appear at every level, from choice of data source to deciding how to process and display the information. Finally, interaction could be with the system itself or with other users, with the system acting as a mediator.

15.3 Designing Ambient Displays

Looking at representation of ambient displays in the introduction, based on Ishii's definition, it is clear that the scope of ambient displays is potentially very large. This is so across a number of dimensions, including the specific setting that is going to be augmented: public, home or work. Additionally, mentioned previously, ambient displays may look at the usage of all human senses to elaborate the information conveyed by the system. On the other hand a few systems have been proposed in recent years that cover just a subset, namely the one addressing the work environment and presenting information through visual means. In the rest of the chapter we will focus on this subset whilst spanning across installations for the public environment and for the work setting.

We believe that the design of these systems is particularly difficult because it requires simultaneously the capacity to understand and create new technology in the field of UC and to understand the behaviour of communities not to mention the usage of locations.

In our research we have been involved with the design of large interactive displays for both work and non-work-related communities, which has given us a broad understanding of how the system features described before can be articulated. This chapter is devoted to presenting two instances of large interactive screens for the work setting and making a critical analysis of what have proved to be the relevant aspects when mapping the features of interactive large screens into work organisations.

15.4 Campiello and the CommunityWall

Our exploration of the possibilities of ambient displays started in the Campiello project (Esprit Long Term Research #25572), a project aimed at promoting and supporting the meeting of inhabitants and tourists in historical cities of art and culture. This overall objective has been undertaken in two main steps: reinforcing the community through collective participation in (1) creating community knowledge and (2) optimising access to it.

To address these steps, at a basic level the Campiello system acted as a Web-based repository of information related to places and events in the city. This information consisted mainly of descriptions and comments. Whilst every user of the system could contribute to the information space, there were some main actors as the cultural managers, who provided an initial set of descriptions and kept on feeding the system with descriptive information. In structuring the information space and designing the access to it, we were faced with the well-known trade-off between structured information systems and unstructured "communication areas". If the system is designed in a structured way, it is easy for users to find information. On the other hand if users are given a lot of freedom in creating new information, they are more likely to contribute, but this usually leads to a loss in structure and so can frustrate people when they try to find something in the network. This is why we decided to model the information services primarily around the recommendation function (Glance et al., 1998) while keeping search and browsing capabilities available for those who choose to use them.

On top of the Web-based repository of information just described, in Campiello a major aim was to offer widespread access to the local inhabitants in order to encourage utmost participation and to ensure that sections of the community were not excluded due to lack of access. For this purpose several additional interfaces to the content have been developed, including paper-based ones and large interactive screens. A complete overview of the system is given in Agostini et al. (2000).

The purpose of our large-screen display, the CommunityWall (CWall) (see Figure 15.2), was to create an environment that fosters social encounters and conversations using topical information and/or news as a trigger. It provided a focus for social activity in a way similar to existing notice boards that display notices (ranging from formal printed notices to handwritten

Figure 15.2 The design of the CommunityWall prototype for the local community in Venice.

scraps of paper) concerning current community activities. Using the CWall, we aimed to provide information on interesting activities, who was actively interested in what and what they were saying. If a topic displayed on the CWall attracted someone's attention he or she could then request that more information be displayed on that item by touching the screen (or using some other means if a touch-sensitive display was not available) or that a News-Card (Agostini et al., 2000) be printed on that topic on a printer/scanner situated near the display. Once the NewsCard is printed it can be used to comment on the topic. In this way the CWall served the project objectives of information discovery and asynchronous communication among members of the community. A detailed analysis of the experimentation is given in Agostini et al. (2000). Furthermore, we hope that if a number of people use the display simultaneously, this will help trigger conversations and allow people to meet others with similar interests (since it will be obvious which topics people are looking at).

In summary, the purpose of the CWall in Campiello was to support information discovery in and across communities and create an environment that fosters social encounters (conversations) using documents and/or news and peoples' opinions on them as a trigger. As the project ended we thought that the same benefits could apply not only to local communities in towns of the kind we addressed in Campiello, but also in work organisations. We had this conviction thanks to studies of work organisations showing that informal communication is important and occurs inside structures such as communities of practices (Lave and Wenger, 1991). Such exchanges have been observed and reported as fully functional to the organisations in several studies (Isaacs et al., 1997).

15.5 Supporting Communities of Practice in Work Organisations

Communities of practice have been widely studied and acknowledged as a major informal organisational structure, pivotal in supporting learning and adaptation to innovation and change (Lave and Wenger, 1991). Because of this central role, many attempts have been made in recent years to foster their creation and people's participation. This includes the effective design of technology in support of them and the deployment of appropriate organisational changes and new processes. These attempts, however, have been troubled in several cases, only resulting in a core set of people participating in community activities and not achieving the broad impact aimed at by the promoting organisation. There could be several reasons for this flop. The first of these, we believe, is due to the top-down approach that is often adopted, imposing community structures from above, instead of aiming to make visible existing activities in the existing areas of expertise whilst soliciting cross-fertilization amongst areas. This contradicts fundamental attributes of community activity such as being spontaneous and bottom-up. Asking people to participate because of management needs, almost certainly ends up in failures because the individual reward is not immediately clear. Community activities flourish in the companies where they do because people participating in them use them as a place for giving and receiving mutual help, or for learning and staying up to date with techniques and problems in a certain field. Another issue, related to this one, is that participation has an additional cost with respect to mainstream daily activities, including deadlines and deliverables. When the effort to participate in the community life is too high, e.g. requiring a learning curve to understand the environment supporting it, or too much time for creating new content, then participation could well be less than expected.

On the basis of these observations, we decided to reconsider carefully every choice that was made for the CWall in Campiello and to adapt it where necessary to the specific requirements coming from its use in a work setting.

As a first choice we decided to have a large-screen display as the main output mechanism for a pre-existing recommender system, Knowledge Pump (KP) (Glance et al., 1999). In this way there would be a display that everyone would see in the course of a normal working day. We were hoping that by using a large screen in a communal space we could at the same time benefit

Figure 15.3 CWall in use.

from existing online exchanges and in addition reach people not currently using KP. By providing a display large enough for several people to view at once, we also hoped to create a social focus and a mechanism for triggering conversations in a more pervasive and systematic way. Initially we used the Campiello prototype and installed the display next to the black and white and colour printers used by members of our workgroup. This also happened to be on the corridor, giving access to the office of the lab manager and the group's assistant – thereby guaranteeing that many people would pass by the display in the course of their working day (see Figure 15.3). After about 10 months of use we moved the display to the entrance foyer of the building used to host visitors to the research centre and later installed a second display in the cafeteria. We found that each site generated different user expectations, as summarised later and fully reported in Snowdon and Grasso (2002).

15.5.1 Lessons Learnt

This first deployment inside a work organisation gave us some insights that we fully reported in Snowdon and Grasso (2002) and that we summarise in the following. Firstly, we learnt that what it takes to prevent people from using a system is surprisingly low. Even though people are accustomed to sharing information and papers by e-mail there is some resistance to using software such as KP. Whilst a core group of users found it of benefit, others said that the effort required to go to the web page and open a form for submitting a document or reading their recommendations was enough to put them off using it. Yet the same users typically make great use of e-mail to recommend and exchange information – this is what led us to create an e-mail interface for the CWall in the first place.

After some initial scepticism the CWall received a favourable response from most users and we received several suggestions for improvement. After the system had been in use for about a year we performed an analysis of the log files and asked both users and non-users to fill in a questionnaire designed to understand why people did and did not use the system.

From the questionnaire it emerged that most of the users who claimed to use the CWall were not using the KP. This was reassuring since it seemed to mean that the CWall was fulfilling a real need and distributing information stored in KP to people who might not otherwise see it. At the same time documents entered via the CWall could become available to KP users. From log analysis we also discovered that there is a correlation between the number of items submitted to the system in a day and the number of items examined on the large-screen display. The correlation was calculated using the formula shown in Equation 1, which indicated a value of 0.276 in this case. We interpreted this to mean that users tend to notice new items relatively quickly and use the screen to find out more information.

$$xi' = \frac{xi}{\max(x)} yi' = \frac{yi}{\max(y)}$$

$$D = \sqrt{\frac{\sum_{0}^{N}(xi - yi)^2}{N}}$$

Equation 1: Calculating the correlation D between two series of values x and y. A result of 0 indicates that the two series are identical and 1 indicates that they are exactly opposite. We treat values in the range 0–0.3 as indicating a reasonable correlation.

We also found that, unsurprisingly, people wanted to be sure that their submission had been seen by others. There was an element of public performance: having submitted a document, it is necessary that it finds an audience. After submitting something to the Cwall, users would pass by the screen in order to see if it was displayed. However, because the rules did not guarantee that new items were displayed immediately, users might not see them initially. We found users less motivated to submit items if they could not see them on the screen, because then they did not know if other people had seen them. To correct this we added a new subrule to the rules for general documents that boosted the priority for new items to ensure that they would get seen at least for a while irrespective of their other merits.

15.5.1.1 The Effects of Location on Users Expectations

One unexpected result was that users' expectations of the system appeared to change according to the location of the display even when location changes were restricted to one workplace. When we installed the first system in proximity to our offices and those of our workgroup, it was our workgroup that mostly used the system. There was a small but loyal group of users, with most submissions being work-related but also some jokes and other sorts of fun items. Comments received about the submissions were generally pertinent. After about 10 months we moved the display to the foyer of a building that was is used to host visitors as well as office space for about 40% of the people on site. This move was done in order to facilitate giving demonstrations to larger groups of people and also to increase the number of potential users by placing the system in a place with more traffic. We found that people were more likely to be curious and experiment with the system. However, we also found that because of the change of building, we lost some of our previous user group. At the same time we got more "junk" comments submitted from the screen, as people would experiment with both the system and the handwriting recognition. For the first time we had to contemplate cleaning the database to remove these spurious comments. It is also possible that, because of the change of location, there was less sense of ownership of the system as it was no longer physically associated with the group that had created it. Three months after the move we installed a second screen in the site cafeteria hoping to capitalise on the large numbers of people using the room during the day for eating lunch, tea breaks and informal meetings. Although this did appear to attract new users, we also found that they too seemed to have different expectations of the information displayed on the screen. Whereas the other two locations had been associated with work, the cafeteria was associated more with leisure, and so we found users requesting that more leisure items be displayed – such as cinema timetables and reviews of current films. In order to respond to this demand we have mined local cinema timetables from the Web in order to automatically generate this information. Since each display has its own rule-set for content selection, we could change the rules for each display so that the one in the cafeteria was more predisposed towards leisure whilst the others were more predisposed towards work.

15.5.1.2 Identification and Trust

Our initial identification mechanism worked by asking users to select their picture from a palette of users whenever they performed an operation that needed their ID (such as rating an item). This worked well initially within our workgroup (membership 15–20 people) where we trusted people to be responsible and everyone knew each other well. An unplanned positive side effect of this was that people could recommend items to others by choosing the e-mail icon and selecting another person's face. However, when the system was moved into the entrance foyer the number of possible users expanded (about 100 potential users) and made this mechanism cumbersome. The appearance of "junk" comments also suggested that it could no longer be assumed that people would be responsible. In further implementations we have investigated both the possibility of using a badge reader compatible with the identification badges used to control access to the building or the possibility of using Bluetooth or Wifi connection to identify users near the CWall through the mobile devices they carry.

15.5.2 Looking for Design Dimensions

Based on this first experience, and on what we learnt about other similar systems like the ones reviewed in O'Hara et al. (2003), we present here a number of dimensions for their success that we believe are crucial when designing these kinds of systems.

15.5.2.1 The Cost/Benefit Barrier

A characteristic of these kinds of systems is to be a "shared" information space, where the goal is to have fresh and highly informative content collected on the basis of the current activities and interests of people. In none of the examples mentioned here was the system a critical element to pursue the work; it was more of an additional layer to make it smoother or faster or more creative. In this sense these systems can be considered as common goods: likely to be happily consumed, but more difficult to be happily maintained and nurtured. This aspect is reported in the experience around the Apple Newspaper (Houde et al., 1998), where they report a drop in the usage after the first wave of enthusiasm. In our own experience with CWall we have noticed drops in the usage that corresponded to periods of time where the unit was experiencing short-term troubles that had an impact on daily life.

This issue can be addressed in a variety of ways, mainly working on the two sides of the cost/benefit ratio. In the following we address the cost side

Table 15.1 The dimensions to reduce usage cost

System smartness	Familiar look and feel	Integration with the working environment
Dynamic profiling based on context sensing	Newspaper	Multiple input methods
Static profile based on editorial rules	Bulletin board	Transition to synchronous support

of it, ranging from how to increase the automatism and smartness of the system to how to make it more similar in its look and feel to other tools already familiar to the users.

15.5.2.2 System Smartness: Group Profile Without Context Sensing

By providing applications with more information about users and their context, systems will be able to be more responsive, the information presented more pertinent and user-machine dialogue a lot less laborious. Context awareness can introduce new possibilities in the improvement of the cost/benefit ratio. For example, the GroupCast project has investigated the system's capability to react to the precise identities of the people in front of the screen (McCarthy et al., 2001). In a first stage the system, comprising an active badge infrastructure, had been designed in such a way as to contain a global profile in which each user would specify their interests. The group of users detected in front of the display was used to compute the intersection of their interests in the global profile and then show something that each of them could talk about in order to sustain the conversation. However, they quickly noticed that such a profile could be very large, therefore a bit impractical and very rarely completed.

In the case of the CWall, since only about 10–15 items can fit on the display at one time, we implemented a mechanism for automatically selecting items from the database. Items can be classified into a number of types (sticky note, paper, meeting, conference announcement, etc.) and each type can be associated with a rule that is responsible for assigning a priority to items of that type. For example the relevance of a scientific paper might be linked to its numeric rating and the number of comments it has received, whilst a meeting announcement might increase in priority shortly before the time of the meeting and then fall to zero afterwards. We implemented a number of simple rules and the possibility to compose them hierarchically, so that administrators of a display can compose composite rules using a simple text configuration format without having to understand how to program. At intervals (currently every 10 minutes) the system reapplies the rules to any active items (those that have been created, rated, commented on or interacted with in the recent past) and selects those with the highest priority for display. This means that the display changes often enough that people who pass by several times in a day should see different items. It also allows us to give priorities to different types of items at different times of day.

Our final rule-set includes a rule that adds a small amount of random noise and a rule that decreases the priority of items according to how much time they have already spent on the display in order to prevent a few items from monopolising the display and to ensure that users see a variety of items if they pass by the display recurrently.

383

15.5.2.3 Familiar Look and Feel: Bulletin Board

When designing the CWall display, we chose the metaphor of a bulletin board to which anyone can post information. The information can be plain text or web pages that include images (including animated GIFs). We wanted to avoid a standard desktop look and feel, so items are placed randomly on the screen to give a more "organic" feeling. However, in order to increase readability, items are not permitted to overlap. The large screen is touch sensitive and people can interact with it using their fingers – the controlling PC's keyboard and mouse is hidden from view. Each item displayed on the screen shows at least a title, a number of stars indicating the average numeric rating, an icon indicating the number of comments on this item, the name of the person who submitted the item and some icons allowing the user to perform specific actions on the item.

Items that the system decides are more relevant at a given moment may also have images selected from a web page displayed together with the first few lines of text from the page and a display of any comments submitted by other users. An example item representation is shown in Figure 15.4.

Interaction with the system is deliberately simple, users can

- touch the item anywhere except for the four action icons to cause the system to expand the amount of space given to the item (up to the limit of about 30 lines of text) in order to find out more about it;
- touch the mailbox icon to e-mail the item to themselves or someone else (a palette of faces is displayed, and the user selects the face of the recipient);
- touch the printer icon to print the document at a nearby printer;
- touch the balance item to record a numeric rating (users are asked to identify themselves by selecting their image);

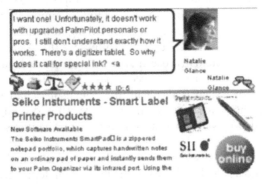

Figure 15.4 Close up on an item displayed on the CWall showing a comment, the picture and name of the person who submitted the comment, some images from the page, the rating (four stars) and the first few lines of text.

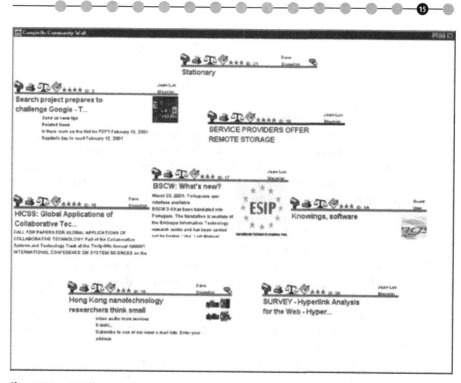

Figure 15.5 CWall screenshot.

- touch the pen and paper icon to write a comment on the screen. We use software delivered with the SMARTBoard to convert the handwritten comment to text.

Figure 15.5 shows a screen shot of a typical display. One point to note is that we make no attempt to display the full text of a document on the large screen – we decided at the beginning that a public space was probably not the ideal location to read possibly long documents and we wanted to keep interaction quick and simple and avoid more difficult interaction such as manipulating a scrollbar. Our idea is that we should give people enough information to decide whether an item is interesting. If they want to read more, they can then e-mail the item to themselves for reading on their desktop, print it or capture it on their Palm Pilot™ (see below).

15.5.2.4 Dynamic Profile Based on Context Sensing

In the case of the Cwall we decided to see if we could improve the utility of the display by differentiating between people passing by and people standing in front of the display and actively reading the content. In this way we hoped to address two criticisms of the existing interface, namely, that

- It was too hard to work out if something interesting was displayed with a passing glance from a distance – people had to approach the display in order to read the titles.
- The display often changed while people were reading it. This occurred when the internal timer caused a refresh and people were reading the contents but not interacting with the display. Since it detected no active usage of its user interface, the CWall had no way of telling that there were people reading the items on display.

For passers-by, the information on the screen should be presented in an eye-catching manner and change frequently, whereas the information shown to an interested reader could be more detailed and stable.

We did not want to force people to carry additional devices such as active badges and so we experimented with two sensing techniques that do not require people to be specially equipped: a grid of infrared movement sensors and face detection using real-time analysis of a video image from a camera mounted on the CWall display.

The face detection sensor was built mixing two image processing techniques that complement each other's weaknesses: motion detection and colour detection. The motion detector allows rapid detection of faces whatever their skin colour and at a wide range of distances. However, it works only indoors; it cannot detect and follow people moving too fast; and finally it cannot distinguish between a face and the back of a head. The skin colour detector is very sensitive to lighting conditions and can produce false detections for background colours similar to skin. However, if applied to the areas already detected by the motion detector, it can be used to refine the results of the motion technique.

We also implemented a module devoted to building a high-level description of context from the basic sensed characteristics. The decision part of the system is basic (user attentiveness to the screen, and activity of groups in the room) but it provides a very strong structure to extend it. We have a working component-based architecture with an embedded inference engine. Rules can be added and tuned easily loading new scripts files through the user interface. The resulting system has the following behaviour:

- It uses bigger titles to attract people's attention when people are in the room but far from the CWall (Figure 15.6).
- It freezes the content of the board when a user is detected to be reading, and communicates the transition to the user. At the same time, it moves from a presentation where only headlines are displayed to one where people can read more contents about the topics (Figure 15.7).

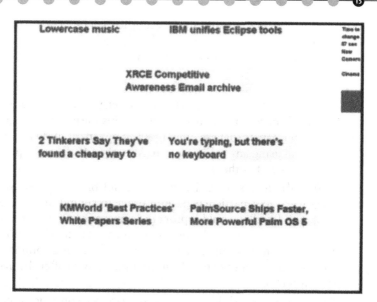

Figure 15.6 No one standing in front of the CWall looking at it.

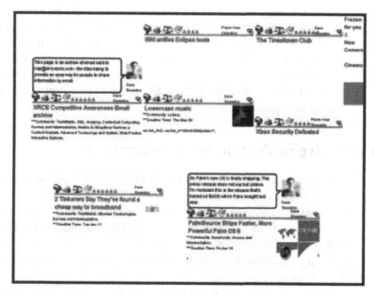

Figure 15.7 Someone looking at the screen (text in upper right corner says "frozen for you").

15.5.2.5 Integration with Current Work Environment: Multiple Input Methods

We formed the hypothesis that one reason why the Web recommender system was felt to be too time-consuming was possibly because the web interface was not always a good match for the medium carrying the information that people wanted to submit. In order to reduce the cost of submitting information, we therefore developed several additional user interfaces that could be used to submit information to the system.

- *E-mail*: Many hi-tech organisations have a strong e-mail culture. Our centre is no exception, and so like Houde et al. (1998) we decided to allow people to submit new items by e-mail. This meant that documents such as calls for papers that are often received by e-mail could be forwarded to the CWall using any mail program. Also, the CWall can be put in copy of messages sent to others.
- *Web bookmarklets*: In order not to disrupt browsing instead of forcing users to go to a particular page in order to submit an item, we created bookmarklets that would capture the current page URL and title and pop up a window allowing the user to confirm the selection and optionally add more information such as a numeric rating or a comment. Cookies are used to prevent the user from needing to login after the first time the system is used.
- *Paper*. Using Xerox's Flowport^{TM} software, we constructed paper forms that allowed users to submit hard copy documents. Using a combined printer/scanner/copier next to the large screen, users were able to place a cover sheet (Figure 15.8) on top of the document and press the start button on the copier. The cover sheet is recognised as introducing a new document, the document is scanned, OCR'd and then the keywords in the OCR'd document are analysed to decide which KP community it is best suited to. Users have the option to override the automatic classification using the normal KP web interface.

15.5.3 A Step Further: The Social Environment

On the basis of the above results and observations, we moved to a second implementation of the system, where we focused on increasing the benefit provided by the system while continuing to build upon our previous findings on design principles to limit usage cost. This second implementation and deployment was done inside the MILK project (Multimedia Interaction for Learning and Knowing (MILK) is a joint research project partly funded by the European Commission – IST Project no. 33165).

The way to support the community exchange in MILK has been similarly based on placing interactive boards where there are activities that are going on in the organisation, as inferred from Document Management Systems (DMS) logs of use, are automatically published in order to make visible community activities. The boards are in semi-public places like the printer room, the entrance hall, the library, etc. Figure 15.9 shows one user site installation of MILK, where a leisure area with magazines is augmented with an interactive large screen.

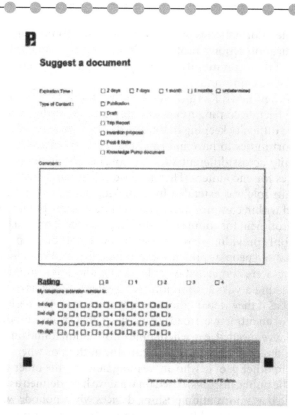

Figure 15.8 Paper from allowing users to submit hard copy documents to the CWall and KP. The form allows users to check boxes corresponding to type of document, a rating and their telephone extension. They can also leave handwritten comments in the box in the middle – these are displayed as images on the CWall and in KP.

Figure 15.9 One of the site installations.

The information is published in order to promote the spontaneous communication among professionals sharing the same interests and profile, even if their community has not yet been given a public space, i.e. is still an emerging community.

However, MILK is grounded on the observation that knowledge workers have been becoming more and more mobile and distributed, thus having more difficulty keeping in touch with their communities. In particular, the opportunities to have informal interactions between members of a community across different sites are considerably reduced due to the lack of spaces for encounters. Therefore we called our system *Social Environment* and its role was extended from making content sharing more straightforward within communities to re-creating in semi-public spaces the required environment for community building and life. The Social Environment was not only providing news or shared documents but also providing information about people's presence and availability. Additionally, these information points were connected by video and audio links. In this way we aimed at recreating a virtual common space where people have the opportunity to meet as if they were in the same physical space. The design idea that we had was of an attractive broadcasting space that could support communication whenever needed, e.g. when some information found on the screen prompts an interest in knowing more from the author, or when we want or need to talk to other people who are reading news on the other site.

The public displays, based on what we have defined as a broadcast model, worked as information pushing devices when nobody was interacting with them, mixing the different channels. As soon as somebody started interacting with the system, it switched to pull mode and the user got access to any information he or she needed. The broadcasting mode in the Social Environment was designed both to give hints of what is going on inside the company to non-interacting onlookers, and to urge them to start interacting with the screen (switching then to pull mode) to go deeper inside the items they found more relevant, or to browse by organisational theme exploring the system content. The alternation between the different types of content in the broadcasting mode depended on a set of rules based on time of day and users' activities, controlled and modified by the system administrator to push certain kinds of data and promote specific information to all fellow workers.

While the entire layout in the CWall was based on the single metaphor of a bulletin board, the Social Environment has to present different types of content with a layout adapted for each of them. In order to ease the comprehension of the interface by users, information to be broadcasted has been clustered in channels. Each channel represents a specific view on the knowledge present in the organisation and is relevant for different communities.

The Thematic channel (Figure 15.10) is the channel providing information about current units of work, i.e. projects and community forums. It is the channel from which new and interesting information can be accessed,

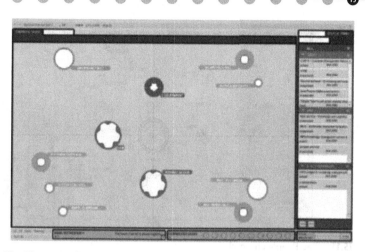

Figure 15.10   The Thematic channel.

read, bookmarked and also enriched with comments and personal notes. Moreover, by using this channel, it is possible to access information across different sites in support of synchronous video contacts. In order to retrieve its content, it monitors the activity in the DMS and prioritises it for providing only live information. It utilises some layout rules in order to represent activity parameters:

- *Colour*: membership of an organisational area of activity.
- *Distance from the centre and colour fading*: the overall amount of recent activity.
- *Size*: overall amount of activity.
- *Thickness*: degree of novelty.
- *Shape*: to differentiate among projects and communities.

The People channel (Figure 15.11) is the channel providing information about people in the organisation, the means to contact each person and their current location and availability. The channel dynamically updates the information about the current location of the users.

The granularity of the information was kept at the level of different organisational locations or else just not in the office. The system integrates functionality from an availability service that can also register at each moment the preferred channel(s) for interactions (e.g. SMS while attending a meeting).

A number of layout rules are used in order to facilitate quick overview:

- *Location tabs*: where each tab represents a site showing the people who have that reference location.

391

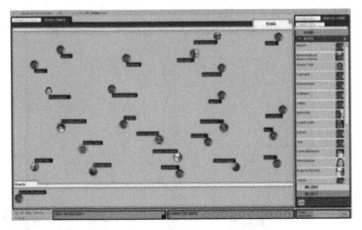

Figure 15.11 The People channel.

- *Colours*: to identify a location, people currently detected in a location are represented with the colour of the site where they are.
- *Guest area*: for people visiting the site.
- *Grey Colour*: for Mobile people.

The News channel (Figure 15.12) was used to broadcast information that the organisation wanted to transmit to everyone at a site or across all sites.

The news had its own channel, but was also overlaid on whatever current channel was displayed if no one was interacting with the screen. It was displayed using a style borrowed from newspapers and was meant to attract attention even from afar.

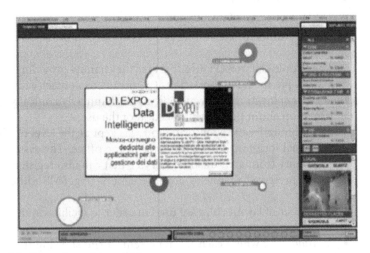

Figure 15.12 The News channel (overlaid).

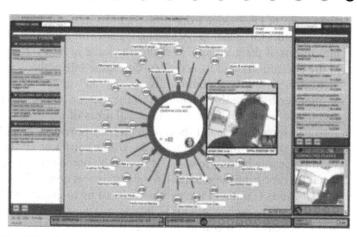

Figure 15.13 The Video channel (bottom right and overlaid windows).

Finally, the video channel was a channel devoted to support synchronous communication. It provided visualisation of the connected sites (Figure 15.13) and supported the possibility of unplanned video-audio sessions. It also included some light support for collaboration; i.e., it was possible to share content on the fly across the various sites in order to support discussion.

15.6 Conclusion

In a report that compares existing systems in support of communities, Wenger (2001) identifies a number of dimensions that are relevant in order to successfully support community life and development. These dimensions include the ease and breadth of spectrum of participation, the provision of value in return for contributions, the visibility of community activity, the support of different levels of membership, openness to the external world and the support for evolving phases of community life. If these are compared to the design principles that guided the design of the CWall and Social View systems – namely, integration of community support functionality with the daily work environment and the development of smart views mapping different work situations – we can see that our approach provides a contribution to most of the relevant dimensions identified by Wenger. Wenger acknowledges that participation in community life is in competition with all the other activities a member is a part of, primarily the work tasks of ongoing projects. Therefore, participation must be easy in order to reduce possible barriers and lower this tension between priorities. In the work presented here we have suggested that participation must not only be easy, but also be fully integrated with the daily work environment, as not to lower just the learning curve for the community system, but also to facilitate the movement of knowledge from projects to communities and vice versa, as opportunities for exchange occur. This is also compliant with Wenger's other indications of the need to provide a clear value in return for participation in the community and its activities. Our large interactive screens also addressed the visibility of community activity, the support of different levels of membership and their openness to the external world by enlarging the visibility of community activity to social areas of the organisation where abstract representations of these activities are provided. It is of course the case that support for evolution of the community was addressed only in an implicit way by us. However, it should be noted on this score that the metadata capturing the activities in the system was nonetheless able to guide the selection of content in some views according to the relative freshness or similarity of the exchange. This facility, in our view, is an important feature in the capacity of any community to evolve.

Matching our own set of requirements to a set that was derived from long observation of community life and activities provides strong confirmation that the design approach we adopted is sound. It remains for the newer design we have discussed here to be validated for us to consider how effectively it further promotes the dimensions of community we have presented in this chapter.

References

Agostini, A., Giannella, V., Grasso, A., Koch, M. and Snowdon, D. (2000) Reinforcing and opening communities through innovative technologies. In Gurstein, M. (ed.), *Community Informatics: Enabling Communities with Information and Communications Technologies*. Idea Group Publishing, pp. 380–403.

Glance, N., Arregui, D. and Dardenne, M. (1998) Knowledge pump: supporting the flow and use of knowledge in networked organisations. In Borghoff and Pareschi (eds), *Information Technology for Knowledge Management*. Springer-Verlag.

Glance, N., Arregui, D. and Dardenne, M. (1999) Making recommender systems work for organizations. In *Proceedings of PAAM'99*, London.

Houde, S., Bellamy, R. and Leahy, L. (1998) In search of design principles for toals and Practices to support communication within a learning community. *SIGCHI Bulletin*, **30**(2).

Isaacs, E., Whittaker, S., Frohlich, D. and O'Conaill, B. (1997) Informal communication re-examined: new functions for video in supporting opportunistic encounters. In Finn, K.E., Sellen, A.J. and Wilbur, S. (eds), *Video-Mediated Communication*. Lawrence Erlbaum, New Jersey, pp. 459–485.

Ishii, H., Wisneski, C., Brave, S., Dahley, A., Gorbet, M., Ullmer, B. and Yarin, P. (1998) ambientROOM: integrating ambient media with architectural space. In *Summary of CHI'98*, pp. 173–174.

Lave, J. and Wenger, E. (1991) *Situated Learning. Legitimate Peripheral Participation*. Cambridge University Press, Cambridge, UK.

McCarthy, J., Costa, T.J. and Liongorasi, E.S. (2001) UniCast, OutCast and GroupCast: three steps toward ubiquitous, peripheral displays. In *UBICOMP-2001*.

O'Hara, K., Perry, M., Churchill, E. and Russell, D. (eds) (2003) *Public and Situated Displays: Social and Interactional Aspects of Shared Display Technologies*. Kluwer Academic Publishers, London.

Snowdon, D. and Grasso, A. (2002) Diffusing information in organizational settings: learning from experience. In *Proceedings of CHI 2002*, ACM Press.

Weiser, M. and Brown, J.S. (1995) Designing calm technology. Retrieved from <http://www.ubiq.com/hypertext/weiser/calmtech/calmtech.htm>

Wenger, E. (2001) Supporting communities of practice: a survey of community-oriented technology. Retrieved from Shareware on the site <www.wenger.org>

16

Serving Visitor Communities: A Mediated Experience of the Arts

Patrizia Marti, Gregory O'Hare, Michael O'Grady, Massimo Zancanaro, Elena Not, Alberto Bianchi and Mick O'Donnell

16.1 Introduction

A 50-year-old man was apparently hesitant carrying such a small and expensive computer around in the museum and in particular he was afraid he would not to be able to use it. As soon as he stepped into the Sala del Mappamondo, wearing a headphone and firmly grasping the device, a deep loud voice started speaking, "*This is the Sala del Mappamondo one of the most important halls of the whole Palazzo Pubblico. The wonderful frescoes, even the more religious ones, impose themselves as the very first examples of laic art involved in history and contemporary political life.*" Just another audio guide, thought the man a little bit relieved about the situation. Yet, when he approached the opposite wall the voice started speaking again, "*This is La Maestà, depicted by S. Martini in 1315. The fresco is located in the main part of the hall, the central point that gave the orientation of the Sala del Mappamondo. By contrast, the Guidoriccio fresco, located just behind you, was a sort of 'poster,' glorifying the power of the Siena Republic.*" The man turned around and the voice continued, "*In front of you, you can see the Guidoriccio, it was also painted by S. Martini [. . .].*" The man was amazed; the audio guide seemed to work by itself. Suddenly, he recalled the small computer that he was still firmly grasping. He looked at it and first of all he noticed a small picture of the wall in front of him, all the frescoes were blurred but one: so that one is the Guidoriccio.

Half an hour later, during the post-experiment interview, the man claimed that the audio guide helped him to get oriented in the space: he actually never discovered that the guide provides a self-oriented map of the museum.

The design of entertainment and edutainment systems is a compelling and challenging adventure. Such systems are not intended to help users perform work-related tasks and, most of the time, they cannot be brought back to clearly stated user requirements: their ultimate mission is to engage the user and to stimulate learning. The nature of this kind of system imposes a mediation between designers' visions and user needs. Within a 3-year project[1] named HIPS (Hyper Interaction within Physical Space) funded by the European Commission in the I3 Programme, we pursued the ambitious goal of designing an electronic tourist guide that transforms the user experience from one of simple consultation (commonly achieved with audio guides, multimedia kiosks, CD ROMs or even books) to an immersion in a rich information environment. We implemented two prototypes, one indoor in the Museo Civico di Siena and one outdoor in the campus of the University College of Dublin. The project finished in October 2000 and the system is currently maintained in laboratory.

Our first objective as designers was to envision new human activities that can be supported by technology and then define the technological requirements needed to support such activities, trying to conceive non-intrusive and non-compulsory instruments. This vision ended up with a number of design concepts that were implemented in the final prototypes. The concepts of *immersive environment* and *environment-sensitive user interfaces* were used to establish the primacy of environment; technology is fundamental only if it supports rather than overcomes the "real" experience – the real interfaces in museums are the rooms and the frescoes themselves. The concept *situation-aware content* highlighted the need to provide information in a way that relates to previously delivered information, for instance, using comparisons to previously seen exhibits, and making frequent use of reference to space and time (e.g. *as you have just seen, in front of you*, etc.). Finally, *social navigation* was the concept that drove the necessity of thinking about a museum visits not just a single person experience but rather as a collective one.

In our vision, the tourist guide should assume the role of a travelling companion that tells stories about what the visitor looks at, without requiring any explicit intervention for accessing information in an attempt to minimise the boundary between the physical space and the related information using the movement of the visitor as mediator.

The technology we used allowed multiple information structures to be overlaid on the physical world in a non-intrusive fashion, opening up new possibilities for creative design. This experience demonstrates that the success of entertainment or edutainment systems, especially those exploiting advanced technologies, strongly relies on a design philosophy that mediates between a deep and continuous focus on the users and innovative visions.

The HIPS audio guide is characterised in five important respects:

---

[1] HIPS Consortium: University of Siena (I) – Project Coordinator, University of Edinburgh (UK), University College of Dublin (IR); IRST-ITC (I), GMD (D), SINTEF (N); ALCATEL (I).

1. It is aware of the visitor's location, orientation and distance from objects.[2] It has knowledge of the physical space and features of the artworks inside it. In presenting information about artworks, appropriate language phrasing is used, along with modified images (e.g. blurring the surrounding space when displaying the location of an artwork), making the user terminal an enrichment of the visiting experience, and helping the visitor discover what is hidden to his eyes.

2. It responds to user movements: the HIPS audio guide interprets visitor movements in context, for example the description of a small painting will be stopped automatically if the visitors walk away, but the description of a wall-size fresco will continue if the visitor moves away to look at the fresco as a whole.

3. The guide stores the history of the visit and of the interaction and uses this to decide what information to present, and thus the user will not be presented with the same information twice: at a second visit to the same object, a richer dynamic presentation will be provided in the new context (e.g. *"you are again in front of [...]"* and following with previously unheard descriptions).

4. By asking for further information, or on the contrary, moving away or explicitly stopping a presentation, the visitor gives strong evidence as to his preferences and interests. The subsequent presentations can be made more effective if the system is able to exploit this knowledge (e.g. *"regarding Simone Martini, you know that the author of this fresco was a strong enemy of his: a legend says that [...]"*).

5. The audio experience follows the user activities dynamically and in a longer term view: the length and the style of the presentations are adapted to the user style of visit (e.g. briefer and more general descriptions where the visitor moves through the art space without becoming particularly involved; or longer, deeper and more exciting explanation styles when they do become involved).

6. While language is the most effective modality for presentation of information and audio is the preferred medium to deliver it, in a situation where artworks have the precedence in using the visual channel, we experimented with the use of images to help the visitor appreciate small details in a fresco and used 3D images to help the user to orient himself or herself in the space.

7. Finally, the HIPS guide gives the user the possibility of annotating the information space for subsequent visitors, offering individual reflections

---

[2] Infrared was used as the basic technology for visitor's localisation in the physical space of the museum. Yet, HIPS actually implemented a three-level representation of the physical space in order to abstract away from the specific technology and to allow symbolic reasoning on the visitor's movements. This representation is briefly introduced in Section 16.3; for further details see Bianchi and Zancanaro (1999).

and comments in the form of *user activated hotspots*; this possibility enriches the space with a sort of personalised track.

But do features such as these actually make a system as useful and enlightening as it was for the 50-year-old man? The effectiveness of the standard usability evaluation methodologies in the field of art, entertainment and leisure is questionable, since the visitor usually does not have a clear goal in mind that can be achieved by a sequence of pre-defined tasks. Rather, it is the context that shapes the relationship between human actions and system outcomes. However, this relationship is not a simple one. Some features of the system may induce errors or confusion, or perhaps a technical fault may occur in the external environment, which the system is unable to handle. Thus it is necessary to analyse the richness of the context of interaction when discussing usability and engagement. We chose to assess the usability of the system by measuring the impact of the HIPS audio guide at four different levels: phenomenological, cognitive, emotive and socio-cultural. The evaluation at the four levels was carried out on the basis of the direct observation of the activity of real users (i.e. English speaking visitors of the Museo Civico in Siena).

The HIPS experience, although limited, demonstrates that the success of entertainment or edutainment systems, especially those exploiting advanced technologies, strongly relies on a design philosophy that mediates between a deep and continuous focus on the users and innovative visions.

16.2 The Museum Experience

Museums[3] are non-competitive and non-evaluative environments where visitors are free to move around and learn concepts through the objects exhibited. A museum visit is a personal experience encompassing both cognitive aspects (e.g. the elaboration of background and new knowledge), as well as emotional aspects (e.g. the satisfaction of interests, the fascination for the exhibits). The optimal tourist guide should support strong personalisation of any offered guidance in an effort to ensure that each visitor be allowed to accommodate and interpret the visiting experience according to their own pace and interests. This is why our tourist guide is able to adapt the presentation of information to the idiosyncrasies of each particular visitor by appropriately selecting:

1. what to tell (taking into account his or her personal interests);
2. the amount of information delivered (providing long or short descriptions of artefacts);

---

[3] In this paper we use the term "museum" in a wide sense to include traditional museums, open-air expositions, tourist sites, zoos, etc.

3. the way in which the information is presented (e.g. deciding whether an audio or textual commentary is appropriate) and
4. making reference to other exhibits in the same museum and in particular to those already presented to the user (in order to reinforce learning).

However, the ideal audio guide should not only guess what the visitors want but also take into consideration what they have to learn, thereby making the museum experience really useful and memorable. Therefore, it is also important for our guide to stimulate new interests and to suggest new paths for exploring the museum. Hence, a system to support visitors in their visit should take into account agenda, expectations and interests as well as the peculiarities of a cultural experience in a physical environment.

16.2.1 Visitor's Agenda, Expectations and Interests

Visitors' own agenda and preferences greatly determine their expectations and behaviour during a visit and should be considered carefully by a flexible guide when planning a situated presentation of information. Usually, visitors have general expectations: they will see valuable objects, they will learn more about what they are interested in, they may possibly buy interesting material at the bookshop and so on. Of course, expectations also vary in accordance with the specific type of museum/cultural site being visited. For example, general art sites, such as historical buildings or cities that were not purpose-fully designed as exhibitions, typically allow the visitor to anticipate a more situated and contextual visiting strategy. Expectations are also a function of how accustomed the users are to museum visits, of their cultural level and of their interest in the specific exhibition topics (for example, the desire not to miss the Gioconda in their first 2-h visit to the Louvre might severely influence the visitors' anxiety and visiting path). The visitor's initial expectations can also change during the visit, possibly as a result of their enjoyment of the exhibition. Typical signals for this change are an increase in interest for the items displayed and slower movement through the items. This scenario requires that the successful digital companion maintain a dynamic model of visitor's expectations, agenda, preferences, interests and knowledge.

16.2.2 Reflective and Experiential Cognition in Exploring Art Settings

In his book, *Things That Make Us Smart*, Don A. Norman (1993) analyses the nature of experiential and reflective cognition. An experiential mode of interacting with the environment is achieved when people assimilate information without apparent effort by just letting the external world drive emotions and perception. Reflective thought takes place in the same "environmental" conditions but requires more conscious effort and initiative in assimilating information. If we apply this notion to the context of a visit to an art setting, we could say that the reflective modality of interaction is related to the deliberate intention of the user to consult information of interest, whilst the experiential modality relies on the capabilities of the environment to attract and stimulate the visitors without requiring an explicit initiative from them. Inspired by this view, we aimed to design both modalities of interaction within our tourist guide. In particular, for mediating the experiential modality, we adopted the concept of *optimal flow* (Csikszentmihalyi, 1990): the absolute absorption in the activity where the experience is guided by external events, which stimulate the visitor and facilitate the assimilation of information. Supporting the optimal flow means embellishing both the physical and virtual environment in such a way as to motivate the visitor through a sense of engagement, thereby enhancing the experience. Our tourist guide supports both the experiential and reflective modalities of interaction, letting the visitor explore content more deeply while experiencing the emotional effect. To elucidate further, the reflective modality of interaction is mostly related to the deliberate intention of the user to ask for information (explicit queries to the system through a Personal Digital Assistant (PDA)), whilst the experiential modality is mediated by a natural input, namely the visitor's physical movement.

In order to design a complete experience for the tourist, it is essential that all the elements that comprise the system be harmonised, ranging from any data that may be manipulated during the interaction to both the contents and the physical layout that cooperate to define the experience.

To reach an optimal flow during the users' interactions, all these elements should be strongly integrated. If the users focus on a single element (e.g. a single data item or a single object) or if they can sense an inconsistency between them, the flow of interaction is damaged. In the following sections, we illustrate how the museum experience was designed, from the initial modelling of the physical space and visiting strategies through the interaction mechanisms and, finally, content design.

16.3 The Vision: Hypernavigation in the Physical Space

16.3.1 Beyond Traditional Audio Guides: Adaptivity and Location-Awareness

Until recently, the state of the art in visitor enhanced user guides was the delivery of content in the form of audio presentations through headphones. There was little consideration as to the individual needs of the tourists or indeed their location within the museum or site.

The technological guide envisaged here is characterised in five important respects:

1. It would be adaptive to the needs of the individual user.
2. It would be aware as to the location of the user.
3. It would respond to user directives and interactions with it.
4. Content would be delivered using a rich media portfolio.
5. Users could annotate the information space for subsequent users offering individual reflections/comments in the form of *user-activated hotspots*.

Many existing systems merely presented prefabricated dialogue to the user resulting in low relevance and user dissatisfaction. In contrast, we seek to track the users movements and orientation, build up a profile of individual user preferences through a user profile and to capture user interventions through the user interface. Collectively this renders an adaptive personalised user experience. In addition, we sought to deploy a rich portfolio of media types in content delivery including audio, text, graphic and video. The incorporation of user augmentation of the content space via hotspots was pivotal in allowing users to take collective ownership of the content, thus enhancing the visitor experience.

However, the most significant disadvantage of the current range of audio guides is often represented by the fixed content of the messages, which does not guarantee that visitors are offered the most suitable information with respect to their current location, interests and needs.

Adaptive hypermedia (De Bra et al., 1999; Chen and Magoulas, 2005) are hypermedia systems in which either the presentation of information or the hyperlink structure (or both!) are dynamically modified according to the specific context of use. They are often compared to dynamic hypermedia. The former exist prior to their use and the user model is employed to hide part of the structure (or to highlight another part). Fully dynamic hypermedia, on the other hand, do not exist until the very moment in which a user explores them; they are dynamically created on the fly using automatic text generation techniques (Oberlander et al., 1998; Isard et al., 2003).

In the digital guide that we envisage, we combine the benefits of location-aware systems with the advances made in parallel in the field of adaptive and dynamic hypermedia; this way we gain more effectiveness in the mobile delivery of information.

Reasoning from localisation information, a model of the current user is collected. For example, a visitor staying in front of an object for a while may, under certain conditions, be interpreted as interest, whereas a quick crossing can show disinterest or boredom. Similarly, the path that the user takes provides a measure of how much that visitor follows the suggestion of the guide. The user's input (both implicit and explicit) is interpreted by the system in order to plan an "intelligent response": simple reactions to user movements are not, in fact, sufficient for this kind of interaction. The movement has to be interpreted in the physical context (e.g. whether the visitor is close to a large fresco or a small painting) and in the semantic context of the presentation (e.g. whether the visitor is listening to the presentation of a painting or to the general presentation of the room).

Indeed, a notion of space that relies on naive geometry rather than the usual Cartesian geometry is needed. Rather than depend on the approximate location and orientation we receive from the location sensing devices, we reinterpret the data *functionally*, in terms of what exhibit(s) the user is near, and which are they looking at. Such an approach allows us to abstract away from the particular localisation system. We use naive notions such as *area* (a closed region of space identified with a name and a type, e.g. a room, a floor, etc.) and *exhibit*. Areas can contain other areas and one or more exhibits.

This representation of space allows the rules of behaviour for the system to be encoded in a more natural and intuitive way. As an example, what should the system do if the visitor moves away during a presentation? Some rules that can be encoded using the above representation of space are:

- If the presentation is about a painting, do not stop it unless the painting is no longer in the visitor's sight.
- If the presentation is about a room, do not stop it unless the visitor leaves the room.

16.3.2 The Tourist Visiting Styles

One of the primary requisites for identifying the most effective presentation strategies for visitors moving in a physical space is understanding visitors' possible behaviour. Direct observation of human activity is one of our primary sources of design inspiration (Hutchins, 1995). In our tourist guide, we exploited the results of ethnographic studies documented in literature and also performed an intense period of observations in two museums in Siena: Museo Civico and Santa Maria della Scala.

Documented ethnographic studies showed that visitors tend to move in exhibition settings in homogeneous modes. In particular two French ethnographers, Veron and Levasseur (1983), classified visitors in four categories defined on the basis of the following variables:

- geometry of pathways;
- time spent in front of each artwork;
- the global time of visit;
- the number of stops.

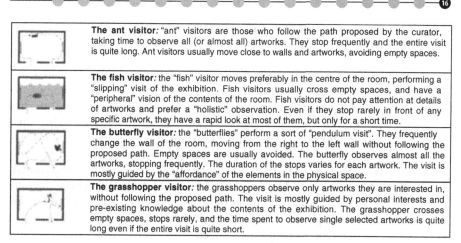

	The ant visitor: "ant" visitors are those who follow the path proposed by the curator, taking time to observe all (or almost all) artworks. They stop frequently and the entire visit is quite long. Ant visitors usually move close to walls and artworks, avoiding empty spaces.
	The fish visitor: the "fish" visitor moves preferably in the centre of the room, performing a "slipping" visit of the exhibition. Fish visitors usually cross empty spaces, and have a "peripheral" vision of the contents of the room. Fish visitors do not pay attention at details of artworks and prefer a "holistic" observation. Even if they stop rarely in front of any specific artwork, they have a rapid look at most of them, but only for a short time.
	The butterfly visitor: the "butterflies" perform a sort of "pendulum visit". They frequently change the wall of the room, moving from the right to the left wall without following the proposed path. Empty spaces are usually avoided. The butterfly observes almost all the artworks, stopping frequently. The duration of the stops varies for each artwork. The visit is mostly guided by the "affordance" of the elements in the physical space.
	The grasshopper visitor: the grasshoppers observe only artworks they are interested in, without following the proposed path. The visit is mostly guided by personal interests and pre-existing knowledge about the contents of the exhibition. The grasshopper crosses empty spaces, stops rarely, and the time spent to observe single selected artworks is quite long even if the entire visit is quite short.

© University of Siena

Figure 16.1   Veron and Levasseur classification.

The patterns of behaviour derived by the combination of these variables resulted in the (metaphoric) visitor categories shown in Figure 16.1. This classification inspired our design and suggested how to isolate significant variables linked to physical movements, and how to relate these movements to the browsing of information spaces.

This classification was confirmed by observations we made in two museums in Siena with the following elaborations:

- The original classification was proposed for exhibitions, that is, physical spaces purposefully designed as exhibition spaces. Since our (indoor) experimental site is a historical museum that was in origin the building of the Sienese municipality, it is neither an exhibition space nor a pure museum. This means that some variables considered extremely significant for the Veron and Levasseur study, such as the concept of "proposed path", acquire a different meaning in our scenario. For us a "path" is any uninterrupted sequence of artworks.

- Visitors move in the space because they are driven both by intentional motivations (personal interests and preferences) and by visiting strategies that are affected by the properties of the environment. In the latter sense, the movements of the visitor mostly depend on natural and contextual affordances of the space (Norman, 1999). Some of these affordances can be used to model the physical movements of the visitors (Marti et al., 2001).

The results of this study were used to develop a "Visiting Style Module" (VSM), a software module that supplies patterns of movements that can be coupled to information contents with the aim of providing the visitor with appropriate content for the specific context of the visit.

16.3.3 The Environment as a Medium

According to Bordegoni et al. (1997), a medium is a physical space in which perceptible entities are realised. Indeed, in a museum (as well as in a cultural city, an archaeological site, etc.) the most prominent medium is the environment itself. If a digital tourist guide is to build coherent and effective presentations that allow the users to appreciate what they are visiting, then the guide has to take into consideration this special status of the environment. The aim is to *integrate* the "physical" experience, *without competing* with the exhibit items for the visitor's attention. From a multimedia point of view, this means that additional uses of the visual channel have to be carefully weighed. In our guide, we prefer to exploit the audio channel (mainly for language-based presentations, although the role of non-speech audio (e.g. music or ambient sounds) has also been investigated). Yet we use images on the PDA to support the visitor in the orientation task (3D or 2D images are used to support linguistic reference to physical objects). In this latter case, the visual channel is shared between the PDA and the environment but the goal is still to provide support to environment-related tasks.[4]

From a multimodal point of view, other modalities are employed to focus the visitor's attention on specific objects or to stimulate interest in other exhibits. For example, in our tourist guide the linguistic part of the presentation (speech audio) makes large use of deictic and cross-modal expressions with respect to both space (such as "here", "in front of you", "on the other side of the wall", etc.) and time ("as you have seen before", etc.). At a deeper level, presentations are planned to contain elaborations on visual details or comparisons to other exhibits.[5]

Ultimately, the goal of a location aware system for cultural visits is to help visitors to adapt their visiting experience to their own interests; but in some cases a visitor should be encouraged not to miss some particular exhibits (for example, you cannot visit the Louvre for the first time and miss the Gioconda). Sometimes this task can be accomplished by direction giving, but there are other ways to promote exhibits: for example, by providing at the beginning of the visit a list of hotspots, or by planning a presentation that, in a coherent way, links the exhibit in sight to others which the visitor's interest model suggests. More generally, further research is needed towards implementing pedagogically motivated systems with meta-goals to pursue, educational strategies to follow and intentions to satisfy. In this respect, the interaction between the visitor and the system must evolve from simple interaction to full-fledged collaboration. (For a discussion on this topic applied to cultural tourism, see Stock, 1999).

---

[4] Actually, the PDA screen is also employed to suggest further navigation in the virtual space of information; an alternative realisation for this would have been the use of auditory icons (Mynatt et al., 1998).

[5] The extent to which the system inserts elaborations and comparisons depends on the user model.

16.4 Interaction and Content Design: Envisioning a Human Experience

16.4.1 Interaction Design

When we started to design the interaction between the visitor and our tourist guide, it was clear that we did not have to design only a tool, but also new human activities that could be performed in the museum, thanks to a new technological infrastructure. Our idea was to not constrain people to follow pre-defined patterns of interactions, confining them to the mere role of "users of technology", but to offer new opportunities for using contents and artistic spaces to build up new experiences. Therefore our first objective as designers was to envision new human activities that can be supported by technology and then define the technological requirements needed to support such activities. This vision concluded with a number of design concepts that were implemented in the final prototypes. These include

Immersive environment

1. The user is immersed in a rich audio environment. Different reading styles characterise the way in which artworks are described from different perspectives (historical, artistic, anecdotal descriptive).
2. The rhetorical styles are tailored to the context (use of deictic expressions) and to the iconographic contents (artworks representing people are described in first person, as if the character presents himself or herself).
3. The rhythm of narration (length, duration) is tailored to the visitor's movement (long and detailed descriptions, for example, are provided to visitors who move slowly and stop in front of each artwork, according to the Veron and Levasseur classification).
4. Experiential cognition is mediated by a natural input: the physical movement. Reflective cognition is allowed by intentional and context-driven interaction (explicit queries to the system).

Situation-aware content

The audio descriptions are segmented into macronodes (Not and Zancanaro, 2000), small blocks of information that are dynamically combined to form an audio presentation. Each of them contains different kinds of contents with explicit deictic reference to the physical position. The flow of narration is made more fluid and harmonised to the context of visit. The use of different reading styles, together with the integration of 3D sounds and music, are means to create rich audio environments. Our tourist guide reproduces a sort of "empathic effect" mediated by human voices and immersive information spaces to engage the user in an intense meeting with art.

Environment sensitive UI

The idea of the environment as an interface is an interesting one, but requires more than just models of visiting strategies. Affordances of cultural settings play a central role in shaping the interaction. These include (a) properties that are "intrinsically" connected to a particular setting, such as the physical dimensions of the artworks, their position, their artistic importance; (b) architectural elements like access points to a room, arches and steps; (c) dynamic and contextual configurations of elements present in the space (crowd, lights). The role of the affordances in attracting the visitor can be hampered when combined in certain configurations (crowd and bad light conditions often drive the visitor to skip important artworks). We envisioned the possibility to design audio triggers to attract the visitor's attention. If the user reacts positively, moving to the mentioned artwork while listening to the description, then the system continues to provide information; otherwise it only mentions the artwork without further elaboration.

Social navigation

Our tourist guide provides some very basic support to the development of a social memory in the community of visitors by "marking" a moment of the visit. By pressing the "hotspot" button on the PDA, the visitor stores into the system the current position, an image of the artwork, the related description and personal comments. This facility may be used to recommend a tour to a friend, to elaborate on contents, to plan another tour and so on.

16.4.2 Content Design

Several issues arise when allowing for the dynamic presentation from a hypermedia repository of information in response to a user moving in a physical space. The most important of these is that the situational context in which the information is presented varies. The main factors determining the situational context are user position and movement (e.g. whether she is in front of an object or whether she is simply walking around a room) and the structure of the surrounding physical space (e.g. whether objects being described are close or not). In order to dynamically adapt the presentations of objects to the particular visitor in a particular context, the HIPS audio guide must have an internal representation of the content that can be expressed.

In the context of the HIPS project, we have developed a formalism to annotate multimedia repositories of data called the *macronodes formalism* (Not and Zancanaro, 2000). In this formalism, the atomic piece of data is called a *macronode*, which, for textual information, typically corresponds to a paragraph. Indeed, a macronode can encode data in different media such as audio, images or a combination of these. A presentation for the HIPS

© University of Siena

Figure 16.2   The PDA interface.

audio guide is the concatenation of different macronodes chosen to fit the visitor's needs.

Macronodes, grouped in networks, have a semantic description that allows the system to decide which macronodes have to be included in the presentation. In addition, the content of each macronode can be expressed in many slightly different ways, the most appropriate one is chosen on the fly to ensure cohesion with respect to other macronodes of the same presentation.

For the HIPS audio guide in the Museo Civico, 170 macronodes were prepared for 31 exhibits. The macronodes encompassed 344 audio files.

Each exhibit has a "welcome" macronode, that is, a macronode used to introduce the exhibit in a simple way when the user approaches it. This type of macronode is very simple and usually consisted of just a welcome statement in different versions (for example, *"this is ..."* or *"in front of you, you can admire ..."*, *"you're back to ..."*). Each exhibit is additionally associated with a number of "caption" macronodes, prepared according to different perspective (for example, historical, anecdotal, artistic, etc.). These kinds of macronodes are used as the core of a presentation and therefore they are quite complex. Usually, each caption macronode consists of a main part (for example, *"The Guidoriccio has been painted in a less complex way than the Maestà but Simone did not renounce to his love for details ..."*) to which some optional parts are added to ensure adaptation and cohesion (for example, after having introduced the Maestà in the previous example, we can add as an optional part *"that you've just seen"*; of course this part will be used only if appropriate).

Finally, other macronodes are provided as "additional info", that is, information that cannot be used by their own but can be added to captions to enrich the presentation with further details. Each macronode in this category has a given perspective but it is also explicitly specified to which caption it can be added and which communicative function it fulfils with respect to the caption (to give background information, to further elaborate one or more details, etc.). For example, a biography of the painter can be linked as background information on the historical caption of a fresco.

Using macronodes, the HIPS audio guide is able to tailor the content of the presentations to the visitor. In particular, the knowledge about the visitor's interests allows it to choose the preferred perspective with respect to which information has to be presented. The visiting style influences the amount of information (i.e. the number of macronodes) as well as the choice of the communicative style of the presentation. Example 1 shows an "elaboration-based" presentation of the Maestà, which is more suited for visitor with the "ant" visiting style and example 2 shows a "comparison-based" description of the same fresco, more suited for "fish" visitors.

Example 1: "elaboration-based" description of the La Maestà
This is the great fresco La Maestà, depicted by Simone Martini in 1315. La Maestà was the first decoration piece in Palazzo Pubblico, therefore it acquired through the centuries a particular value for the Sienese population. It's not surprising that the very first decoration of Palazzo Pubblico (the site of political power) was a religious artwork. Only four years before, in fact, the "Fabbrica del Duomo" the other civic power of Siena influenced by the bishop, commissioned the famous "Maestà" to Duccio di Boninsegna. The traditional spirit of competition between the two great "factories" of the city demanded an adequate reply.

Example 2: "comparison-based" description of the La Maestà
In front of you, you can admire the great fresco La Maestà, depicted by Simone Martini in 1315. The fresco is located in the main part of the hall, the central point that gave the orientation of the Sala del Mappamondo. On the contrary the Guidoriccio fresco, on the opposite side of the room, was a sort of great "poster", glorifying the power of the Siena Republic. It was a sort of historical documentation more than an artwork to be judged for its artistic value

16.5 Walkthroughs

16.5.1 San Bernadino and the Monogram

This example illustrates that two visitors who access information on the same frescoes in a different order get presentations that are similar[6] in the content yet "tailored" to their physical movements in the museum.

---

[6] Of course, the presentations could have been different on other respect as well, depending on their interest models and the visiting styles. This example has been simplified in order to clearly illustrate the effect of the previous interactions with the system on a presentation.

A prominent goal of any intelligent audio guide in a museum should be to help the visitor in locating the frescoes presented in the physical space while avoiding confusion when referring to different frescoes in the same presentation.

Let us consider a visitor in the Sala del Mappamondo. After having listened to the general introduction to the room, the visitor walks towards the fresco representing the Bernadian Monogram. As soon as he or she approaches the fresco, the HIPS audio guide starts the presentation: *"This is the Bernardian Monogram. Batista di Niccolò da Padova painted it in 1425. The portrait of San Bernardino is located behind you."*

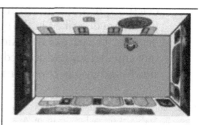

As you can note, HIPS uses spatial expressions such as *"This is"* and *"behind you"* to help the user to easily recognise the topic of the presentation and the other relevant fresco involved in the description.

When the visitor follows the system's implicit suggestion and moves towards the opposite side of the room, the HIPS audio guide starts the following presentation: *"On the column in front of you, there is the portrait of San Bernardino. It was painted by Sano di Pietro in 1460–1461. The saint can be easily identified by the monogram he holds in his hand. You just saw the bernardian monogram depicted on the opposite wall."*

As before, the system makes use of spatial expressions, but openly remarks that the other mentioned fresco has already been seen to avoid confusing to the visitor.

Now, let us consider a different visitor who just after entering the room turns his or her attention to the San Bernardino fresco. The presentation produced by HIPS is quite similar to the one presented to the previous visitor for the same fresco, but with a notably exception: *"On the column in front of you, there is the portrait of San Bernardino. It was painted by Sano di Pietro in 1460–1461. The saint can be easily identified by the monogram he holds in his hand. The bernardian monogram is also depicted on the top of the opposite wall."*

Of course, when the visitor then moves towards the Monogram fresco, the presentation will be similar to the one presented to the first visitor for the same fresco, but with the explicit mention of the fact that the San Bernardino fresco has already been seen: *"This is the Bernardian Monogram. Batista di Niccolò da Padova painted it in 1425. You just saw the portrait of san Bernardino on the opposite wall."*

16.5.2 The Maestà and Santa Caterina da Siena

These examples illustrate how different reading styles are used in the HIPS guide to immerse the visitor in a rich audio environment and to produce engaging presentations.

The rhetorical styles of the presentations are tailored to the context (use of deictic expressions) and to the iconographic contents (artworks representing people are described in the first person, as if the character presents him- or herself). The pace of narration (length, duration) is tailored to the visitor's movement (long and detailed descriptions are provided to visitors who move slowly and stop in front of each artwork).

Let us suppose that the visitor now moves towards the big fresco La Maestà at the end of the room. As soon as he or she approaches the fresco, the HIPS audio guide starts presenting the fresco with a polite male voice: *"In front of you is the Maestà, one of the absolute masterpieces of Sienese art, painted by Simone Martini in 1315."* And after a small pause, a new male voice with a strong Italian accent continues:

"The Virgin is depicted as the Sienese people's protector, and as a symbol of municipal justice: this particular devotion to the Virgin derived from the famous Battle of Montaperti in 1260, when Siena defeated the army of Florence and preserved its freedom."

The visitor laughs because of the strange Italian accents and walks away from the fresco.

As soon as he turns to the left a creepy female voice starts talking into the headphones: *"I'm Santa Caterina da Siena . . . ."* The visitor is astonished but soon, looking at the PDA display, he or she sees that the portrait of Santa Caterina is depicted on the column just in front.

After he or she has listened to the life of Santa Caterina in her own voice, the visitor decides to move out of the Sala del Mappamondo.

But while he or she is going to exit, the HIPS guide with a male voice and a 3D sounds effect says *"Behind you, there is another important fresco of Simone Martini: Guidoriccio da Fogliano."* Is it worth a look?

16.6 The Evaluation

The evaluation of the impact of enabling technologies in the field of art, leisure, learning and entertainment is not a trivial task. It is evident that the subject of evaluation is not "usability" per se but a wider and fuzzy notion of engagement and intellectual comfort that the user may experience.

From the very early stages of the HIPS development process, we applied traditional user-centred design to conceive and refine initial concepts, to learn from users' visiting strategies and needs, to observe and develop user profiles and to evaluate efficacy and effectiveness of intermediate solutions. These research methods provided indispensable input into the development process, but they were insufficient to answer some of the most crucial questions which were more global in nature, such as

- Will the users have to change their practices to exploit the new capabilities of the system?
- How motivated will they be to do so?
- Will they discover new ways to use the system to improve their experience?
- Will they use the system in a creative way to learn about art?
- Will the system adapt to the learning process of the users who will continue to explore and use additional functionality over time?

For each of these questions we wished to assess the complex relationships between artistic contents, context of use, physical environment, cognitive, cultural, motivational and emotional components at stake – in short, the overall experience of use.

However, the evaluation of such experience was not focused on in just the final prototype, but evolved through the entire life of the project. From the initial phases of *collection and inspiration,* we learned from previous works including field studies and surveys, and we defined the theoretical and technological framework. We observed and interviewed people, consulted experts, studied the environmental features of the museum (artistic, physical, social, economical, cultural) and shared ideas and concerns in every stage of design. During *concept generation,* we explored ideas, composed concepts, developed mock-ups and evaluated alternative solutions with users. During the *sharing and demonstration of the final system,* we presented results and asked different users to try out and assess the final product. The outcomes of this phase were fed back into the design process to refine and consolidate the system.

To provide the reader with a broad view of the evaluation process of HIPS, we will consider the history of the project from the evaluation of specific aspects of the system carried out on intermediate prototypes to the description of a full scale assessment carried out in a museum.

Part of the initial evaluation was inspired by a comparison with existing audio tourist guides, in particular, tape cassette guide and "telephone-like" guide. Whilst the tape cassette guides show evident limitations for a successful visit (content-centred organisation, fixed contents, constrained paths), the telephone handset exhibits a more interesting behaviour.

413

What sets this kind of audio guide apart from traditional tape cassette-based tours is the freedom from fixed routes. Presentation is self-paced, and the user can browse through the museum at will. Location-centricity enhances the access to the contents since the code organisation is not structured and very easy to access. Indeed to hear a description for a particular piece, the user must enter on the keypad the code displayed alongside the works of art.

However, these audio guides provide commentaries for a limited number of the items on display in the museum, and the information is fixed (the same description is available for each work of art, the visitor cannot ask for more/different information). Another feature that seems particularly relevant in the case of these guides is that they are "pull" only: information is "pulled" by the users, where, when and if they desire. The opposite case, of a "push" only guide, is the traditional tape tour – the user has no input and passively receives information. From the beginning of the project, it was clear to us that what could distinguish HIPS was its potential as both a pull and a push system. On the one hand we thought about a system that could be used to facilitate browsing: the user could pull information simply by standing in front of an exhibit and walking around; on the other hand, it could be used to push information, by guiding the user along predetermined routes.

Another distinguishing feature of HIPS with respect to traditional guides was the auditory interface of the system. We put specific effort into designing an auditory space that could support the visitor not only in getting information about a piece of art but in sustaining a quality of experience based on the perception of the physical space around. Indeed, we considered the role of hearing in sensing and interpreting the outside world and designed informative auditory events for perceiving the space and interacting with it.

During the project development, we produced some evidence concerning the relationship between the behaviour of people in physical spaces and auditory information content (Marti et al., 2000).

We developed "enriched commentaries", that is, varying descriptions in terms of voice used, pace of narration, reading styles and the use of 3D sounds, each associated with different contents (episodes from the Saints biographies, anecdotes regarding characters and events of Siena, general and artistic descriptions, etc). These commentaries turned out to be very effective in supporting the memorability of contents and were highly effective in keeping the attention and encouraging their interest in the artworks. The different voices and reading styles were easily perceived by the listeners, and they helped them in recognising different topics and types of contents: in particular, the longer the commentary, the more recognisable the type of content. The listeners' perceptions of them and their ideas and expectations about their "way to work" contributed to form a coherent mental model of the system's behaviour.

In terms of a qualitative appraisal of the *enriched* comments, such comments were preferred to the plain ones; moreover, some of the users who listened to the plain comments suggested that they could be made more appealing by introducing different voices.

These results formed an extremely rich basis for the design of a texture of sound that could be endowed with cues for information recognition and interpretation, without interrupting the user's condition of flow in listening to the information. The spatialisation of sound was used in the final prototype to match the perceptual features of the physical environment seen by the user with a corresponding information space heard by the user. In this respect, within such a texture of auditory space, the delivered descriptions were endowed with further perceptual features able to mark different levels of content.

Another aspect of the system that received a systematic evaluation in the course of the project was the adaptation of contents to the physical movement of the visitors. Our guess was that different types of users (ants, butterfly...) would require presentations that differed in both quality and duration. For instance, we thought that short descriptions were probably suitable for *fish* visitors, whereas longer ones were more suitable for *ants*. We conducted experiments administering presentations that were not adapted to user movement strategies (Marti et al., 1999).

The results showed interesting behavioural differences between those users who were assigned a path matching their visiting style versus those who were not: the degree of information skipping and of explicit requests for more information significantly increased if the spatial user model was altered. These results produced evidence on the relationship between visitor behaviour in physical spaces and information content. In particular, we showed how visitors' behaviours can be used as a heuristic principle for content organisation and specification of spatial user models.

A prototype of the final system was installed in the museum for 2 months and evaluated by visitors who volunteered to try it and provide feedback. The subjects who took part in the evaluation were recruited in the museum on the day of the test. All of them were English or English-speaking tourists.

© University of Siena

Figure 16.3   Trials in the Museo Civico

415

The user performance was assessed on four levels: phenomenological, cognitive, emotive and socio-cultural (Marti and Lanzi, 2001). At the phenomenological level, the performance measure concerned

- user's perception of the adaptation to the visiting style (personalisation of the information, pauses, pace of narration) and the physical movement as a primary means for accessing information;
- effectiveness of auditory comments (deictics, pronouns, etc.) in supporting the user's orientation and recognition of artworks;
- tool flexibility (tailoring to the user's changes of path or visiting style).

At the cognitive level, the performance measure concerned the cognitive effort associated with the use of the tool, the comprehension of the contents and the user's conceptual model. At this level, scenarios were used when questioning the design of the system. Norman's cycle of cognition based on goals, intentions, planning, execution, perception and evaluation was used to generate questions such as "How does the artefact evoke goals in the user?" or "How does the artefact make it easy or difficult to carry out the activity?" or "How does the artefact support the user when a shift in her or his goals occurs?"

At the emotive level, the performance measure mainly concerned aspects of experiential cognition including observation of frustration or confusion and expressions of satisfaction and engagement.

At the socio-cultural level, the performance measure concerned the social aspects of group activity mediated by the system (communication, knowledge sharing, collective memories), appraisal/dislike of contents and the impact of narrative styles (male/female voices, accents, music, reading styles).

The evaluation at the four levels was carried out on the basis of the direct observation of the activity (video-recorded) during free exploration and scenario execution. Scenarios were used as a means to create a context for the activity whilst a series of exceptional circumstances or constraints were artificially provoked in order to evaluate the system under specific conditions. At the end of each session, the visitors were involved in a debriefing and encouraged to comment, analyse and interpret events that occurred during the test. The subjects were asked to describe their experience by commenting on the video recording of the test.

The most enthusiastic comments regarded the possibility of freely moving during the visit whilst being simultaneously assisted by the new guide. Visitors felt comfortable listening to descriptions without interacting too much with the PDA interface, which was mainly used in case of poor performance of the system (delay in loading a presentation, lack of information, etc.). Another feature that was truly appreciated was the tailoring of information to the context. They recognised an original capability of the tourist guide to follow their movements and offer appropriate information at a specific moment of the visit. Essentially, the capability of the system to activate a relationship between the user, the guide and the surroundings, with the aim of facilitating and improving the visit, was recognised.

The use of different voices and unusual stories about the museum made the visit even more pleasant and enticing. All visitors appreciated these features and declared an interest in using the system for exploring open spaces. Some of them asked for a direct connection to their PC at home so as to prepare the visit and elaborate on it afterwards. Some visitors were amazed to discover unknown interests when the guide presented unexpected information. We could directly observe the capability of the guide to catch the visitor's attention since they liked to spend quite a long time listening and looking for details. Tourists who tried the system were very aware of how to handle the prototype. Some remarkable problems were noticed that did not hamper the positive effect on the global experience: slow performance and heaviness of the hardware used for the trials, a rare but still possible fragmentation in the narration. We believe that the perception of robustness of the system by the intended audience depended crucially on the quality of interaction. Speed of response was not the only issue, but information presentation, gradual degradation and effective feedback were some other crucial elements. Whereas issues such as data integrity, reliability, security, etc., were significantly less important, quality of interaction never was, and the users appreciated the effort of centring design on human needs and characteristics.

16.7 The Outdoor HIPS Prototype

As a proof-of-concept of the validity of the design concepts that drove the HIPS project, a second prototype has been implemented in a completely different scenario. The new HIPS audio guide has been designed with the purpose of guiding a visitor on an outdoor walk on the campus of University College Dublin (UCD) and describing the various buildings' architecture and history. With few exceptions, the system architecture behind of the outdoor prototype was the same as the indoor one.

Being designed for outdoor visits, this prototype cannot employ infrared to localise the visitor physical position: global positioning system (GPS) has been used instead. Moreover, the adaptivity component, prominent in the indoor prototype, has been replaced with a simpler module that just triggers canned presentations when the visitors reach certain zones. By contrast, the hotspot functionality plays a central role.

An evaluation of the outdoor implementation of HIPS was carried out on the campus of the university. Each subject was given a brief introduction to the goals of HIPS. They were then asked to follow a short route around the campus that included four exhibits. On completion, each user was interviewed and a short questionnaire was filled in. The whole procedure took approximately 30 minutes. Weather conditions at the time were dry but overcast. As the evaluation took place during late summer, students had not yet returned and there were still some visitors around the campus. A number of these agreed to participate in the evaluation.

Results of the evaluation were encouraging. Users grasped the concept reasonably quickly and, in general, considered the system usable, though some expressed some reservations about the perceived lack of user input. One issue that caused some concern was the discrepancy between perceived position and actual position. Even though the position fell within the standard GPS error (± 20 metres), some users seemed somewhat perturbed when they discovered that their actual position did not coincide exactly with that shown on the map. A solution to this might be the use of a more course-grained map. While users were generally satisfied with the presentations, some concern was expressed with the actual responsiveness. Ideally, a presentation would be triggered at the exact place from which the accompanying photograph was taken. However, this can be difficult to achieve and, at best, one can only assume that the presentation will be triggered within a certain zone. Clearly, such issues need to be considered carefully when designing information spaces.

16.7.1 Current Developments: Gulliver's Genie

Development of the HIPS outdoor prototype has continued after the project formally terminated and it has morphed into a system that we eventually termed Gulliver's Genie (O'Grady and O'Hare, 2002; O'Hare and O'Grady, 2003). In principle, the objectives are identical; however, the architecture of the Genie differs radically from HIPS in that its constituent entities consist of intelligent agents. Historically, the use of such agents on mobile devices would have been computationally prohibitive. Ongoing developments in PDAs and so-called smartphone technologies have rendered such concerns obsolete. In summary, the Genie may be regarded as a multi-agent system (MAS) that encompasses fixed network servers and databases as well as mobile devices. Agent Factory (O'Hare, 1996; Collier et al., 2003), an integrated environment for the design and fabrication of intelligent agents, provides the delivery engine around which the Genie is constructed.

In the case of outdoor users, the availability of a wireless data networking facility is essential. The somewhat protracted deployment of 3G networks represents a significant development for people planning the deployment of services for mobile users. However, it must be remembered that mobile computing users will always be at a computational disadvantage in comparison to traditional fixed workstation users. This issue is particularly pertinent when it is considered that users' expectations are shaped by their normal experiences; thus mobile computing applications have ample scope to disappoint! In the case of the Genie, a particular challenge is to dynamically

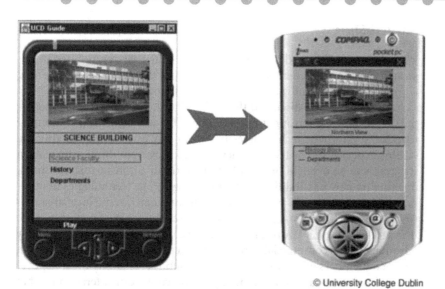

Figure 16.4 The evolution of the HIPS outdoor prototype to Gulliver's Genie.

build presentations and make them available to the tourist in a just-in-time basis. The key obstacle is of course the limited bandwidth available to subscribers. Downloading content with a rich multimedia component over a standard 2.5G connection (in this case, general packet radio service (GPRS)) takes minutes, not seconds. To overcome these limitations, the Genie adopts a strategy that we have termed *intelligent pre-caching* (O'Grady and O'Hare, 2004). In brief, a detailed model of the information space is constructed. This model is continuously interrogated in light of the tourist's ongoing movement. Presentations are assembled dynamically and stored on a cache on the server. When the tourist passes a certain threshold, a presentation is downloaded to the cache on the user's device such that it is available when the tourist encounters the attraction. The unpredictability of the available bandwidth and the inherent inaccuracy of GPS both conspire to make this process somewhat error-prone. However, ongoing development in 3G networks as well as satellite-based augmentation systems (SBASs) offer considerable opportunities for improving and fine-tuning the process.

Naturally, it was necessary to conduct user trial, the results of which are described elsewhere (O'Grady et al., in press). The results suggest that users see the potential for services such as the Genie, and would, in principle at least, be prepared to pay for them. However, they have high expectations, and delivering services that meet their expectations will continue to challenge both service providers and software engineers for the foreseeable future.

16.8 Conclusion

We believe that, in the development of HIPS, the user-centred design approach played a fundamental role in generating original ideas and combining them with the use of advanced enabling technologies.

In this chapter, we discussed the design and the prototype implementation of an innovative tourist audio guide. This was conceived combining together several new technologies in such a way as to deeply modify the visiting experience. The technologies we used allowed multiple information structures to be overlaid on the physical world in a non-intrusive fashion, opening up new possibilities for creative design. The technologies in question include both hardware (localisation technologies, radio local area networks, handheld computers) and software (adaptive text generation, GIS, user modelling, etc.). The result, while not a commercial system, offers many advantages over commercially available systems. It allows communication between members of a community via the public availability of "hotspots". The presentations are adapted, firstly to the context of interaction (what have they heard/seen before, etc.), and secondly in regards to the very way they appear to use the museum (visiting style modelling). This experience demonstrates that the success of entertainment or edutainment systems, especially those exploiting advanced technologies, strongly relies on a design philosophy that mediates between a deep and continuous focus on the users and innovative visions.

A further valuable result of our work, and all I3 projects, is the sharing of expertise between the consortium members – the connected community of research. We each came away with knowledge held elsewhere in the consortium and have established strong research links between the institutions.

16.9 References

Bianchi, A. and Zancanaro, M. (1999) Tracking user's movements in an artistic physical space. In *Proceedings of I3 Annual Conference*, Siena, October 1999.

Bordegoni, M., Faconti, G., Maybury, M.T., Rist, T., Ruggeri, S., Trahanias, P. and Wilson, M. (1997) A standard reference model for intelligent multimedia presentation systems. *Computer Standards and Interfaces*, **18**, 477–496.

Chen, S. and Magoulas, G. (2005) *Adaptable and Adaptive Hypermedia Systems*. IRM Press.

Collier, R., O'Hare, G.M.P., Lowen, T.D. and Rooney, C.F.B. (2003) Beyond prototyping in the factory of agents. In *Proceedings of Third International Central and Eastern European Conference on Multi-Agent Systems (CEEMAS)*, Prague, Czech Republic.

Csikszentmihalyi, M. (1990) *Flow, The Psychology of Optimal Experience*. Harper Perennial, New York.

De Bra, P., Brusilovksy, P. and Houben, G. (1999) Adaptive hypermedia: from systems to framework.

Hutchins, E. (1995) *Cognition in the Wild*. MIT Press, Cambridge, MA.

Isard, A., Oberlander, J. and Matheson, C. (2003) Speaking the user's language. *IEEE Intelligent Systems*, **18**(1), 40–45.

Mann, W.C. and Thompson, S. (1987) Rhetorical structure theory: a theory of text organization. In Polanyi, L. (ed.), *The Structure of Discourse*. Ablex Publishing Corporation.

Marti, P., Ciolfi, L. and Gabrielli, F. (2000) Designing auditory spaces. In *Proceedings of the Annual Conference i3AC2000*, Joenkoeping, Sweden, September 2000.

Marti, P., Gabrielli, L. and Pucci, F. (2001) Situated interaction in art. *Personal Technologies*, **5**, 71–74.

Marti, P. and Lanzi, P. (2001) "I enjoyed that *this* much!" A technique for measuring usability in leisure-oriented applications. In Bawa, J. and Dorazio, P. (eds), *The Usability Business: Making the Web Work*.

Marti, P., Rizzo, A., Petroni, L., Tozzi, G. and Diligenti, M. (1999) Adapting the museum: a non-intrusive user modeling approach. In *Proceedings of UM99*, Canada.

Norman, D.A. (1993) *Things That Make Us Smart*. Addison-Wesley, Reading, MA.

Norman, D.A. (1999) *The Invisibile Computer*. MIT Press, Cambridge, MA.

Not, E. and Zancanaro, M. (2000) The macronode approach: mediating between adaptive and dynamic hypermedia. In *Proceedings of the International Conference on Adaptive Hypermedia and Adaptive Web-Based Systems, AH'2000*, Trento, August 2000.

Oberlander, J., O'Donnell, M., Knott, A. and Mellish, C. (1998) Conversation in the museum: experiments in dynamic hypermedia with the intelligent labelling explorer. *The New Review of Hypermedia and Multimedia*, **4**, 11–32.

O'Grady, M.J. and O'Hare, G.M.P. (2002) Accessing cultural tourist information via a context sensitive tourist guide. *Information Technology and Tourism Journal*, **5**(1).

O'Grady, M.J. and O'Hare, G.M.P. (2004) Just-in-time multimedia distribution in a mobile computing environment. *IEEE Multimedia*, **11**(4), 62–74.

O'Grady, M.J., O'Hare, G.M.P. and Sas, C. (in press) Mobile tourists, mobile agents: a user evaluation, Gulliver's Genie. *Interacting with Computers*.

O'Hare, G.M.P. (1996) The agent factory: an environment for the fabrication of distributed artificial systems. In O'Hare, G.M.P. and Jennings, N.R. (eds), *Foundations of Distributed Artificial Intelligence, Sixth Generation Computer Series*. Wiley, New York.

O'Hare, G.M.P. and O'Grady, M.J. (2003) Gulliver's Genie: a multi-agent system for ubiquitous and intelligent content delivery. *Computer Communications*, **26**(11), 1177–1187.

Petrelli, D., Baggio, D. and Pezzulo, G. (2000) Authors, adaptive hypertexts, and readers: supporting the work of authors for user's sake. In *Proceedings of the International Conference on Adaptive Hypermedia and Adaptive Web-Based Systems, AH'2000*, Trento, August 2000.

Stock, O. (1999) Was the title of this talk generated automatically? Prospects on intelligent interfaces and language. In *Proceedings of IJCAI99, the 16th International Joint Conference on Artificial Intelligence*, Stockholm, pp. 1412–1419.

Veron, E. and Levasseur, M. (1983) Ethnographie de l'exposition Bibliothèque publique d'Information, Centre Georges Pompidou, Paris.

Index